Genetics for the Animal Sciences

Genetics for the Animal Sciences

L. Dale Van Vleck

E. John Pollak

E. A. Branford Oltenacu

Department of Animal Science
Cornell University

W. H. Freeman and Company
New York

Cover photograph by Grant Heilman,
Grant Heilman Photography

Library of Congress Cataloging-in-Publication Data

Van Vleck, L. Dale (Lloyd D.), 1933–
Genetics for the animal sciences.

Includes bibliographies and index.
1. Animal genetics. 2. Animal breeding.
I. Pollak, E. John (Emil J.)
II. Oltenacu, E. A. Branford (Elizabeth A.) III. Title.
QH432.V36 1987 636.08'21 86-9819
ISBN 0-7167-1800-6

Printed in the United States of America

1 2 3 4 5 6 7 8 9 0 MP 4 3 2 1 0 8 9 8 7 6

CONTENTS

Preface

An appreciation of the relationship between chance and genetic expectations based on precise laws of inheritance is the key to the study of genetics and to the application of genetics to animal improvement. *Genetics for the Animal Sciences* covers three main areas of genetics, which a breeder of animals will encounter. First, the inheritance of many characteristics that arouse the curiousity of the breeder, such as color and hair length, is explained through the basic principles that Gregor Mendel outlined. Chapters 2, 3, and 4 provide a thorough exposition of Mendelian genetics. Chapter 5 emphasizes a fact of genetics that puzzles inexperienced breeders: the phenotypic frequencies that Mendel's laws predict for matings are not always realized with small numbers of progeny. Although Mendel's laws have a precise mathematical basis, the randomness of segregation and recombination may lead by chance to unexpected frequencies of phenotypes.

Second, Chapters 6 through 9 extend the concepts of Mendelian genetics from individual matings to matings in large populations. In populations, the form of gene action interacts with selection, with the size of the population, with migration and mutation, and with systems of mating to lead to fixation of desirable genes or to elimination of undesirable genes. As with Mendelian genetics, many examples and applications reinforce the presentation of population genetics. Although the mathematics of genetics is not usually beyond basic algebra, following through the simple mathematics of genetics is the best way to understand and apply both Mendelian and population genetics.

The third area, quantitative genetics, is more statistical because it involves both populations of animals and traits influenced by many genes with small effects and with each allele having a relatively low frequency. Traits having a quantitative mode of inheritance are generally those with the greatest importance to the animal breeder, such as milk yield, growth rate, and litter size. This section, Chapters 10 through 15, is supplemented by boxed material, which presents the mathematics and complexities of quantitative genetics in considerable detail.

We are well aware that nearly all teachers of animal genetics organize their courses differently. *Genetics for the Animal Sciences* has been written to allow considerable flexibility. Most sections can stand alone. The more quantitative chapters are organized so that the simpler material precedes more complicated developments. Some chapters, such as the one on relationships, provide alternative ways of teaching the same material.

The book is geared to the introductory-level student with little or no background in genetics. By including the more detailed information found in the boxes, it can also be used for advanced courses in quantitative genetics. The principles of genetics are stressed throughout the book, but considerable emphasis is also placed on breeding goals and the procedures necessary to achieve them.

Chapters 2 through 5 emphasize the principles of classical genetics. Selected parts of chapters on population and quantitative genetics would provide an overview to those important areas of genetics. For students with a solid basis in Mendelian genetics, thorough study of Chapters 5 through 11 provides a detailed foundation in population genetics. Remaining time could be used to preview quantitative genetics. Chapters 5 through 15 could also be used for a quantitative genetics course with equal emphasis on population genetics and quantitative genetics.

The number of people who should be thanked for their help is too great to be listed. Three special groups, however, must be acknowledged. The efforts of our unsigned colleagues, Susan H. Herbert and Judith E. Sherwood, are especially appreciated. Comments and suggestions by the reviewers, particularly Tom Cartwright, Texas A & M University; Larry Chrisman, Purdue University; Tom Famula, University of California, Davis; Bill Hohenboken, Oregon State University; Rodger Johnson, University of Nebraska; Max Rothschild, Iowa State University; and Bill Vinson, Virginia Polytechnic Institute have resulted in substantial improvements for which the authors are grateful. A most important group to be thanked is the many students and graduate teaching assistants who have tested versions of this text at Cornell University and at the University of California at Davis.

Ithaca, New York
July 1986

L. Dale Van Vleck
E. John Pollak
E. A. Branford Oltenacu

Genetics for the Animal Sciences

CHAPTER 1

Genetics and Animal Breeding: An Overview and a Short History

Animal genetics is the study of the principles of inheritance in animals. *Animal breeding* is the application of the principles of animal genetics with the goal of improvement of animals. The two terms are often confused and for all practical purposes are closely linked. Thus, the terms, animal genetics and animal breeding, are often used interchangeably.

The study and application of animal genetics falls naturally into three main areas: Mendelian genetics, population genetics, and quantitative genetics. The principles of transmission of genetic material from one generation to the next are the basis of *Mendelian genetics,* the laws of particulate inheritance first formulated in 1865 by the Austrian monk Gregor Mendel from results of his experiments with the common garden pea.

Three European botanists, Correns, deVries, and Tschermak, are given varying degrees of credit for the rediscovery of Mendel's laws in 1900. An Englishman, William Bateson, in 1901 provided the first evidence from experiments with chickens that Mendel's particles are the basis of inheritance in animals as in plants. In 1906 Bateson also provided the classical definition of genetics as a field of study: "Genetics is the science dealing with heredity and variation seeking to discover laws governing similarities and differences in individuals related by descent" (Hutt, 1964). He was the leading promoter of Mendelian principles, in opposition to the biometricians (biological mathematicians) who were the main opponents of these new ideas for the first two decades of the twentieth century. In addition, Bateson coined such technical words of genetics as homozygote, heterozygote, and allelomorph that are now in common use. A botanist, Wilhelm Johannson, introduced in 1906 the even more commonly used terms gene, genotype, and phenotype.

Although Mendelian genetics has relatively little direct importance in animal improvement, the principles of Mendelian genetics are the basis for two specialized areas of genetics with major implications for animal improvement —population genetics and quantitative genetics. The second section of the book deals with population genetics.

Population genetics, in simplest terms, is the study of Mendelian genetics in populations of plants or animals. The basic foundation of population genetics, the Hardy-Weinberg law, was formulated in 1908 by the English mathematician G. H. Hardy and the German physician, W. Weinberg. Population genetics usually is limited to the inheritance of qualitative characters which are influenced by only a small number of genes. Population genetics is therefore important to the understanding of why characteristics, desirable or undesirable, can either become fixed or continue to exhibit variation in natural populations. Even more important, principles of population genetics can be applied to the design of selection strategies to increase the frequencies of desirable genes or, more likely, the elimination of deleterious genes.

Quantitative genetics is conceptually the most difficult of the three areas because the effects of individual genes can seldom be seen or measured, and because many genes are hypothesized to contribute to the expression of traits such as milk yield, growth rate, or litter size. The theory of selection for such traits is further complicated by random influences of the environment and other nongenetic factors that tend to mask the combined effects of the many genes influencing the trait. Nevertheless, from a practical point of view, quantitative genetics is the most important of the three areas. Response to selection for quantitative traits generally has much more potential monetary value than has selection for simply inherited traits. Although the theory of selection for quantitative traits is often thought to be difficult to understand, the underlying principles lie in Mendelian and population genetics. These principles of genetics combine with relatively simple statistical concepts to form the basis of quantitative genetics. The early leaders in this field were R. A. Fisher in England and Sewall Wright in the United States, who bridged the intellectual gap between the early Mendelians and the followers of the English biometrician Francis Galton and later, Karl Pearson. Even before the rediscovery of Mendel's principles, Galton and Pearson, using the statistical tools of regression and correlation, had discovered the principle that the similarity between an animal and its descendants decreases by one-half from one generation to the next. For example, using actual records they found the correlation between records of parent and offspring to be twice the correlation between records of grandparent and grandprogeny. For several years neither the Mendelians nor the biometricians would concede the validity of the others' principles. The main difficulty was that the Mendelians expressed their results in terms of frequencies of genotypes and phenotypes, whereas the biometricians' results were correlations and regressions. Finally, however, Fisher and Wright demonstrated that Mendelian frequencies were the basis of biometrical correlations.

The history of animal breeding shows that the empirical principles of the biometricians had been applied successfully to change animals in desired ways for many years — even thousands of years — before either Mendel or Galton. The history of animal breeding probably started before recorded history with the domestication of animals. Although domestication may have been accidental in some cases, in other cases there must have been intentional selection for more friendly and tractable animals. Behavioral traits, as an overall term for the traits needed for initial domestication, probably are quantitative traits involving many genes. Selection for affinity to humans has been more important with the dog than with any other species. The dog was likely the first domesticated species, with distinct breed types arising as long ago as 12,000 years. The dog, probably derived from the wolf, was the only animal to have been domesticated independently in both Europe and North America.

Selection in other domestic animals probably progressed very slowly for better performance, primarily through better adaptation to the environment. Most performance traits are quantitative. The best example of selection for performance traits may be the various types of horses. The horse was developed for draft — pulling and hauling large loads — and for transportation of people. Multipurpose horses also were developed. The power of selection is clearly shown by the difference between the modern draft horses — the Percheron, Belgian, and Clydesdale, which descended from the "cold-blooded" horses of northern Europe — and the Shetland pony, also a descendant of the cold-blooded horses, which was developed as a small pack animal and later imported to England for use in the coal mines. Horses of the Middle East, the ancestors of the modern Arab and other "hot-blooded" horses, trace back at least 2500 years. These horses were bred for speed with endurance under a rider. Selection for such diverse types and different performance traits would have taken many hundreds of years with the primitive identification and records systems available during those times.

Records of performance and reliable identification are now universally accepted as the necessary foundation for progress in selection for quantitative traits. Records provide the basis for designing optimum breeding strategies, such as progeny testing (selection based on progeny records) or performance testing (selection based on own record).

An English livestock breeder of the eighteenth century, Robert Bakewell, is known as the father of animal breeding. His success as a breeder is attributed to his care in keeping records and his use of inbreeding to fix a desired type. He laid the foundations for the Shire breed of horses, the old Longhorn breed of beef cattle, and the Leicester breed of sheep. Many sayings attributed to Bakewell still have meaning today and, as Bakewell probably well understood, are true only "on the average" and not every time. "Like begets like" simply means superior parents are more likely to produce superior progeny than are inferior parents. But, as Jay Lush of Iowa State University, the father of modern animal breeding, stated, "like does not always beget like." In other words, some progeny of the best parents may be inferior to some progeny of the worst parents. Chance plays a part in the success of a particular mating. Similarly, the axiom "breed the best to the best," although basically sound, requires the additional phrase "and hope for the best," because the animal that appears to be best may not be genetically the best. The goal of the animal breeder is to make the best use of the available records to maximize the probability of selecting the best animals.

The first herdbook, with herd referring to a national breed, is associated with English Thoroughbred horses. *An Introduction to the General Stud Book* was started in 1791 to record the pedigree of *performers,* meaning those that raced. Unfortunately most other herdbooks such as the Coates herdbook for Shorthorn cattle, in 1822 the first herdbook for cattle, emphasized only pedigree information—names of ancestors—and not performance. Most herdbooks eventually become closed, meaning that only animals with parents in the herdbook are eligible for registration in the herdbook and thus are called purebreds. At the beginning of a herdbook, and for a period of time, the book is open to outstanding performers. Unfortunately, performance often is not required as a criterion for registration after the herdbook is closed. Both pedigree information and records of performance are important for genetic evaluation. Pedigree information without records is essentially useless.

Outstanding examples of the value of records and identification are the programs of the Dairy Herd Improvement (DHI) Associations which started as cooperative Cow Test Associations in 1906 in Michigan and in 1908 in New York. Although the records were originally intended for herd management, they provide the data resource for the most advanced methods of genetic evaluation of dairy sires. The best sires can, by artificial insemination, produce thousands of offspring each year. The combination of DHI records, computer evaluation of the records, and dissemination of the genes of the best bulls by artificial insemination led to one of the most successful stories in animal breeding. The availability of records for evaluation and of artificial insemination for taking advantage of the best bulls provided the incentive for animal breeders to design breeding programs to increase the rate of genetic gain and to develop improved methods of predicting genetic merit. Between 1958 and 1980, in the northeastern United States, the average milk production per lactation increased from 12,500 pounds to 17,500 pounds. Of the total increase, nearly 40 percent is attributed to increased genetic merit.

Using the theories of Wright, Lush in the 1930s established the foundations of modern methods of estimating the merit of animals for breeding purposes, that is, breeding values. Following Lush, C. R. Henderson of Cornell University and Sir Alan Robertson of the University of Edinburgh developed in the early 1950s methods for the computer evaluation of dairy sires which have resulted in much of the genetic improvement enjoyed by the dairy industry. Henderson, during the following 25 years, developed most of the increasingly more advanced genetic evaluation systems in use in dairy and beef cattle breeding. These systems became possible to implement because of the breakthroughs in computer speed and capacity.

Genetic gain, in contrast to gain made by improved management, is permanent. The following chapters describe the basic principles developed by many students of genetics, including such notable pioneers as Mendel, Hardy and Weinberg, Fisher and Wright, and Lush and Henderson, that lend an understanding of how knowledge of Mendelian, population, and quantitative genetics can be used for permanent improvement of farm animals.

Further Readings

Bateson, W. 1902. *Mendel's Principles of Heredity: A Defense.* Cambridge University Press, Cambridge, England.

Galton, F. 1889. *Natural Inheritance.* Macmillan, London.

Hutt, F. B. 1964. *Animal Genetics,* Ronald Press, New York.

Johansson, I., and J. Rendel. 1968. *Genetics and Animal Breeding.* W. H. Freeman and Company, New York.

Lush, J. L. 1945. *Animal Breeding Plans.* Iowa State College Press, Ames, Iowa.

Lush, J. L. 1951. Genetics and animal breeding. In *Genetics in the 20th Century.* Edited by L. C. Dunn. Macmillan, New York.

Pearson, K. 1903. Mathematical contributions to the theory of evolution. XI. On the influence of natural selection on the variability and correlation of organs. *Phil. Trans. Royal Soc.,* London, Series A, **200:1**.

Peters, J. A., ed. 1959. *Classic Papers in Genetics.* Prentice-Hall, Englewood Cliffs, N.J.

Provine, W. B. 1971. *The Origins of Theoretical Population Genetics.* The University of Chicago Press, Chicago.

Stern, C., and E. R. Sherwood, eds. 1966. *The Origin of Genetics, A Mendel Source Book.* W. H. Freeman and Company, New York.

Zirkle, C. 1951. The knowledge of heredity before 1900. In *Genetics in the 20th Century.* Edited by L. C. Dunn. Macmillan, New York.

CHAPTER 2

Mendelian Genetics

2-1 Historical Perspective

Gregor Mendel entered the Augustinian monastery in Brno in 1843 to prepare for the priesthood. The monastery was a place of learning, which encouraged the study of the natural sciences alongside religion. Having been ordained, Mendel temporarily left the monastery in 1847, his scientific interests leading him to further study at the University of Vienna. He then returned to a teaching career within the monastery.

Mendel's interests lay in the field of *hybridization,* or the interbreeding of purebreeding strains with one another. He chose plants as his experimental tool. Mendel's approach to his research was different from that of any predecessor. He chose to work with a single species of plant rather than deal with the complexities of interspecific hybridization. He chose a frequently cultivated species, the garden pea. Before experimenting with any variety, Mendel took time to verify that it was indeed purebreeding for the traits of interest. He concentrated on a few clearly defined characteristics in the plants, produced and raised large numbers of progeny in each of several generations, and repeated his experiments several times before publishing his results. Finally, Mendel applied his mathematical training to the large data base that he had accumulated and sought to explain his results in statistical terms. Consequently, he was able to deduce much about the inner workings of the hereditary material within the cell long before any knowledge of the physical basis of heredity existed. His methods remain a superb example of the scientific approach to studying inheritance.

In some ways, inheritance in plants is easier to study than in animals. Plants are often capable of self-fertilization, plants generally produce many more progeny per mating than do domestic livestock, and *generation turnover,* the time between parents producing offspring and those offspring reaching reproductive age, is rapid. Nevertheless, the conclusions Mendel drew from his work are as applicable to animals as to plants. Mendel worked with simply inherited characteristics, but his deductions provide the basic foundation on which knowledge of the inheritance of complex traits has been built. It is appropriate, therefore, to investigate the conclusions Mendel drew from his studies, using animal traits that parallel the traits with which Mendel worked, as a foundation on which to develop an understanding of the genetics of domestic livestock.

There is now so much genetic terminology that it is difficult to imagine the conditions under which Mendel worked. The development of specific terminology has an important function, as it permits single, descriptive words to replace cumbersome explanations. To enable the student to become familiar with this terminology, appropriate terms have been introduced in the following discussion, even though Mendel did not have them at his disposal when he reported his results.

2-2 The Law of Segregation

Mendel's first law is also known as the *law of segregation* (or separation) of gene pairs, where a *gene* is defined as a unit of inheritance. The law of segregation states that genes exist in pairs in the cells of the individual, and that members of a gene pair separate into the reproductive or sex cells formed by that individual, such that half carry one member of the gene pair and half carry the other. The reproductive cells are called *gametes,* and are the vehicles that pass the hereditary material from parents to progeny.

Suppose that Mendel had taken a strain of purebreeding black animals, and had mated them with individuals of a purebreeding red strain. *Purebreeding* animals produce progeny like themselves when they are mated inter se, that is, to one another. The mating of two purebreeding strains that differ for a single characteristic is called a *monohybrid* cross, as it involves just one pair of genes. The two purebreeding types chosen as parents are referred to by geneticists as the *parental generation.* Their offspring, called the *first filial generation,* F_1, appear as all black individuals. This first cross readily dispenses with the ancient theory that inheritance was a "blending" process such that the F_1 should be a combination of the two parental colors. In reality, one color has totally *dominated,* or concealed, the other.

Matings among F_1 individuals give the F_2, or *second filial generation.* These progeny also contradict the predictions based on the blending theory of inheritance that colors, once blended, could not be separated back into their component parts. When large numbers of F_2 progeny, not just the offspring of one or two matings, are classified, there are about three black offspring to every red one. The red color, identical to that in the parental strain, has reappeared, thus requiring the development of a theory that better fits the results than does the blending theory. This testing of hypotheses is an essential process in the study of genetics. A simple theory is useful to start with, but the researcher must be prepared to dispense with it, or modify it, should the experimental evidence not agree with its predictions. A new theory should be developed and then tested with further experimental results, which was precisely the approach taken by Mendel.

Mendel rejected the blending theory of inheritance in favor of a *particulate* theory. He postulated that traits were controlled by factors—now called genes—that remain discrete in each generation, allowing parental characters to reappear without having been changed by their passage through intervening generations.

Mendel tested the F_2 with further matings before he drew his conclusions. To continue with the example, the red individuals, mated inter se (among themselves), produce only red offspring. These individuals carry only the genes for red color, as did the purebreeding parental strain. Such animals are *homozygous,* having a "matched" pair of genes (see Figure 2-1). Each individual, or *homozygote*—from homo meaning "like" and zygote meaning "fertilized egg"—produces only one kind of gamete for this particular gene pair.

Test matings of black F_2 individuals are needed to determine which of

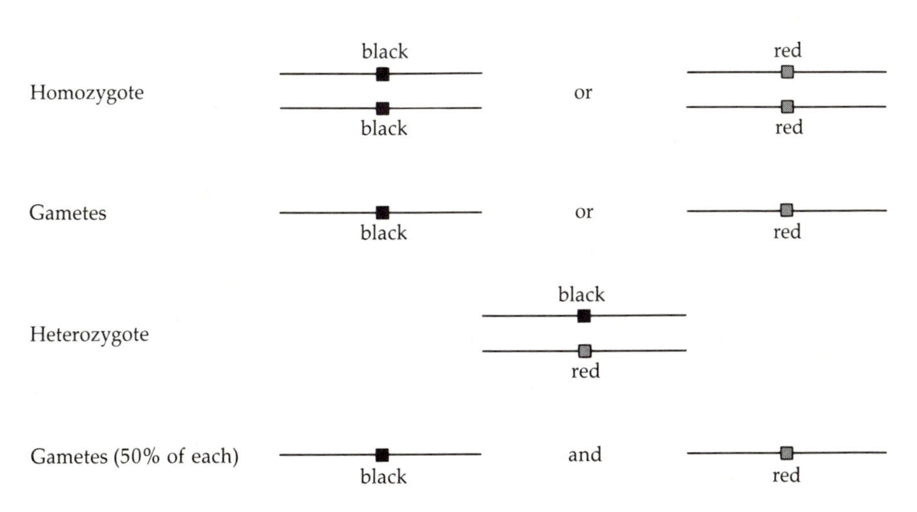

Figure 2-1 Diagram of the genetic makeup of a homozygote and a heterozygote.

them are the homozygotes. The reds are used for mates as the genes they transmit will not conceal the genes transmitted by the black individuals being tested. As the F_1 indicated, black is *dominant* to red, completely concealing the effect of the *recessive* red gene in an animal having an unmatched pair of genes. The *heterozygous*—hetero meaning "different"—black animal produces two kinds of gametes, half with the red gene and half with the black (see Figure 2-1). A *heterozygote* therefore does not breed true. Homozygous black animals would be of no use in test matings, as the effect of their dominant genes would appear in all their progeny.

Test matings of black F_2 animals show that one-third are homozygous, producing only black progeny, and two-thirds are heterozygous, producing black and red offspring in the ratio of 1 : 1. The number of homozygous blacks approximates the number of reds, so the overall F_2 genetic ratio is 1 : 2 : 1, homozygous black to heterozygous black to homozygous red.

The results of test matings of heterozygotes were Mendel's evidence that genes occur in pairs in the individual; this is the only way to account for the heterozygote's ability to produce two kinds of offspring in equal proportions. Mendel deduced that the genes segregate during gamete formation, so that a parent passes on only one member of each gene pair to each of its offspring. The identity of the gene received by each offspring is determined randomly.

Mendel thus concluded from the first part of his study that the F_1 produces offspring with one or other of the two differentiating characters, and, of these, one-half develop the hybrid (that is, heterozygous) form, while the other half yield progeny that remain constant and receive the dominant or recessive characters in equal numbers (Peters, 1959).

2-3 Terminology

Some further terminology is useful before the complexities of Mendel's second law are discussed. The location of a gene on the genetic material is referred to as a *locus* (plural, loci). The members of the gene pair at a given locus may take different forms—for example, the genes for black or red—and these forms are called *alleles* or *allelomorphs.* As already indicated, an allele that conceals the effect of its pair member is said to be dominant, and the hidden allele is recessive.

During gamete formation, *gametogenesis,* pair members segregate into the reproductive cells so that the male sperm, *spermatozoön* (plural, spermatozoa), and female egg, *ovum* (plural, ova), each contain only half the genes of the individual. Gametes are therefore said to be *haploid.* The body cells of the individual are *diploid;* that is, they contain the full, paired complement of genetic material. The process of gametogenesis is discussed fully in Chapter 3.

A form of genetic shorthand can be used to designate the different alleles at a locus. The most dominant allele is assigned an uppercase letter, while recessive alleles are written as lowercase letters. The same letter is used for all alleles at the same locus. In the black *(B)* and red *(b)* color example, an animal with a red *phenotype,* or appearance, has a *genotype,* genetic makeup, written *bb.* Phenotypically black animals are either *BB* or *Bb,* with a breeding test or knowledge of the parents or both necessary to distinguish between the two genotypes. The notation *B*_ is often used to denote an animal with the dominant phenotype and unknown genotype. The process of gamete formation is diagrammed in Figure 2-2, while Figure 2-3 outlines the succession of generations.

A problem begins to be apparent. Without a breeding test to determine what genes an animal transmits to its offspring, the observer must guess its genotype from its phenotype. There is no difficulty with a recessive homozygote, such as red in the example, but individuals with the dominant phenotype may be deceptive as some are heterozygous and will not breed true. This problem is always present in genetic studies involving dominance, and is one reason why several generations must be studied before conclusions can be drawn. The *test cross,* as a procedure for distinguishing between homozygous

Parents	*BB* (black) and *bb* (red)	
Gametes	*B*	*b*
F_1	*Bb* (black)	
Gametes	1*B* : 1*b*	

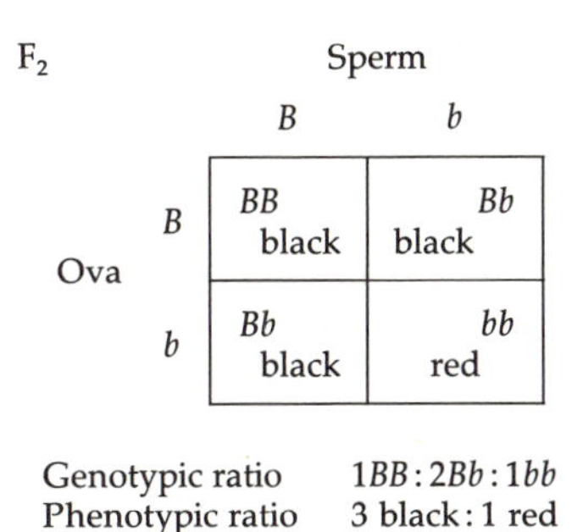

Figure 2-2 Gamete formation and union for a monohybrid cross (two alleles, simple dominance) in cattle. The allele for black color is designated *B*, and the allele for red color *b*.

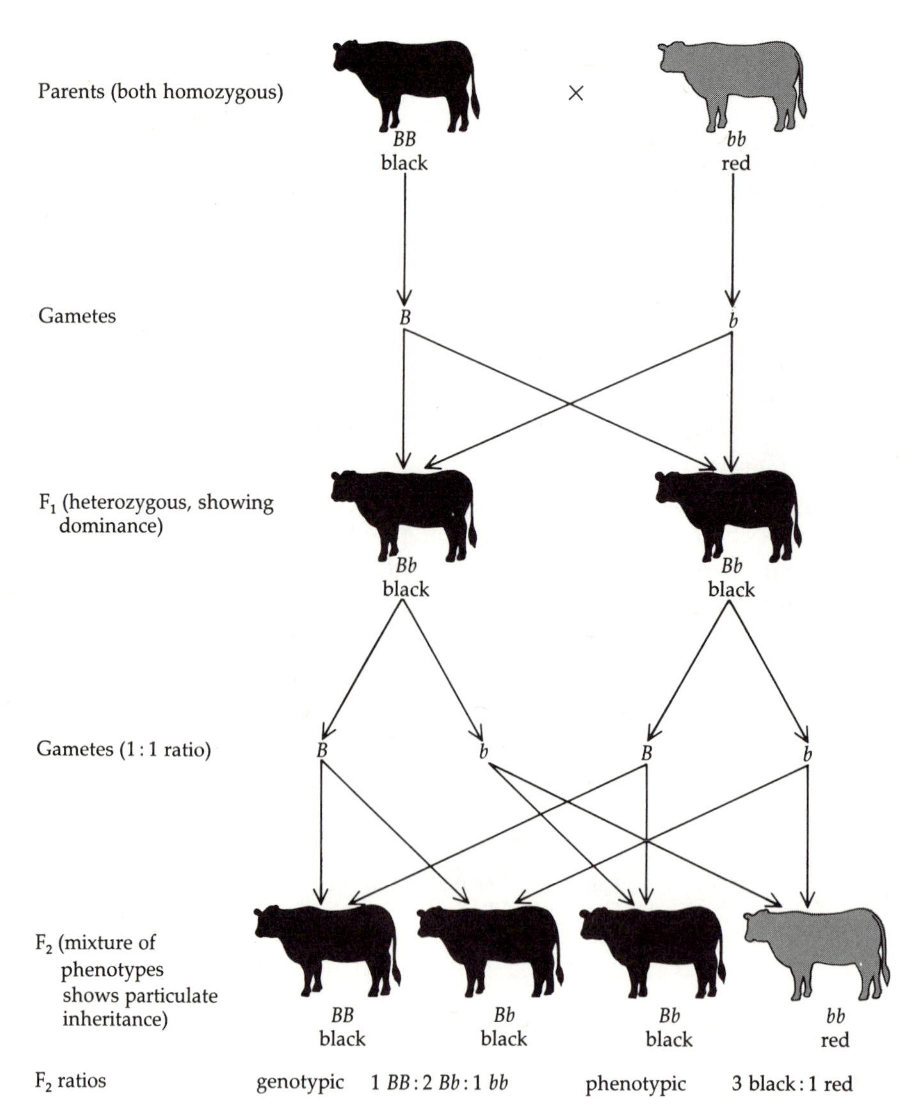

Figure 2-3 Diagrammatic representation of allele segregation in a monohybrid cross with simple dominance. The allele for black color is designated *B*, and the allele for red color *b*.

and heterozygous animals sharing the dominant phenotype, is an essential tool in this process. Recessive homozygotes are used in test matings because they can transmit only recessive alleles to their offspring, thus allowing expression of all genes transmitted by the individual with the dominant phenotype.

2-4 The Law of Independent Assortment

Mendel next sought to explain the behavior of genes controlling separate characteristics. His conclusions led to his second law, the *law of independent assortment,* which states that genes controlling separate traits segregate *independently;* that is, segregation at one locus does not influence segregation at another.

There exists in the Hampshire breed of swine a dominant allele causing a white belt around the body. This allele can be designated *W*, with the allele for a self, or solid color, coat designated *w*. Suppose that purebreeding red, belted swine, *bb WW*, are mated with purebreeding black, solid-colored individuals, *BB ww*. The F_1 are genotypically *Bb Ww*, and phenotypically black and belted. What should be expected in the F_2? If these traits are truly independent, each gene pair will segregate according to Mendel's first law and give a 1:2:1 genotypic ratio. Figure 2-4 illustrates the expansion of the multiplication of the ratios for the two loci to derive the F_2 ratio.

The algebraic approach may also be used with the expected F_2 phenotypic ratios of 3 black:1 red and 3 belted:1 solid. Multiplication of these two ratios yields an F_2 phenotypic ratio of 9 black belted:3 black solid:3 red belted:1 red

F_1 mating *Bb Ww* × *Bb Ww*

By locus

Bb × *Bb* 1 *BB*:2 *Bb*:1 *bb*
Ww × *Ww* 1 *WW*:2 *Ww*:1 *ww*

F_2 Progeny

Genotypic ratio: multiply the ratios for the two loci:
(1 *BB*:2 *Bb*:1 *bb*) × (1 *WW*:2 *Ww*:1 *ww*)

1	*BB WW*	black, belted
2	*BB Ww*	black, belted
1	*BB ww*	black, solid
2	*Bb WW*	black, belted
4	*Bb Ww*	black, belted
2	*Bb ww*	black, solid
1	*bb WW*	red, belted
2	*bb Ww*	red, belted
1	*bb ww*	red, solid

Phenotypic ratio

9 black, belted
3 black, solid
3 red, belted
1 red, solid

Figure 2-4 Algebraic derivation of F_2 progeny ratios for a dihybrid cross. *B* is the allele for black color, *b* the allele for red, *W* the allele for white belt, and *w* the allele for no belt.

solid, which could also be expressed as 9 double dominant:3 of each single dominant:1 double recessive. The separate gene pairs segregate independently for the two traits such that progeny may inherit any combination of alleles with an equal probability.

The inheritance of color illustrates the case of separate loci controlling the same general aspect of the phenotype. Coat color is an example of a "trait" that is actually a composite of several genetically controlled traits. An illustration is coat color in Doberman dogs. Two color patterns are most familiar, the black-and-rust and the red-and-rust. In fact, this "color" can be subdivided into color and pattern. The rust markings on legs, muzzle, eyebrows, and chest are a *fixed* pattern in the breed; that is, they are always present, with only minor variations. A study of color inheritance in this breed is therefore concerned only with the major body color, either black or red. In addition, there are two rarer colors, blue and fawn (or Isabella).

The inheritance of black and red has already been discussed. The blue and fawn are actually a dilute black and dilute red, respectively. Observations of the results of actual matings provide the information to deduce the combined mode of inheritance of the four colors.

Fawns mated inter se give only fawn offspring, and are suspected therefore of being the recessive homozygotes. Reds are known already to be *bb* homozygotes. When mated inter se, some reds can produce both red and fawn pups, and hence must be heterozygous for the allele pair influencing fawn color. Using *d* to represent the allele that dilutes color, and *D* its partner for full color,

$$\text{red} = bb\ D\_$$
$$\text{fawn} = bb\ dd$$

Reds that produce fawn pups must be *bb Dd*, purebreeding reds *bb DD*.

The data relating to matings of reds and fawns can be summarized as

Mating		Possible progeny (depending on mating)*	
		Red	**Fawn**
red × red	($bb\ D\_ \times bb\ D\_$)	x	x
red × fawn	($bb\ D\_ \times bb\ dd$)	x	x
fawn × fawn	($bb\ dd \times bb\ dd$)		x

* The phrase "depending on mating" is used to indicate that actual progeny results depend on the specific alleles transmitted by each parent. For example, some of the red × fawn matings will produce all red pups, some red to fawn in a 1:1 ratio, depending on the genotype of the red parent. Hence, ratios have little meaning in this type of table, which is based on phenotypes alone.

What prediction does the preceding information suggest for matings involving blacks and blues? Black color is known to be $B\_D\_$, and it is assumed that blue (dilute black) is $B\_\ dd$. Matings could give

Mating		Possible progeny (depending on mating)			
		Black	Blue	Red	Fawn
black × black	($B\_D\_ \times B\_D\_$)	x	x	x	x
black × blue	($B\_D\_ \times B\_\ dd$)	x	x	x	x
blue × blue	($B\_\ dd \times B\_\ dd$)		x		x

These are the outcomes observed over a number of matings. Again, the results of specific matings will depend on the genotypes involved. Because blacks are the result of two dominant alleles they potentially can produce pups of any color.

Using the preceding information, a complete table of possible mating outcomes can be compiled:

Mating		Possible progeny (depending on mating)			
		Black	Blue	Red	Fawn
black × black	($B\_D\_ \times B\_D\_$)	x	x	x	x
black × red	($B\_D\_ \times bb\ D\_$)	x	x	x	x
black × blue	($B\_D\_ \times B\_\ dd$)	x	x	x	x
black × fawn	($B\_D\_ \times bb\ dd$)	x	x	x	x
red × red	($bb\ D\_ \times bb\ D\_$)			x	x
red × blue	($bb\ D\_ \times B\_\ dd$)	x	x	x	x
red × fawn	($bb\ D\_ \times bb\ dd$)			x	x
blue × blue	($B\_\ dd \times B\_\ dd$)		x		x
blue × fawn	($B\_\ dd \times bb\ dd$)		x		x
fawn × fawn	($bb\ dd \times bb\ dd$)				x

Such a table permits the prediction of the possible outcomes of specific matings without knowledge of the actual parental genotypes. Observation of the pups born will allow possible parental genotypes to be narrowed down more specifically.

Suppose that Dobermans of a purebreeding black strain are mated to a purebreeding fawn strain; that is, each strain has been tested for several generations and never produces any other color offspring:

$$BB\ DD \text{ (black)} \times bb\ dd \text{ (fawn)}$$

This mating involving two gene pairs is called a *dihybrid cross.* The F_1 will be uniform, *Bb Dd* in genotype, and phenotypically black. When the F_1 reproduces, each gene pair will segregate yielding four possible types of gametes: *BD, Bd, bD,* and *bd* in a 1 : 1 : 1 : 1 ratio. Mating the two genotypes, *BB DD* and *bb dd,* gives the results outlined in Figure 2-5. The *Punnett Square* (named in

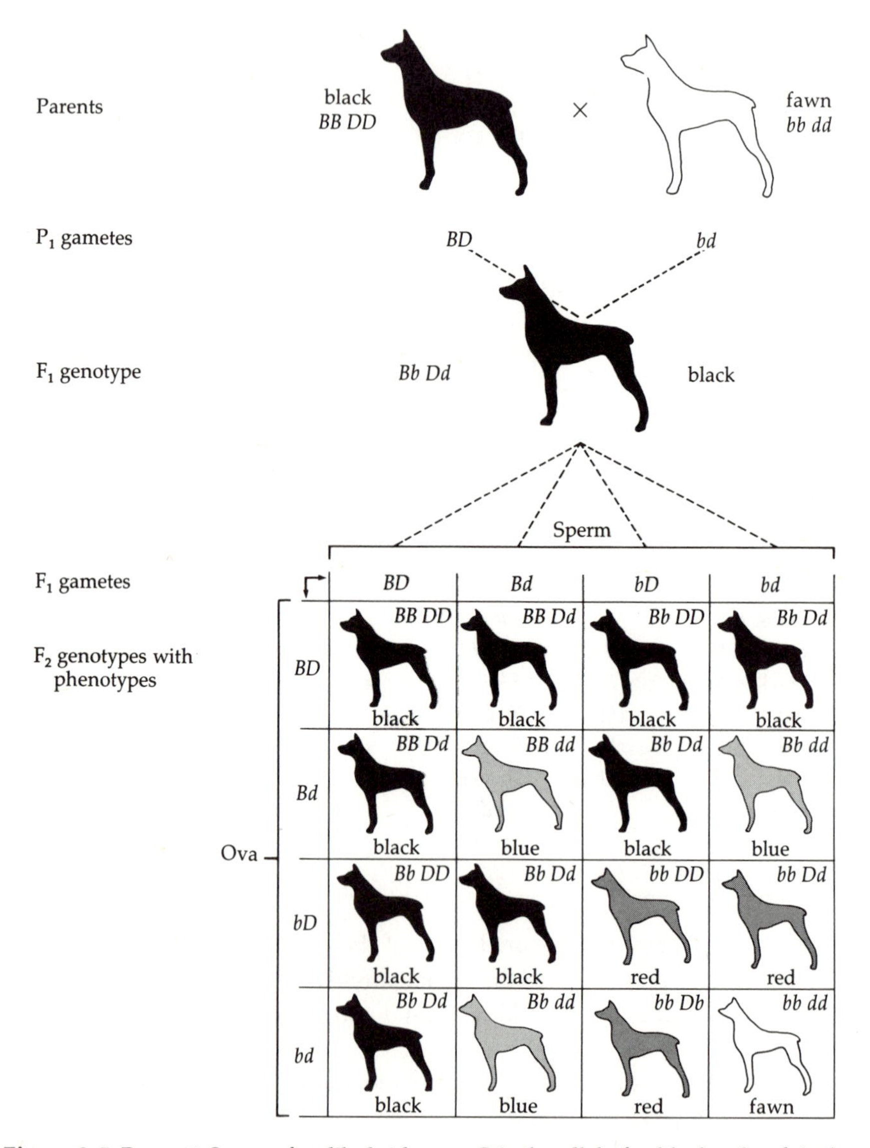

Figure 2-5 Punnett Square for dihybrid cross. *B* is the allele for black color, *b* is the allele for red, *D* is the allele for full color, and *d* is the allele that dilutes color.

honor of British geneticist R. C. Punnett) illustrated in Figure 2-5 is a diagrammatic representation of the union of gametes in all possible combinations, and is an alternative to the use of the algebraic derivation shown in Figure 2-4. There are nine possible F_2 genotypes but only four phenotypes. When large numbers of progeny are obtained, the F_2 ratio approximates 9 black : 3 red : 3 blue : 1 fawn, or 9 double dominants : 3 of each single dominant : 1 double recessive.

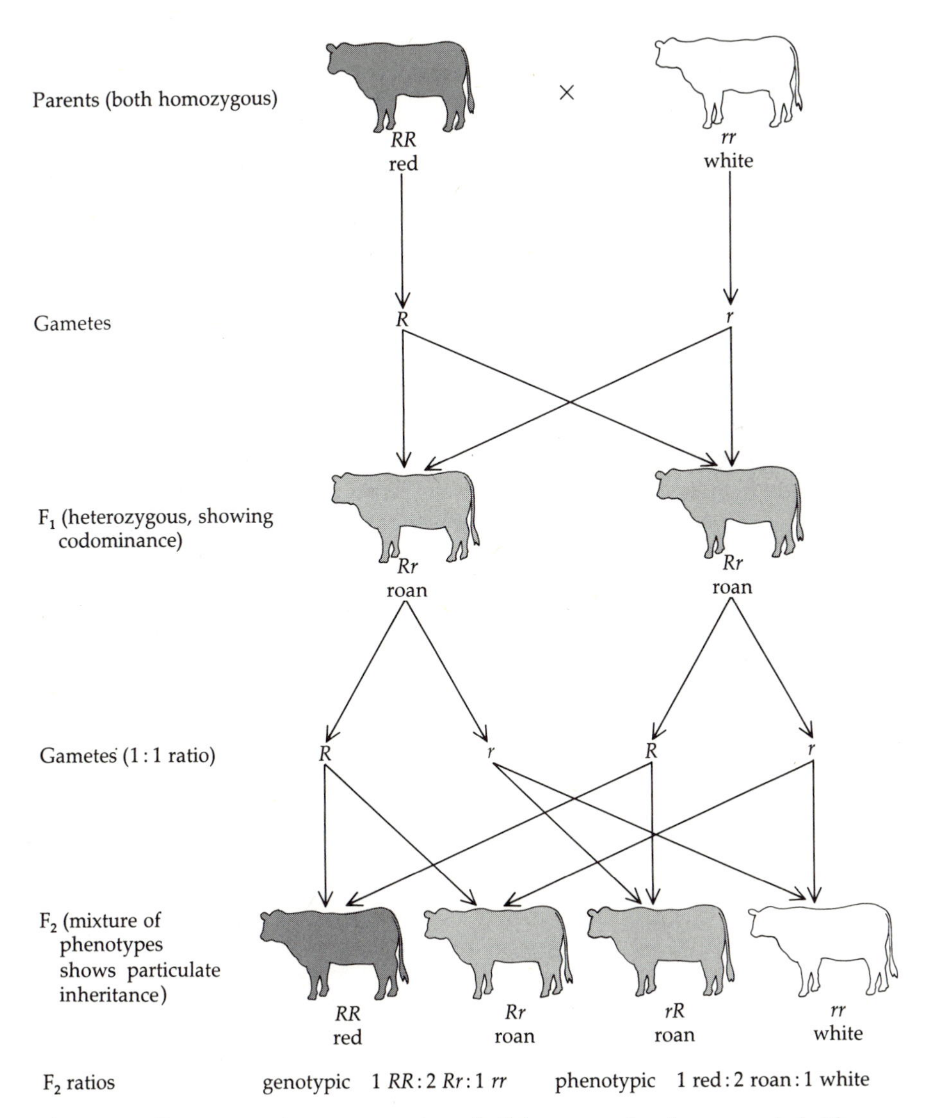

Figure 2-6 Diagrammatic representation of allele segregation in a monohybrid cross with codominance. *R* is the allele for red color, and *r* is the allele for white.

Mendel, of course, approached the problem from the opposite end. He started with the progeny results and, using his mathematical expertise, deduced the underlying behavior of the genes from the progeny ratios. He concluded that "the relation of each pair of different characters in hybrid union is *independent* of the other differences in the two original parental stocks" (Peters, 1959; italics added).

2-5 Codominance: An Apparent Exception to Mendelian Ratios

Had Mendel chosen to work with the color variants of Shorthorn cattle, he might have believed, at least for a time, in the blending theory of inheritance. Purebreeding red Shorthorns mated to purebreeding whites give a uniform F_1 that is unlike either parent. The F_1 is a red roan, which is most accurately described as a blending of red and white hairs in the coat. However, further study of the F_2 restores faith in Mendel's conclusions (see Figure 2-6). The F_2 progeny appear as 1 red : 2 roan : 1 white, indicating that this is not an exception to Mendel's law but rather a situation in which there is not complete dominance. The results illustrated in Figure 2-6 should be compared with those shown in Figure 2-3 for a trait showing dominance. The heterozygote has its own phenotype, different from either homozygote. This situation is known as *codominance*, no dominance, lack of dominance, or incomplete dominance. All these terms tend to be used interchangeably for situations in which both alleles express their effects on the phenotype, although more precise definitions will be found in the section on population genetics (Section 7-4).

The mating table for possible progeny from the various matings of red, roan, and white parents (that is, with no black allele present) is

Mating		Possible progeny (depending on mating)		
		Red	**Roan**	**White**
red × red	(*RR* × *RR*)	x		
red × roan	(*RR* × *Rr*)	x	x	
red × white	(*RR* × *rr*)		x	
roan × roan	(*Rr* × *Rr*)	x	x	x
roan × white	(*Rr* × *rr*)		x	x
white × white	(*rr* × *rr*)			x

2-6 Multiple Alleles

Thus far in this discussion, the existence of only two alternative alleles at each locus has been postulated. In fact, there may be many alleles at a single locus in a population of animals. However, each individual may have only one pair of genes at a locus and the individual must be either homozygous or heterozygous. An example of multiple alleles is found at the pattern locus referred to in dogs. Rust (also called tan) points and solid (also called self) color result from the action of two alleles at the pattern locus. Self color (A^s), or uniform dark pigment, is dominant to tan points (a^t). The third allele at the locus is a^y—restriction of dark pigment—which is dominant to a^t but recessive to A^s. In order of dominance, the alleles can be written as $A^s > a^y > a^t$. The a^y allele, when homozygous, gives a uniform tan color called sable by dog breeders (Searle, 1968). Note that the most dominant allele in the series is generally assigned the uppercase letter, with superscripts used to distinguish between alleles.

A phenotypic mating table can be built for a multiple allele situation just as it was for the dihybrid example of color in Dobermans. Because the allele for tan points is recessive to all the others, such dogs should be purebreeding when mated inter se. (Dobermans almost never produce offspring without tan points; the rare situation of a new mutation [see Chapter 3] would be the exception.) Matings between sable dogs and dogs with tan points should never give self-colored offspring, as neither parent has the A^s allele, the most dominant in the series. Matings involving self-colored dogs ($A^s\_$) potentially can give any kind of offspring, depending on the second allele involved in each parent. For example, if the parents were A^sa^y and A^sa^t, they could produce self ($A^s\_$) progeny and sable (a^ya^t) progeny, but none with tan points. Again, the table predicts all possible outcomes, and the results of actual matings will narrow down the possible genotypes of the parents.

		Possible progeny (depending on mating)		
Mating		Self	Sable	Tan points
self × self	($A^s\_ \times A^s\_$)	x	x	x
self × sable	($A^s\_ \times a^y\_$)	x	x	x
self × tan points	($A^s\_ \times a^ta^t$)	x	x	x
sable × sable	($a^y\_ \times a^y\_$)		x	x
sable × tan points	($a^y\_ \times a^ta^t$)		x	x
tan points × tan points	($a^ta^t \times a^ta^t$)			x

Table 2-1 Possible outcomes of matings involving the *C* multiple allelic series in cats

Mating		Possible progeny (depending on mating)					
		Color	Burmese	Siamese	Tonkanese	White	Albino
color × color	($C\_ \times C\_$)	x	x	x	x	x	x
color × Burmese	($C\_ \times c^b\_$)	x	x	x	x	x	x
color × Siamese	($C\_ \times c^s\_$)	x	x	x	x	x	x
color × Tonkanese	($C\_ \times c^sc^b$)	x	x	x	x		
color × white	($C\_ \times c^a\_$)	x	x	x		x	x
color × albino	($C\_ \times cc$)	x	x	x		x	x
Burmese × Burmese	($c^b\_ \times c^b\_$)		x			x	x
Burmese × Siamese	($c^b\_ \times c^s\_$)		x	x	x	x	x
Burmese × Tonkanese	($c^b\_ \times c^sc^b$)		x	x	x		
Burmese × white	($c^b\_ \times c^a\_$)		x			x	x
Burmese × albino	($c^b\_ \times cc$)		x			x	x
Siamese × Siamese	($c^s\_ \times c^s\_$)			x		x	x
Siamese × Tonkanese	($c^s\_ \times c^sc^b$)		x	x	x		
Siamese × white	($c^s\_ \times c^a\_$)			x		x	x
Siamese × albino	($c^s\_ \times cc$)			x		x	x
Tonkanese × Tonkanese	($c^sc^b \times c^sc^b$)		x	x	x		
Tonkanese × white	($c^sc^b \times c^a\_$)		x	x			
Tonkanese × albino	($c^sc^b \times cc$)		x	x			
white × white	($c^a\_ \times c^a\_$)					x	x
white × albino	($c^a\_ \times cc$)					x	x
albino × albino	($cc \times cc$)						x

Alleles in a multiple allelic series do not always fall into a neat pattern of dominance. Some alleles in the series can show complete dominance, others codominance, while others are fully recessive. The C series of alleles in cats is an example. Full color (C) is dominant to all other alleles in the series. The Burmese (c^b) and Siamese (c^s) alleles are recessive to C, but codominant to one another. The c^bc^s cat has a color pattern intermediate between the Burmese and Siamese, and is called a Tonkanese. The allele for blue-eyed white color (c^a) is recessive to C, c^b, and c^s, while the albino (c) allele is recessive to all others. Thus, $C > (c^b = c^s) > c^a > c$. Table 2-1 shows all possible mating outcomes. A useful exercise is to list the possible genotypes for some of these matings and verify the potential outcomes. A population may have many alleles at a locus, but any normal individual in that population will have at most two of these alleles as its gene pair.

2-7 Lethal Genotypes

Certain alleles prove fatal to the individual inheriting them in the homozygous condition. The loss of the offspring may occur at some time during gestation, at birth, or between birth and the offspring's reaching reproductive age. An example of early embryonic loss is associated with dominant white color in horses. The two alleles at this locus determine whether the horse will be white or colored. *Which* color would be specified by genes at other loci. Horses with the dominant white phenotype will produce two white foals for every one colored foal when mated inter se. White horses mated to colored (recessive) horses produce one colored foal for every white foal. Tabulated, the possibilities are

Mating		Possible progeny (depending on mating)	
		White	**Colored**
white × white	($W\_ \times W\_$)	x	x
white × colored	($W\_ \times ww$)	x	x
colored × colored	($ww \times ww$)		x

This illustrates a limitation in the use of such a table, as the indication of lethality comes not from the possible outcomes, but from the actual progeny ratios, these ratios being different from what would be expected with a pair of alleles and simple dominance.

The actual ratios indicate that all white horses are heterozygotes, showing the dominant phenotype, but carrying the recessive allele for color. Why then do they produce a progeny ratio of two white to one colored instead of 3 : 1? The conclusion is that one-third of the expected whites do not appear as live progeny. Since all test matings show that white animals breed as heterozygotes, the missing foals must be the homozygous whites. They are apparently lost in utero, so that the genotypic ratio of 1 : 2 : 1 is never expressed phenotypically. Figure 2-7 illustrates the outcome of the mating of heterozygous white horses.

The allele for dominant white is an example of what is called a *lethal* gene. In this case, the allele must exist in the homozygous condition before expressing its lethal effect. In a single dose, such genes may have a marked effect on the phenotype of the heterozygote as in the suppression of color formation in the white horse. The lethal genotype has a drastic effect on the development of the individual, causing considerable abnormalities and death of the embryo. Early embryonic losses mean that not all the results of conception are seen, so that a 2 : 1 phenotypic ratio of progeny occurs at birth from matings of heterozygotes for this type of allele.

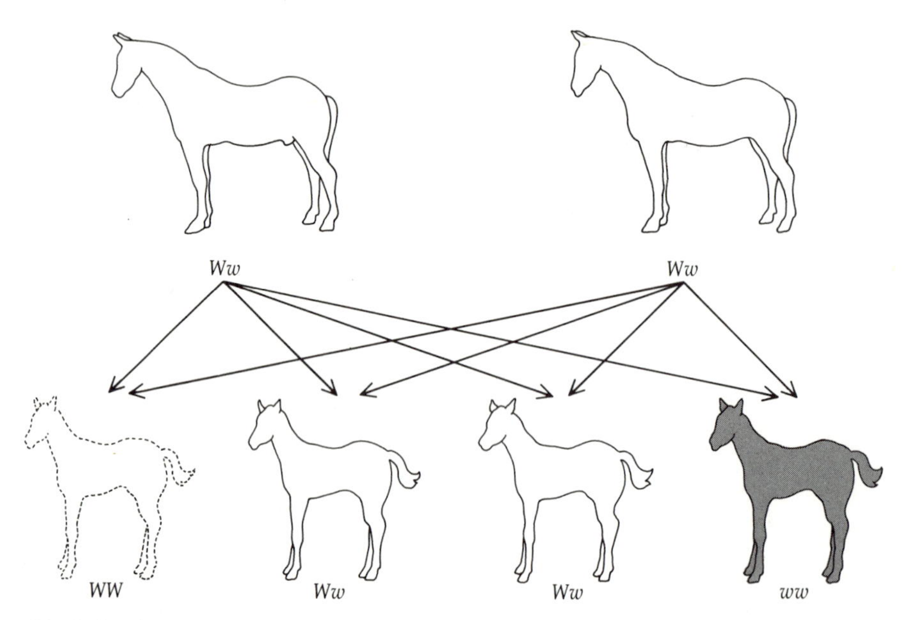

Figure 2-7 Outcome of mating of two dominant (lethal) white horses. *W* is the allele for white color, and *w* is the allele for colored.

Some lethal genotypes produce a conceptus that is delivered but which is so abnormal that it is born dead or dies soon after birth. The phenotypic ratio at birth from mating heterozygotes for such loci is 1 homozygous abnormal : 2 heterozygotes : 1 homozygous normal. An illustration of such a trait is the "amputated" lethal found in cattle and sheep. This allele affects development of the extremities such that the heterozygote has abnormally short legs. In cattle, the Dexter breed is such a heterozygote; in sheep, it is called the Ancon. The homozygous effect of this allele is to cause severe skeletal abnormalities that result in the calf or lamb being born dead with an almost legless (hence "amputated") phenotype. Some lethals are recessive, such that a 3 normal : 1 abnormal progeny ratio is produced instead of the 2 : 1 and 1 : 2 : 1 ratios discussed. If the recessive abnormality acts early in gestation, no abnormal offspring will be observed. The presence of the lethality must be inferred from the lowered fecundity of matings able to produce the lethal genotype.

Lethal genotypes are obviously undesirable in a population. How then does the breeder of dominant white horses or short-legged cattle approach the problem? Mating white with white or short with short yields two-thirds of the desired phenotype in the live progeny and one-third undesired, the undesired phenotype being colored or long-legged. One-quarter of all conceptions are wasted because the result is a dead fetus that received the homozygous abnormal genotype. The herd is more productive if all matings are between animals of the desired phenotype and those of the undesired, but normal, type. The phenotypic ratio of the progeny for these matings will be 1 desired : 1 undesired, but no dead progeny will be produced as a result of lethal homozygosity. In either case, the breeder must settle for a less-than-perfect situation, as the ideal is the heterozygote, which by definition is not a purebreeding type. It should be noted that a lethal dominant allele, denoted as A, will be eliminated from the population in the generation in which it occurs, as all $A\_$ individuals will die.

2-8 Testing for Carriers of Recessive Alleles

An animal heterozygous for a recessive allele is said to be a *carrier* of that allele. Detection of such carriers may be an important part of a breeding program. The horned versus polled, or hornless, trait in cattle, sheep, and goats is controlled by alleles at a single locus. Polledness is dominant to horns, so horned individuals are homozygous for a recessive allele and should be purebreeding. Polled individuals require a breeding test before their genotype is known with any degree of certainty. If the polled animal had a horned parent, it must be heterozygous, as the horned parent could transmit only the recessive allele for horns. To test the genotype of a polled animal that had two polled parents it could be mated with horned individuals. The first horned offspring produced would confirm that the polled animal transmitted the allele for horns and thus was heterozygous. If only polled offspring are born, how many must the breeder see to be reasonably confident that the polled individual does not carry the allele for horns?

If the animal is a carrier, one-half of its gametes will have the allele for horns and one-half the allele for polledness. Each progeny, therefore, has a 50 percent chance of receiving the allele for polledness from a carrier parent. The chance that two offspring will both receive the polled allele is $1/2 \times 1/2 = 1/4$. (Chapter 5 discusses the basic rules for determining probabilities.) Thus there is a 25 percent chance that a truly heterozygous parent would produce consecutively two polled offspring when mated with horned individuals. This probability decreases with each successive polled offspring:

$$\text{chance of 2 polled} = (1/2)^2 = 25 \text{ percent}$$
$$\text{chance of 4 polled} = (1/2)^4 = 6.25 \text{ percent}$$
$$\text{chance of 6 polled} = (1/2)^6 = 1.5 \text{ percent}$$

If the tested parent had really been heterozygous, and with six polled and no horned offspring produced by test matings, there is a less than 2 percent chance the heterozygosity would remain undetected. At this point, the breeder might conclude that this individual is most likely homozygous for the polled allele.

Chapter 8 discusses procedures for testing for carriers of recessive alleles in further detail.

2-9 Pleiotropic Effects

The polled versus horned condition in dairy goats provides an example of another genetic phenomenon known as pleiotropy. Dairy goats are most desirable without horns as they are less likely to cause injury to one another or their handlers. Therefore, it would be logical to prefer polled individuals in a breeding program to avoid the necessity of disbudding kids. However, matings between polled goats in many breeds produce an unexpected result, as shown in Figure 2-8. Male offspring are produced in a phenotypic ratio close to the Mendelian 3 : 1 ratio of polled to horned, indicating that the polled parents are, in fact, heterozygotes. However, there are fewer female progeny than the expected 50 percent, and their ratio is 2 polled : 1 horned. An extra class of offspring, polled intersexes, is produced in a number approximately equal to

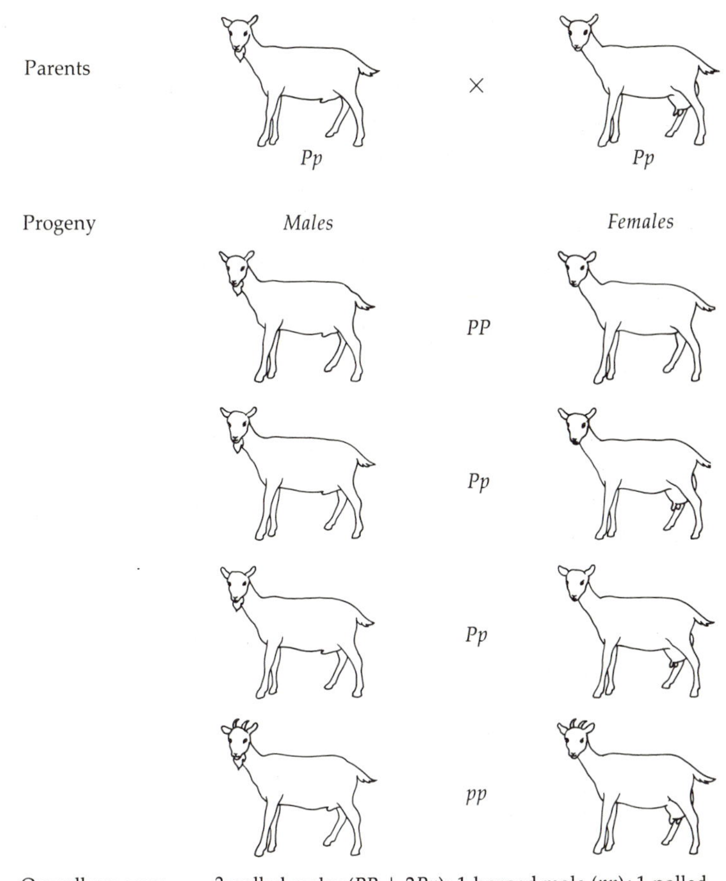

Figure 2-8 Outcome of mating of two heterozygous polled goats. *P* is the allele for polled, and *p* is the allele for horned.

that of horned females, that is, one-eighth of all progeny. An *intersex* is an animal that has phenotypic characteristics of both male and female but cannot function as either sex.

The explanation of the results is that the polled intersexes represent the missing females, and that homozygosity for the polled allele leads to abnormal development of the female, producing a sterile intersex. Thus, the phenotypic ratio of genetic females is 1 polled intersex : 2 polled females : 1 horned female.

This phenomenon, *pleiotropy,* is defined as a single allele influencing more than one distinct trait. Pleiotropy probably occurs more often than is realized since the product of a single gene is likely to affect more than one metabolic body process during development. Pleiotropic effects are probably associated with the lethal effects of homozygosity for the white color allele in horses, for example.

Purebreeding polled male goats are rarely found among the males remaining after selection in the breeding population. Therefore, it is suspected that the homozygous condition in males leads to abnormalities that cause total, or perhaps partial, sterility. The homozygous males are phenotypically normal, unlike the homozygous females, but they are reproductively abnormal and thus would soon be eliminated from a breeding program.

Polledness in goats is thus another situation where the desired phenotype is not purebreeding. An increase in the polled allele in a herd would also lead to increased fertility problems: matings between polled goats are matings of heterozygotes that produce an average of 25 percent infertile offspring. The breeding alternative is to mate polled with horned to give a progeny ratio of 1:1 polled to horned. The horned kids must then be disbudded. Neither solution is ideal for the breeder, whose decision on which program to use is based on the breeding objectives and a knowledge of the genetic consequences.

2-10 Variable Expressivity

Another example of a complication to the one-locus, two-allele situation is the "mulefoot" trait in swine and cattle. These two species are naturally cloven-hoofed, as compared to the solid hoof of equids. Hence the name given to this inherited condition, which gives affected animals hooves that resemble those of a mule. Breeding tests have shown that the gene causing mulefootedness is probably a dominant allele in swine but recessive in cattle. When an animal is referred to as mulefooted, what is meant exactly? Does the reference indicate one mule foot or more? Does this trait in fact vary in its expression? Breeding tests show that animals having the mulefoot genotype may indeed have from one to four affected feet. The *variable expressivity* of the same genotype is yet another complication in the study of inheritance. The question of how many feet are affected is really of secondary importance in this case, as the geneticist actually is interested in mulefoot or mulefeet versus normal cloven hooves.

2-11 Incomplete Penetrance

Variable expressivity of the genotype leads to confusion in the study of the inheritance of such a trait since the phenotype is not consistently indicative of the genotype. In the previous example, cattle with the recessive homozygous genotype did show some degree of mulefootedness. In contrast, there are some traits where the genotype for the trait may not be expressed in the phenotype at all; the animal would appear as a normal, rather than an affected, individual. An example is polydactyly in the fowl. A polydactylous bird has an extra digit on the foot. The allele for polydactyly usually behaves in a dominant manner in breeding tests. A dominant allele should express itself in the phenotype of every bird that has it. However, examples have been found of birds that produce almost all polydactylous progeny when mated to normal hens, indicating that they are homozygous for this allele even though their own feet are perfectly normal. The gene for polydactyly is indeed dominant over the gene for normal feet, but it is said to have *incomplete penetrance;* that is, the genotype for polydactyly is not always expressed in the phenotype. The percent penetrance can be calculated using the progeny of matings which should result in 100 percent polydactylous offspring, and is defined as the percent of individuals with the genotype for the trait that actually show the trait in their phenotypes. For example, if, from a mating of a homozygous polydactylous male to various females, 70 out of 100 offspring actually show polydactyly in their phenotypes, penetrance of the allele is said to be 70 percent. To add to the confusion, polydactyly also tends to have variable expressivity.

Incomplete penetrance is possibly the result of some kind of "threshold" effect, where the product of the gene has to build up to a certain point before it can affect the development of the trait. Another theory is that it is due to some delay in the switching on or off of the gene so that the gene is not active at the time when the trait is sensitive to change. Whatever the cause, incomplete penetrance further complicates the work of the animal breeder who seeks to understand the inheritance of a trait in order to develop an effective selection program for the desired phenotype.

2-12 Variations in Dihybrid Ratios

Lethality

The complications of codominance, lethality, variable expressivity, and incomplete penetrance also may affect the outcome of dihybrid crosses. If one trait has a lethal associated with it, then any phenotype having the lethal genotypic combination would disappear from the F_2, thus changing the ratio observed. In the preceding example of white horses, *WW* is lethal. An allele for curly hair (*C*) exists in horses, and is dominant to the allele (*c*) for normal straight hair. Mating of heterozygous *Ww Cc* parents gives ratios in the progeny of 2 *Ww* : 1 *ww* for the color locus, and 1*CC* : 2 *Cc* : 1 *cc* for the hair type locus. These ratios, when multiplied to find the overall genotypic ratio, give

2 *Ww CC*	white, curly
4 *Ww Cc*	white, curly
2 *Ww cc*	white, straight
1 *ww CC*	colored, curly
2 *ww Cc*	colored, curly
1 *ww cc*	colored, straight

which condenses to the phenotypic ratio

6 white, curly
3 colored, curly
2 white, straight
1 colored, straight

Such an F_2 ratio may be initially confusing. However, if it is broken down into the constituent loci, the results are more readily interpreted. Looking at the color locus first, there are 8 white : 4 colored, which is the 2 : 1 ratio that suggests lethality. The hair type locus gives 9 curly : 3 straight, or the 3 : 1 ratio typical of simple dominance. Mendel's second law suggests this treatment of loci independently of one another.

Epistasis

Epistasis is the term describing the relationships between loci. The genotype at one locus may influence the effect of the genotype at a separate locus and may actually prevent the expression of genes at that second locus in the phenotype. Following are examples of recessive and dominant epistasis, and epistasis with alleles that have mimic effects on the phenotype:

Recessive epistasis The albino phenotype is the result of a recessive homozygote which prevents pigment formation anywhere in the body. Hence, all other color loci of an albino individual will be prevented from being expressed in the phenotype. The albino animal will, however, still transmit alleles at all these loci to its progeny.

The albino (*c*) allele in cats has already been introduced, together with one of its alleles, *C*, for full color. Cats also have alleles at the *B* locus, with the *B* allele giving black color and *b* a chocolate brown color (called red in other species—the description of color is very subjective, another problem for the geneticist). The interaction of these loci is such that a cat must have the *C_* genotype to be colored, with the alleles at the *B* locus determining what the color will be. A tabulation of the possible outcomes of matings would show

Mating		Possible progeny (depending on mating)		
		Black	Brown	Albino
black × black	($C\_ B\_ \times C\_ B\_$)	x	x	x
black × brown	($C\_ B\_ \times C\_ bb$)	x	x	x
black × albino	($C\_ B\_ \times cc\_\_$)	x	x	x
brown × brown	($C\_ bb \times C\_ bb$)		x	x
brown × albino	($C\_ bb \times cc\_\_$)	x	x	x
albino × albino	($cc\_\_ \times cc\_\_$)			x

Albinos mated inter se can produce only albinos, as there is no *C* allele involved. However, in matings with other colors, albinos can produce a variety of offspring colors because they transmit alleles at the *B* locus, which are revealed once a *C* allele is introduced, from the mate's gamete, into a progeny genotype.

In general, the term dominance is used to refer to the masking effect of one allele over another at the same locus, while epistasis describes the relationship between combinations of alleles at different loci. The genotype that is epistatic may actually be recessive within its own locus—as in the case of the albino genotype—even though it masks the effects of alleles at other loci.

A mating between heterozygous (*Cc Bb*) cats would give an F_2 progeny ratio of

$$9 \text{ black } (C\_ B\_) : 3 \text{ brown } (C\_ bb) : 4 \text{ albino } (cc\_\_)$$

Note that this is not greatly different from the 9 : 3 : 3 : 1 ratio expected in the F_2 of a dihybrid cross where there are two loci segregating, with a simple dominant at each locus.

It is instructive to compare the tabulation of possible outcomes from the matings of black, brown, and albino cats with the tabulation earlier in the chapter of self, sable, and tan point dogs. The tables of all possible matings and their possible progeny phenotypes are almost identical in the two examples. There is only one mating that enables the geneticist to determine whether the inheritance of the trait is by way of a series of three alleles at one locus, or is controlled by two loci with two alleles at each. This mating involves the phenotype resulting from the postulated middle allele of the series with the recessive homozygote. For example, compare the results of the mating of brown cats to albino cats with those of mating sable dogs to tan point dogs.

Offspring phenotypes can only be the same as those of the parents if the colors result from a multiple allelic series. The appearance of offspring of three phenotypes indicates that the multiple allelic model is inadequate; however, this outcome is possible if two loci are involved as the genetic basis for the trait.

Dominant epistasis An example of dominant epistasis is that of white, black, and brown color in sheep. White is a dominant allele in sheep, as in horses, but

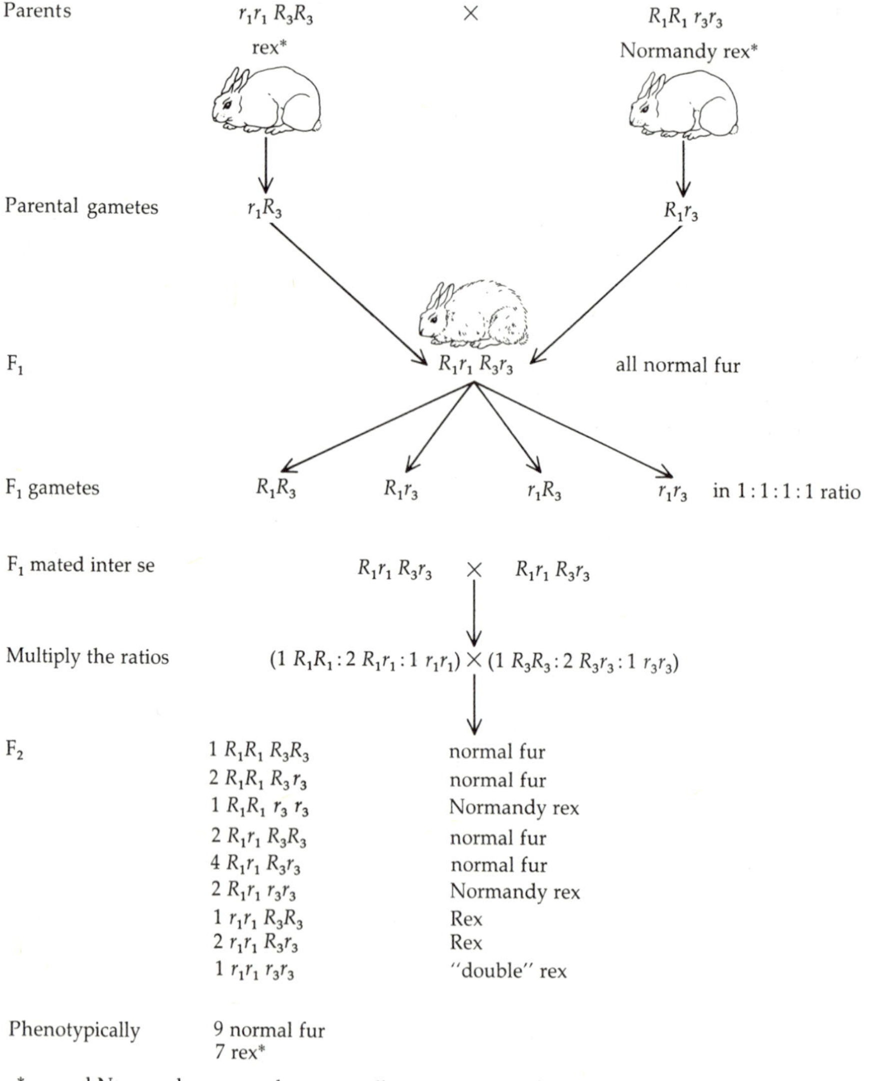

Figure 2-9 Mimic loci in rabbits—rex fur type.

is not associated with deleterious effects. Hence, both *WW* and *Ww* whites exist. Sheep that are *ww* are colored, and may then express the effect of alleles at the *B* locus. $B\_$ sheep are black, and *bb* sheep are brown.

Heterozygous sheep (*Bb Ww*) are white, and, when mated together, will give an F_2 progeny ratio of

$$12 \text{ white } (\_\,\_W\_) : 3 \text{ black } (B\_\ ww) : 1 \text{ brown } (bb\ ww)$$

Mimic effects Alleles at different loci can have the same effect on the phenotype; they are called *mimics.* There are variations in the actual gene action at each locus, as the following three examples show.

Example 2-1 Both alleles recessive; 9 : 7 F_2 ratio

Rex rabbits have a dense, velvetlike fur which breeding tests show to be recessive to normal fur. There are different rex breeds which were developed in different places. Interestingly, when rex rabbits of different breeds are mated with one another, the rex fur does not appear in the F_1. All the F_1 progeny have normal fur, despite the fact that rex is known to be recessive in each breed. This pattern of inheritance is a key indicator that there are two loci with recessive alleles at each that mimic one another's effect on the phenotype; that is, the alleles at different loci produce similar phenotypes. *R* will designate the normal fur allele, *r* the rex fur allele, and the subscripts, 1 and 3, will be used to indicate two different loci affecting fur type. The F_2 rabbits are produced in a ratio of 9 normal fur : 7 rex fur (see Figure 2-9). There are three ways to obtain rex fur in the F_2: with an r_1r_1 genotype, r_3r_3, or $r_1r_1\ r_3r_3$. Hence the observed ratio.

Assuming that rex rabbits of one breed do not have the rex allele of the other breed in their genotypes, the following mating tabulation can be made:

	Possible progeny (depending on mating)		
Mating	**Example**	**Rex**	**Normal**
rex × rex (of same breed)	$(r_1r_1\ R_3R_3 \times r_1r_1\ R_3R_3)$	x	
rex × rex (of other breed)	$(r_1r_1\ R_3R_3 \times R_1R_1\ r_3r_3)$		x
rex × normal (of same breed)	$(r_1r_1\ R_3R_3 \times R_1\_\ R_3R_3)$	x	x
rex × normal (of other breed)	$(r_1r_1\ R_3R_3 \times R_1R_1\ R_3\_)$		x
normal × normal (of same breed)	$(R_1\_\ R_3R_3 \times R_1\_\ R_3R_3)$	x	x
normal × normal (of other breed)	$(R_1\_\ R_3R_3 \times R_1R_1\ R_3\_)$		x

---

Example 2-2 Both alleles dominant; 15 : 1 F_2 ratio

Both Hereford and Simmental cattle have a characteristic white facial color. Either breed, when mated to Angus cattle that are self color (that is, no white markings), produces an F_1 that is white-faced. Hence, the whiteface pattern appears to be dominant over self color. However, the white facial areas of the progeny of Hereford crossed with Angus are larger than those on the progeny of Simmental crossed with Angus. Thus, the suspicion arises that the white pattern may be genetically different in the two breeds. Progeny of crosses of Hereford with Simmental are white-faced. When these F_1 Hereford and Simmental crosses are mated to Angus—a test cross situation, as self is recessive—the progeny appear in a ratio close to 3 white-faced to 1 self. Hence, it is certain that the two whiteface patterns are genetically distinct, as the expectation from the test cross would be all white-faced calves if only one locus with two alleles controlled this trait. Multiple alleles at one locus do not account for the test cross results, as the F_1 heterozygote would have carried both white-face alleles, and all their test cross progeny would have had one or the other whiteface pattern. How do the data fit the theory that there are dominant alleles as two loci for whiteface pattern?

Purebreeding Hereford cattle can be represented as *HH ss*, and Simmental as *hh SS*, where *H* represents the dominant allele for the Hereford face pattern and *S* the dominant allele for the Simmental pattern. Their F_1 cross would then be *Hh Ss* with a "double" whiteface because of the presence of both dominant alleles. The genotype of Angus cattle must be *hh ss*. The test cross of F_1 with Angus is therefore

$$Hh\ Ss \times hh\ ss$$

The progeny will be

1 *Hh Ss*	double whiteface (like F_1)
1 *Hh ss*	Hereford whiteface
1 *hh Ss*	Simmental whiteface
1 *hh ss*	self color (no white face)

This 3 whiteface : 1 self color ratio is the clue that there are two loci, each with a dominant allele.

When F_1 cattle (*Hh Ss*) are mated together, the F_2 phenotypic ratio will be 9 double whiteface : 3 Hereford type : 3 Simmental type : 1 self color; that is, 15 whiteface : 1 self color.

Example 2-3 Dominant at one locus crossed with recessive at other; 13:3 F_2 ratio

Chickens of several breeds have a white color. White color is interesting to work with, as it can have many genetic bases. White Leghorns mated to white Wyandottes give an F_1 that is predominantly white, although F_1 birds do show a few flecks of color. The whites in the two breeds are known to be different genetically because they give different results when mated to birds of a colored breed. White Leghorns give white progeny flecked with a little color, indicating that their white allele behaves as a dominant. White Wyandottes give colored progeny in matings with a colored breed, indicating a recessive white allele. The Wyandotte white is recessive to color (C), so Wyandottes can be represented as *cc*. Is the Leghorn white an allele at the same locus? The F_1 from Leghorn mated with Wyandotte is an all white bird with a few flecks of color. This could support the theory of an allele at the same locus, although the appearance of the flecks of color is disturbing. The F_2 generation should be all whites of one kind or another if the theory is correct. Figure 2-10 shows the actual ratio of 13 white birds, some with color flecks, to 3 colored birds. The single locus theory must be discarded since it does not allow for the colored birds. Instead, the possibility of a second locus should be considered. The recessive (Wyandotte) white allele is at one locus, and the dominant (Leghorn) white allele at another. The mating of heterozygotes should give a ratio of 3 Leghorn whites : 1 Leghorn colored at one locus and 3 Wyandotte colored : 1 Wyandotte white at the other. Multiplying these:

(3 Leghorn white : 1 color) × (3 color : 1 Wyandotte white) gives

9 Leghorn white/color
3 Leghorn white/Wyandotte white
3 color/color
1 color/Wyandotte white

White at either locus overrides the color development in the plumage, so 13 of these birds will be white to every 3 that are colored, the observed 13 : 3 F_2 ratio.

If the observer singles out the white birds with flecks of color, the ratio is seen to be 7 pure white : 6 flecked white : 1 colored. Figure 2-10 shows the origin of the flecked birds, which are like the F_1. They are in fact the double heterozygotes *Ii Cc*, where *I* represents the dominant pigment inhibitor allele that produces white in the Leghorn breed. These birds can form pigment ($C\_$), although the *I* allele almost completely shuts down pigment production. Because its epistatic effect is not quite complete in the heterozygous combination a little pigment if formed.

It must be remembered that alleles at different loci may produce genotypes that are phenotypically indistinguishable. The various genetic white colors found in many species are excellent examples. The white individuals all look alike to the observer; only breeding tests can reveal the true genetic situation.

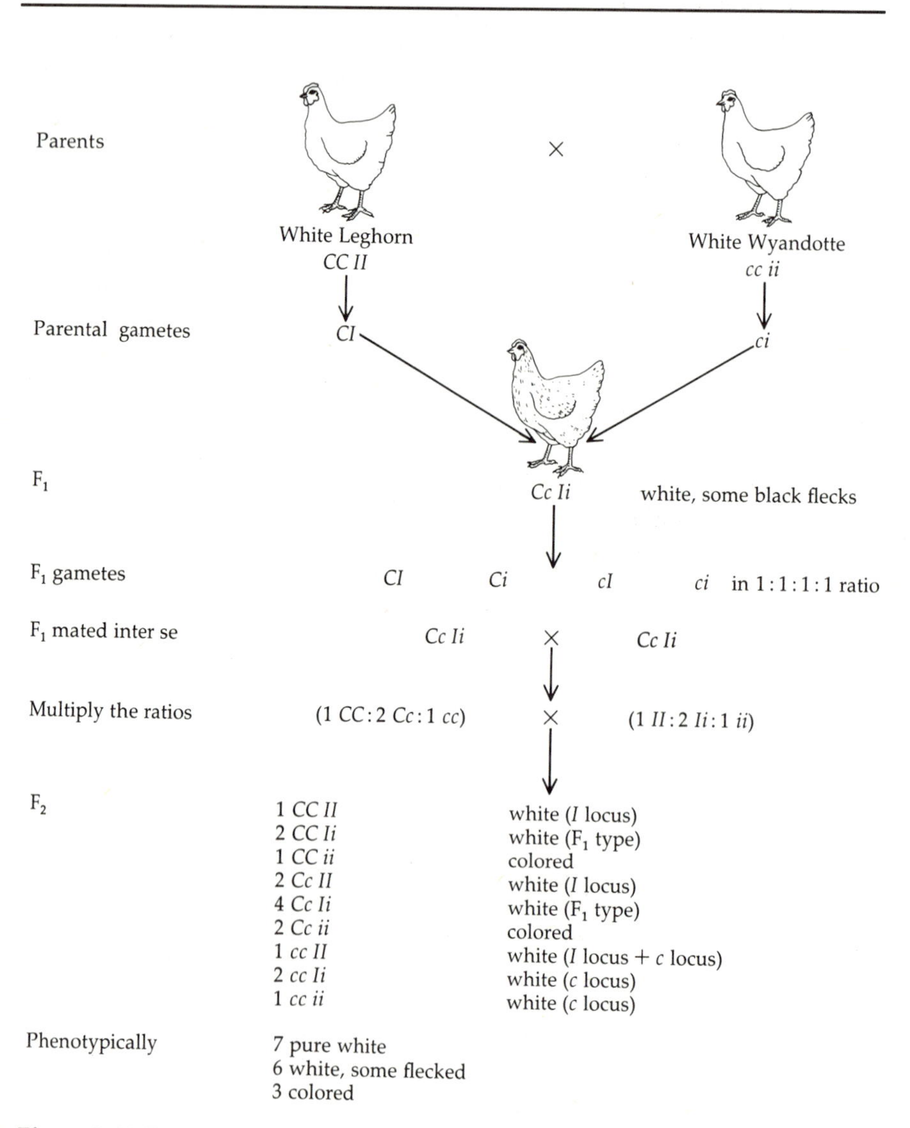

Figure 2-10 Epistatic white color in poultry.

Table 2-2 lists established examples of dihybrid F_2 ratios for a variety of species, which depend on various combinations of the types of gene action and locus interaction discussed in this chapter. Table 2-3 summarizes the different dihybrid F_2 ratios.

2-13 Modifying Genes

It is known that the spotting pattern found in Holstein cattle is determined by a recessive allele. This allele is fixed in the Holstein breed: all Holsteins are spotted. However, the spotting patterns of all Holsteins are not alike. Some Holsteins are almost all white, some almost all black, with the majority in between. Variation in white spotting is an example of genes at additional loci *modifying* the effect of genes at the first locus. One locus determines whether there will be spots or not; this is the *major* gene. Other loci determine how many and where on the body these spots will be; these are the modifying genes. There is a gradation of spottedness from one end of the scale to the other, which hints that there are not just one or two, but rather many loci determining the degree of spotting. The interplay of large numbers of loci on a single trait will be developed in depth in Chapter 11. Traits that are controlled by few loci, such that each allele has a marked effect on the phenotype and individuals can be phenotypically classified into one of a few groups, are termed *qualitative* traits. In cases where there are many loci, a gradation of phenotypes, and small effects of single alleles, the geneticist is working with *quantitative* traits. Most traits such as color, hair type, and major abnormalities are qualitative. Most production traits, such as milk yield, growth rate, and feed efficiency are quantitative in nature. For both types of traits, Mendel's laws of inheritance hold true.

2-14 Phenocopies and Environmental Effects

Not all changes in the phenotype are the result of genetic causes. The environment, for example, temperature, may influence the development of an embryo. If the environmental effect occurs at a stage of development that can also be affected by an allele causing a particular abnormality, then the phenotypic result mimics the effect of that allele. The result of an environmental influence which mimics the effect of a specific allele is referred to as a *phenocopy.* The phenocopy obviously is not transmitted to offspring in the way that a genetic

Table 2-2 Some variants and examples of the F_2 progeny ratio with two loci and two alleles* per locus

Mating	F_2 ratio	
1. dominant × dominant: (3 : 1) × (3 : 1)		
Dogs: black/red, nondilute/dilute	9 : 3 : 3 : 1	black : red : blue : fawn
Cattle: Hereford face/solid, Simmental face/solid	15 : 1	white face : solid face color
Chickens: inhibitor/colored, colored/white	13 : 3	white : colored
Swine: white/colored, black/red	12 : 3 : 1	white : black : red
2. dominant × lethal: (3 : 1) × (2 : 1)		
Horses: curly coat/straight hair, white/colored	6 : 3 : 2 : 1	white curly : colored curly : white straight : colored straight
Horses: Tobiano spots/no spots, white/colored	8 : 3 : 1	white : Tobiano : colored no spots
3. dominant × codominant: (3 : 1) × (1 : 2 : 1)		
Cattle: polled/horned, solid/roan/white	3 : 6 : 3 : 1 : 2 : 1	polled solid : polled roan : polled white : horned solid : horned roan : horned white
Chickens: inhibitor/colored, black/blue/white	13 : 2 : 1	white : blue : black
Cattle: black/red, solid/roan/white	3 : 6 : 4 : 1 : 2	black : blue roan : white : red : red roan
4. lethal × lethal: (2 : 1) × (2 : 1)		
Sheep: Ancon/normal leg, lethal grey/white	4 : 2 : 2 : 1	Ancon grey : Ancon white : normal grey : normal white

Mating	F_2 ratio	
5. lethal × codominant: (2 : 1) × (1 : 2 : 1)		
Cattle: Dexter/normal leg, red/roan/white	2 : 4 : 2 : 1 : 2 : 1	Dexter red : Dexter roan : Dexter white : normal red : normal roan : normal white
Horses: white/colored, cremello/palomino/chestnut	8 : 1 : 2 : 1	white : cremello : palomino : chestnut
6. codominant × codominant: (1 : 2 : 1) × (1 : 2 : 1)		
Chickens: black/blue/white, frizzle/moderate frizzle/normal feather	1 : 2 : 1 : 2 : 4 : 2 : 1 : 2 : 1	black frizzle : black moderate frizzle : black normal : blue frizzle : blue moderate frizzle : blue normal : white frizzle : white moderate frizzle : white normal
7. recessive × recessive: (3 : 1) × (3 : 1)		
Rabbits: normal fur/rex_1, normal fur/rex_3	9 : 7	normal fur : rex fur (either kind)
8. dominant × recessive: (3 : 1) × (3 : 1)		
Cats: black/brown, colored/albino	9 : 3 : 4	black : brown : albino

* The dominant allele is listed first, and the recessive second. In cases of codominance, the heterozygote is in the middle.

Table 2-3 Summary of gene action and locus interaction resulting in the F_2 ratios discussed in Chapter 2

F_2 ratio:	F_2 ratio by locus	Types of allele	Locus interaction	Page reference
9:3:3:1	(3:1) × (3:1)	Simple dominant at each	None	13
6:3:2:1	(2:1) × (3:1)	Lethal at one locus, simple dominant at other locus	None	27
9:3:4	(3:1) × (3:1)	Simple dominant at both	Recessive epistasis from one locus	28
12:3:1	(3:1) × (3:1)	Simple dominant at both	Dominant epistasis from one locus	30
9:7	(3:1) × (3:1)	Simple dominant at each, recessives mimic	None	31
15:1	(3:1) × (3:1)	Simple dominant at each, dominants mimic	None	32
13:3	(3:1) × (3:1)	Simple dominant at each, dominant at one locus mimics recessive at other	None	33

abnormality is. A phenocopy is a one-time event affecting the phenotype but not causing any transmissible change in the genotype.

The environment influences the phenotype for many traits, and it may be considered to consist not only of external factors such as temperature, but also of internal factors influenced by genes at other loci, such as the sex of the individual. Subsequent chapters discuss the concept of environmental influences on the phenotype in more detail.

2-15 Practical Implications

This chapter has expanded on the basic foundation of Mendelian genetics to include many of the complexities beyond the single locus, simple dominance situation. The student of genetics now has the tools required to attempt to analyze the modes of inheritance for many traits controlled by a few loci with major effects. Using illustrations drawn from a variety of species, the application of these tools has been demonstrated.

In general, the approach to a study of the inheritance of a new trait should be made in a stepwise, organized fashion. The steps should be

1. data accumulation;
2. hypothesis development, always starting with the assumption that simplest is best;
3. plan for further matings, with prediction of results based on hypothesis;
4. more data accumulation;
5. hypothesis testing; and
6. hypothesis acceptance or rejection. Rejection indicates the development of a new hypothesis, returning then to step 3.

The tabulation of results is a major step in understanding their meaning. Tabulation based solely on phenotypes of the parents involved in a mating and of the progeny produced is extremely useful in the initial stages of a study. It enables the geneticist to identify recessive phenotypes and begin to assign genotypes. Possible genotypes for other phenotypes can be developed along with the genetic hypothesis. Once the geneticist has established possible genotypes, predictions of expected progeny ratios from specific matings can be made and tested. The appearance of progeny phenotypes not predicted by the hypothesis invalidates the hypothesis, as do ratios that deviate excessively from expectations.

2-16 Summary

Mendel's work on the inheritance of traits in crosses of purebreeding varieties of garden peas resulted in the two basic Mendelian laws of inheritance. Mendel's first law states that genes are particulate and exist in pairs in each individual. Members of a gene pair separate into the gametes produced by that individual, such that half the gametes carry one member of the gene pair, and half carry the other.

Mendel's second law applies to the inheritance of characters controlled by separate gene pairs. It states that genes controlling separate traits segregate independently; that is, segregation at one locus does not influence segregation at another.

Gene action within a single locus may be of several types:

1. simple dominance, where *AA* and *Aa* show the dominant phenotype and *aa* the recessive;
2. codominance, where *AA*, *Aa* and *aa* will have distinct phenotypes;
3. dominance, with or without codominance, within a series of multiple alleles at one locus;
4. lethal, in which one or more of the genotypes—*A*_, *AA*, or *aa*—may be lethal. Offspring with the lethal genotype may die at any stage during gestation or between birth and reproductive age. The stage at which death occurs may influence the observed progeny ratio. The dominant lethal *A*_ is eliminated from a population as soon as it appears, as all individuals with the *A* allele will die;
5. pleiotropic, in which the effect of a certain genotype influences more than one observable trait in the individual;
6. variable in expression, in which the genotype for a trait is expressed to a greater or lesser degree in the phenotype;
7. incomplete penetrance in which the genotype for a trait may not always show its expression in the phenotype.

Testing procedures for detecting carriers of undesired recessive alleles in a population exist and are discussed in detail in Chapter 8.

Interactions between loci can influence the expected F_2 progeny ratio in two ways:

1. dominant epistasis, where, for example, the A_ genotype conceals the effect of all alleles at a second locus; or
2. recessive epistasis, where the *aa* genotype conceals the effect of all alleles at the other locus.

Alleles at one locus may produce an effect on a characteristic similar to alleles at a second locus; that is, they mimic one another. Mimic alleles will also give rise to variants of the F_2 progeny ratio:

1. recessive mimics, 9:7 F_2 ratio;
2. dominant mimics, 15:1 F_2 ratio; and
3. dominant at one locus mimics recessive at other, 13:3 F_2 ratio.

Table 2-2 lists some established examples of F_2 progeny ratios from a variety of species and Table 2-3 presents a summary of all F_2 ratio variations discussed in the chapter. Minor genes that modify the effect of major genes provide an introduction to quantitative (continuous) inheritance. Most traits considered in Chapter 2 are qualitative (discontinuous) in nature.

The phenotype may be influenced by the environment as well as by the genotype of the individual. Phenocopies occur when an environmental effect mimics that of a particular gene not possessed by the affected individual.

References

Peters, J. A., ed. 1959. *Classic Papers in Genetics.* Prentice-Hall, Englewood Cliffs, N.J.

Searle, A. G. 1968. *Comparative Genetics of Coat Colour in Mammals.* Logos Press, distributed by Academic Press, New York.

CHAPTER 3

Chromosomes and Their Behavior

3-1 The Discovery of Chromosomes

Mendel had no knowledge of the physical nature of the genetic material when, in 1865, he promulgated his laws which were based on years of data collection and analysis. The scientific world's intense interest in the recent publications of Charles Darwin was at least partly to blame for the oblivion into which Mendel's report disappeared. By 1900, when Mendel's results were rediscovered, scientists had succeeded in observing the inner workings of the cell, although they did not understand the functions of the structures they saw through the microscope.

W. Flemming, in 1880, observed a collection of threadlike bodies within the nucleus of the cell. He had prepared his slides using a stain which was taken up by these intranuclear structures. Because of their brightly stained appearance, Flemming assigned these structures the collective name *chromatin.* In 1888, W. Waldeyer renamed them *chromosomes.*

In 1908, Walter S. Sutton, in an elegant argument, identified the chromosomes as the bearers of the hereditary material. The deductive process by which Sutton pieced together the available evidence was:

1. Only sperm and eggs link one generation to the next, hence they must carry all the information needed to produce the new individual.
2. Sperm cells are composed almost totally of nuclear material, yet they contribute as much to the next generation as does the egg with its mass of cytoplasm. Hence, the nucleus must be the location of the genetic material.

3. Chromosomes have been observed to show precise division behavior, which is essential for the transmission of genetic information. They reside in the nucleus.
4. Both chromosomes and Mendelian factors occur in pairs.
5. Both chromosomes and Mendelian factors segregate into different cells at gametogenesis.
6. Chromosome segregation appears to be independent, meeting the requirements of Mendel's second law.

Taken together, these facts led Sutton to the conclusion that chromosomes are the location of genetic information.

3-2 Chromosomes of Domestic Species

The early studies of chromosomes were done on laboratory species, primarily insects. As cell preparation techniques became more sophisticated, researchers could observe the chromosomes of more species. In essence, cells must be arrested during their division cycles, stained with specific chemicals, and placed on a microscope slide such that the chromosomes of a single nucleus can be observed. (It is important in the slide preparation to avoid spreading out the cells to the extent that the chromosomes of one cell spill over and mix with those of another.) Each species requires a slightly different preparation technique, so it was some time before chromosomes had been studied in all common species. Not until 1956 did Tjio and Levan succeed in establishing the precise chromosome number of the human species. The illustration of a species' chromosome complement is called a *karyotype.* Figure 3-1 shows the karyotypes of a boar and a cockerel.

Table 3-1 lists the characteristic number of chromosomes for some familiar species. Chromosome number is not related to an organism's level on the evolutionary scale, as some relatively simple organisms have large numbers of chromosomes and others small numbers. Bacteria and viruses are exceptions to the rule that chromosomes exist in pairs; they generally have a single ring-shaped strand of genetic material.

3-3 Chromosomes and Sex Determination

As more species were studied and more karyotypes developed, researchers found a frequent exception to the rule that all chromosomes exist in matching pairs. Within each species, one sex has all matching pairs whereas the other sex has one chromosome pair that does not match. In mammalian species, the male

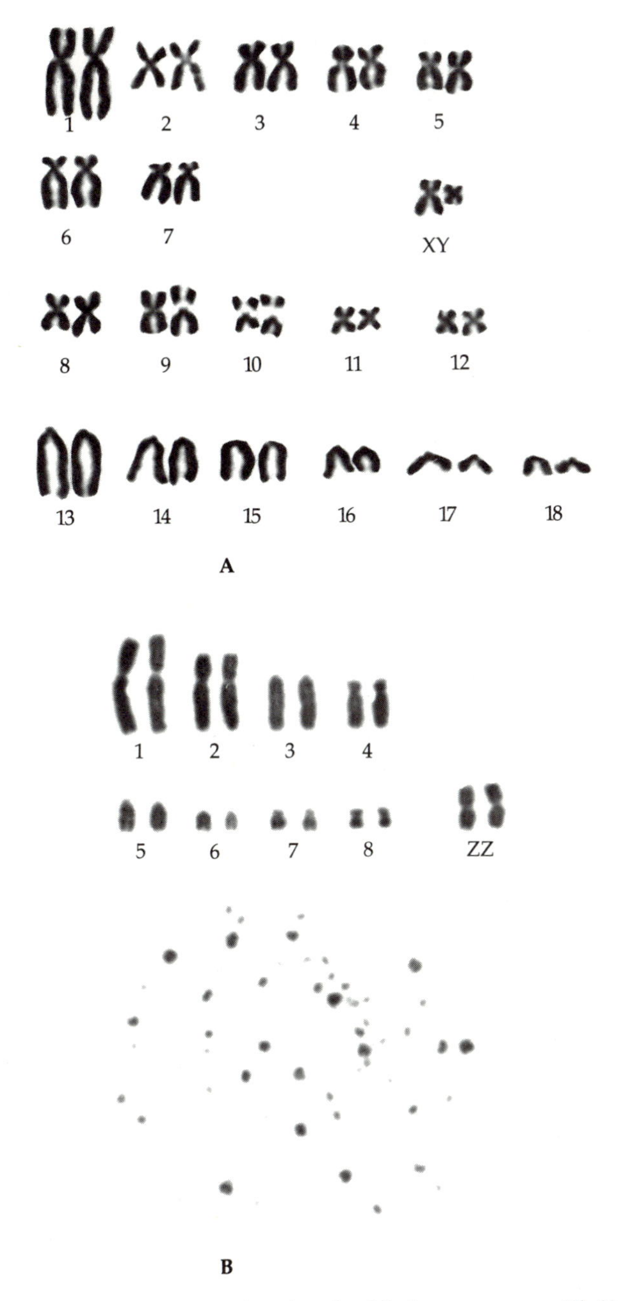

Figure 3-1 Karyotype of boar *(A)*, showing the 38 chromosomes with X and Y on the right end of the second row, and partial karyotype of cockerel *(B)*, depicting the 8 largest pairs of autosomes and the 2 set chromosomes. The 60 microchromosomes (shown unpaired) make up the full complement of 78 chromosomes. (Photographs courtesy of N. S. Fechheimer, Department of Dairy Science, Ohio State University.)

Table 3-1 Chromosome numbers of familiar species

Species	Diploid number of chromosomes
Swine	38
Cat	38
Rabbit	44
Human	46
Sheep	54
Goat	60
Cattle	60
Bison	60
Donkey	62
Horse	64
Dog	78
Chicken	78*
Duck	80*
Turkey	80*

* Avian species have a group of microchromosomes that are difficult to distinguish using normal karyotyping methods. However, researchers are extending their knowledge of avian species' chromosome complements using electron microscopy.

has an unmatched pair; in avian species, the female. There are some insect species where one chromosome exists as a singleton, lacking even an unequal partner with which to pair.

In 1896, E. B. Wilson had stated that "the determination of sex is not by inheritance, but by the combined effect of external conditions" such as temperature and nutrition. By 1909, he had reversed his stance completely and stated that sex was undoubtedly an inherited trait, like so many others. The correlation of the sex of individuals with the unmatched pair of chromosomes was convincing evidence.

In fact, the inheritance of sex is one instance of a trait controlled by such a large piece of genetic material that it can be observed through the microscope. As shown in Figure 3-2, for a mammalian species such as cattle, the ova of the female are all alike with respect to sex chromosomes. All her eggs contain one X chromosome, so called because of its shape in many species. She is therefore called the *homogametic* sex, gamete meaning "reproductive cell." The male is *heterogametic* in that half his sperm carry an X chromosome and half a Y

XX: homogametic sex
XY: heterogametic sex
Gametes and fertilization

	X	X
X	XX Female	XX Female
Y	XY Male	XY Male

Offspring ratio 1 male : 1 female

Figure 3-2 Sex determination in mammalian species.

chromosome, the Y chromosome often being small and having a different configuration. Hence, the male gamete determines the sex of each offspring; Y-bearing sperm will produce males, and X-bearing sperm females. The sex chromosome complement of females is designated XX, that of males is XY.

In mammalian species, an individual lacking a Y chromosome will develop as a female. The presence of a Y chromosome influences development as a male in a normal individual. However, this effect can be overridden, for example, by the gene for testicular feminization in humans. This gene causes a genetic male (XY) to develop the external genitalia of a female, such that only a karyotype will reveal that the individual is a male.

In some species, for example the familiar laboratory fruit fly Drosophila melanogaster, it is the *balance* between the sex chromosomes and the *autosomes*—the remainder of the chromosome complement—that determines sex. In this species, an XXY individual would be female (in humans, an abnormal male). An XO, where O denotes the lack of a second sex chromosome, would develop as a male (an abnormal female in humans). The section on the Imbalance of Sex Chromosomes, later in this chapter, explains how XO and XXY individuals can arise.

In avian species, in contrast to mammals, the male is homogametic and the female heterogametic. To avoid confusion, the sex chromosomes in birds are usually referred to as Z and W chromosomes, so that males are ZZ and females ZW. The W chromosome may be one of the larger chromosomes, or one of the microchromosomes characteristic of avian species.

Some exotic species have more unusual multiple sex chromosome arrangements, and many fish lack clearly defined sex chromosomes. They exhibit labile sex determination in that they can change functional sex during their lives. Such organisms are believed to have "sex genes" distributed on the autosomes instead of being located on specific sex chromosomes.

3-4 Chromosome Morphology

The naming of the sex chromosomes reflects their morphology in some species, although their shape varies with species. The remainder of the chromosome complement, the autosomes, also have characteristic morphologies that differ from species to species. The karyotype of each species shows this clearly as illustrated in Figure 3-1.

A karyotype illustrates the chromosomes during their division cycle. The cycle is arrested at the point at which the chromosomes have replicated themselves, but before the replicates—called *sister chromatids*—have split apart. They are still held together at the *centromere*, which is the constriction observed in each chromosome as pictured in the karyotype.

Chromosomes, like the X chromosome, which have the constriction located centrally are referred to as *metacentrics.* A *submetacentric* chromosome has the constriction located a little off center. The restriction is further off center in a *telocentric* chromosome, and an *acrocentric* has the constriction

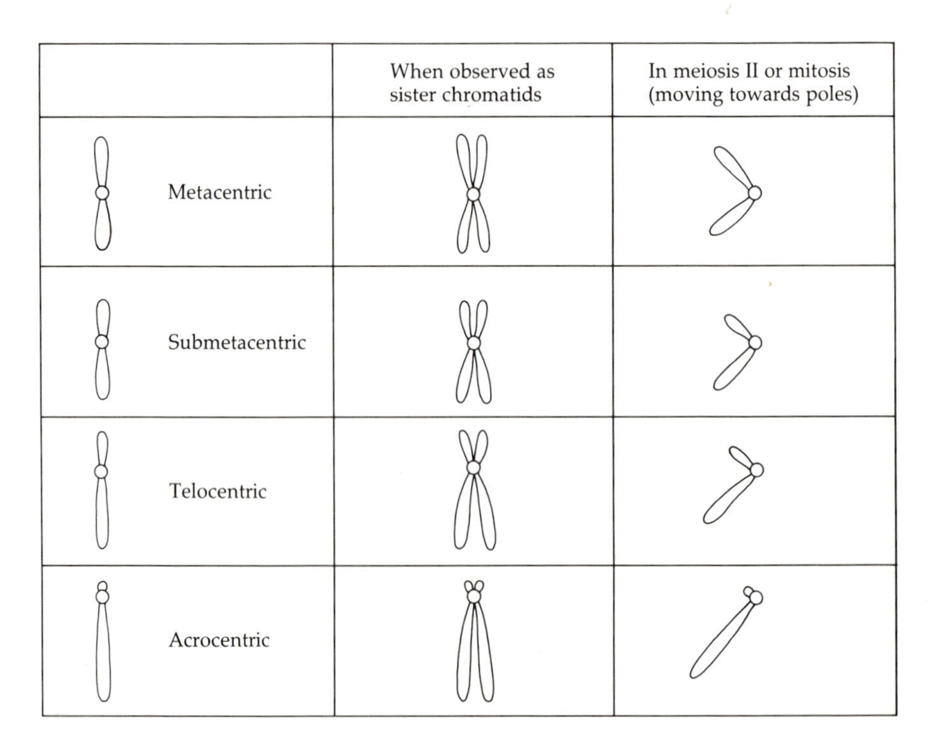

Figure 3-3 Chromosome morphology.

apparently in a terminal position on the chromosome; in fact, there is always a small piece of chromosomal material beyond this constriction. Figure 3-3 illustrates the various chromosome morphologies.

3-5 Somatic Cell Division—Mitosis

To comprehend fully the behavior of the genetic material, the student must understand the process of cell division. This process will be described first for the body, or *somatic,* cells, then for the reproductive cells formed during gametogenesis.

The entire process of *mitosis* is diagrammed in Figure 3-4. Essentially, mitosis is the ongoing process of body cells replicating themselves as part of normal growth and body maintenance.

During *interphase,* the resting phase of the cell and the longest phase of the cellular cycle, the chromosomes are diffuse and virtually invisible. They condense as cell division progresses, becoming clearly visible. By the time they can be observed, they have duplicated themselves and appear as a pair of sister chromatids held together at the, as yet unduplicated, centromere. The chromosomes line up on the equator of the cell, then the centromere divides and separates, allowing one of each pair of sister chromatids to move to each pole. At this stage, a "spindle" is observed in the cell, composed of *spindle fibers* that run from each centromere to a common point, called the *centriole,* one of which is found at each pole of the cell. The nuclear membrane has also disappeared by this point in the division cycle. The spindle fibers appear to draw the chromatids away from the equator and towards the poles of the cell, whereupon the fibers disappear, two nuclear membranes reform, and two daughter cells take the place of the parent cell.

3-6 Reproductive Cell Division—Meiosis

One additional step distinguishes reproductive division, or gametogenesis, from normal somatic cell division. This step is the process by which the diploid condition, maintained during normal mitosis, is changed into the haploid condition of the sperm or ovum. *Meiosis* is the process of reduction division, and, as diagrammed in Figure 3-5, is composed of two stages: *meiosis I,* which is the reduction division of diploid to haploid, and *meiosis II,* which is essentially a mitotic division occurring in a haploid cell.

In gamete formation, the chromosomes come out of interphase with their

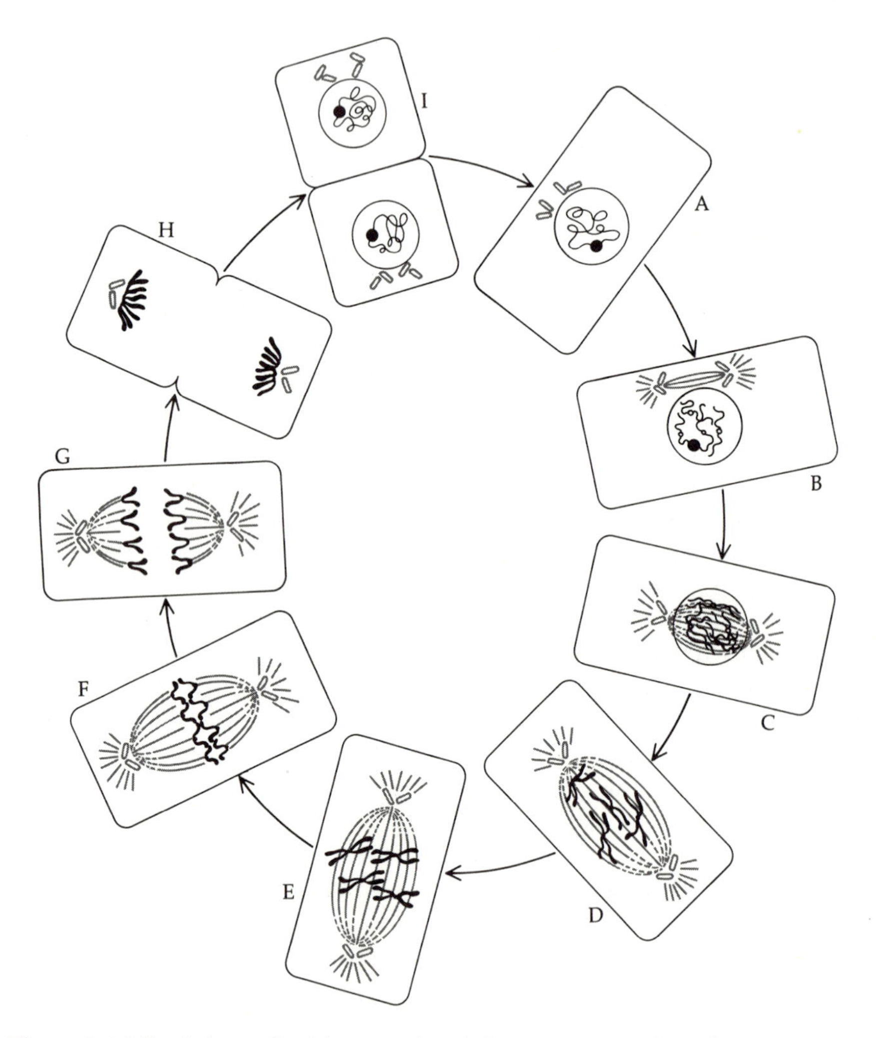

Figure 3-4 Mitosis in a cell with two pairs of chromosomes A, *interphase*: the resting nucleus, chromosomes hard to distinguish; B, C, D, *prophase*: chromosomes condensing, reveals that they have already duplicated themselves; E, *metaphase*: duplicated chromosomes line up on spindle; F, *anaphase* : duplicates (sister chromatids) separate, one to each pole of the cell; G, H, *telophase*: spindle fibers disappear, nuclear membranes reform, and new cell membrane defines the daughter cells; I, *interphase*: two new daughter cells. (After L. W. Sharp, *Fundamentals of Cytology*, McGraw-Hill, New York, 1943.)

duplication process completed so that they comprise pairs of sister chromatids. These line up on the equator of the cell in a manner different from that during mitosis. In meiosis I, the members of each chromosome pair line up alongside one another such that sister chromosomes line up together, their four chromatids forming what is known as a *tetrad* (see Figure 3-5). In the first meiotic division, the sister chromosomes separate into the two new cells being formed; that is, one member of each pair goes to each new cell. This is in contrast to

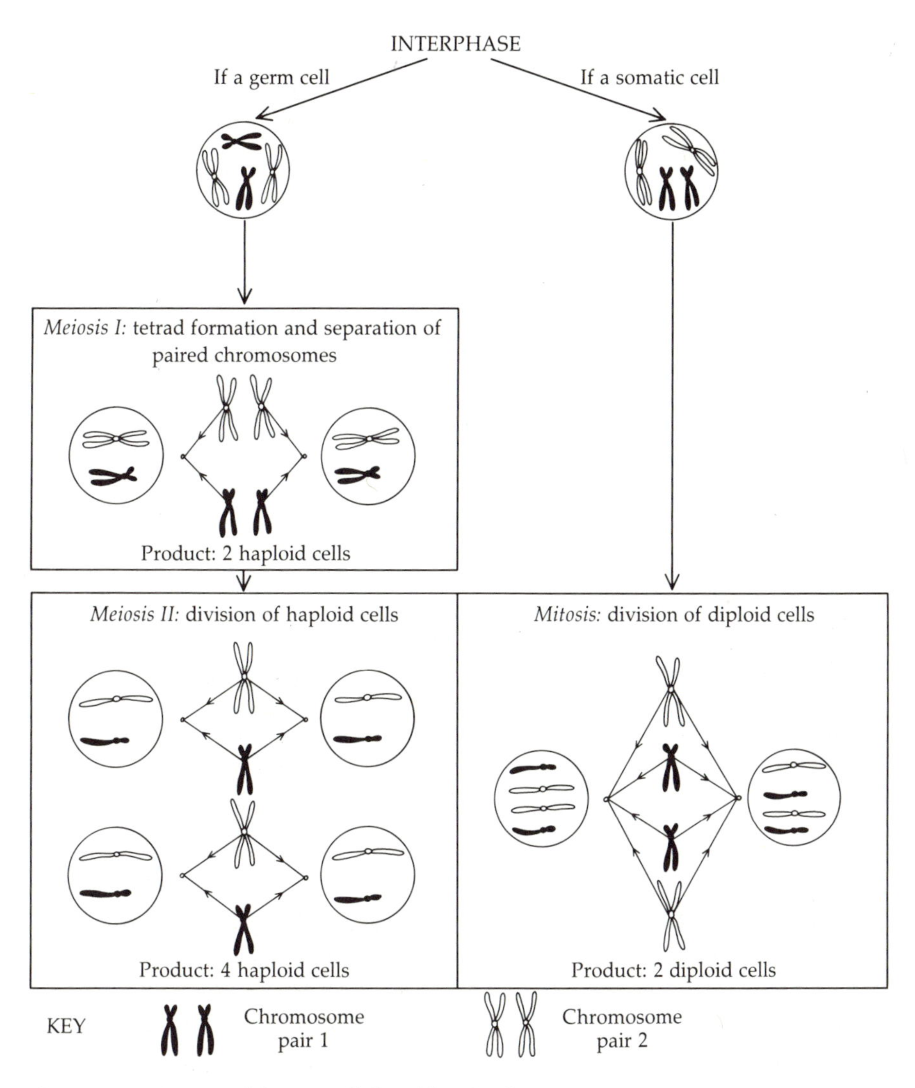

Figure 3-5 Meiosis and its parallels with mitosis.

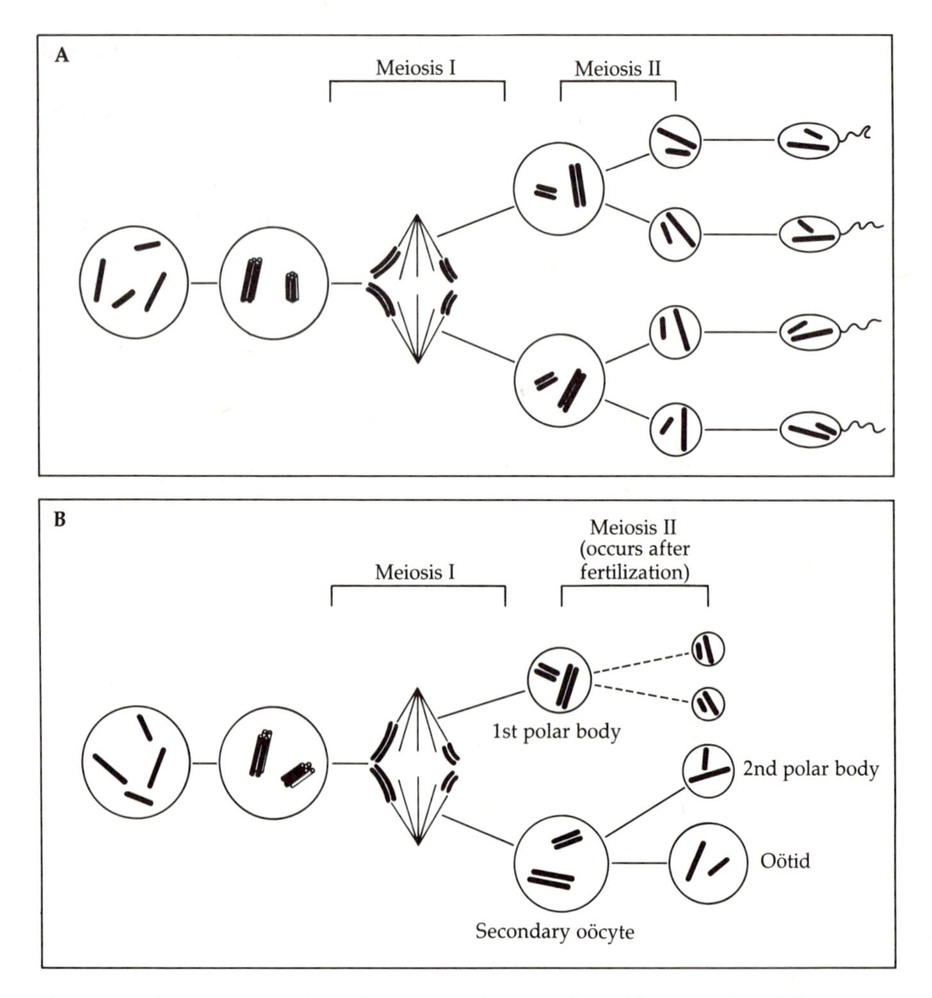

Figure 3-6 Spermatogenesis and oögenesis in a species with two pairs of chromosomes. *(A)* In the formation of sperm, duplicated members of each pair of chromosomes come to lie side by side in four-strand configurations. Two successive nuclear divisions then result in the formation of four sperm, each with one member of each pair of chromosomes. *(B)* Meiosis in a female animal gives rise to only one functional egg from each primary oöcyte. (After L. W. Sharp, *Introduction to Cytology,* McGraw-Hill, New York, 1934.)

mitosis, in which the chromatids separate. At the end of meiosis I the chromatids of each chromosome are still held together at the centromere, so the net result is a pair of cells, haploid in chromosome composition but with each chromosome still made up of a chromatid pair. The second meiotic division serves only to separate the chromatids, as occurs during mitotic division, the difference being that this is a division of haploid rather than diploid cells. The final product in meiosis is therefore a group of four cells, as compared to the two daughter cells produced during mitosis.

Figure 3-6 shows the difference between meiosis in the female, called *oögenesis,* and meiosis in the male, called *spermatogenesis.* In the male, four equal products, the *spermatozoa,* result from meiosis in the testes, and all of them function as reproductive cells. In the female, one product becomes the reproductive cell, the *ovum,* which contains most of the cytoplasm of the cell. The other three products of meiosis do not normally function as reproductive cells. The first meiotic division in the female is completed when the secondary oöcyte is shed from the ovary. The smaller product of this division is the first polar body, which usually degenerates, but which may go on to divide again while retained in the zona pellucida that surrounds the mammalian egg. The second meiotic division does not occur until after a sperm penetrates the egg. Following this division, the chromosomes of the sperm and egg coalesce, and the zygote begins its development.

3-7 Events at Fertilization

Fertilization is the process which restores the diploid constitution of the individual. The oöctye is penetrated by one sperm cell which stimulates it to undergo the second meiotic division (see Figure 3-6). The nuclei of the ovum and sperm then fuse to produce a zygote which then commences mitotic cell divisions to produce a new individual. The male and female gametes that actually produce the next generation are randomly determined from the gametes of each parent. Each gamete contains a random reassortment of the parental genes such that each fertilization event is virtually guaranteed to produce a unique individual.

In some species the female usually ovulates a single egg at a time and hence, should that ovum be fertilized, produces a single young. Humans, cattle, and horses are examples of such *monotocous* species. Other species, such as swine, dogs, and cats, are *polytocous,* producing litters of young at each birth. In these species, the female routinely ovulates a number of ova, and the reproductive system is capable of nourishing several growing embryos at once.

Multiple births can occur in monotocous species and are referred to as twins, triplets, quadruplets, and so on. If these multiple births result from the ovulation and fertilization of several ova, the resulting progeny are no more alike genetically than are brothers and sisters produced at successive births, except that they share a common prenatal environment which may make them appear more alike. Twins of this kind are called *fraternal* or *dizygotic* twins. Occasionally, *identical* twins appear. They are also called *monozygous* twins, which indicates their origin from the division of a single zygote to produce two genetically identical individuals. Differences in their environment will cause them to differ to some extent, especially for the more complex quantitative traits. Identical twins are useful in studying the effects of genetics and environment, as genotypes are the same in these rare pairs of individuals. The armadillo cooperates mightily in the study of the same genotype in different environments in that it routinely produces identical quadruplets (octuplets in some species) at a parturition. These quadruplets are a naturally occurring example of a *clone*, which is defined as a group of genetically identical individuals.

3-8 Errors of Cell Division

To transmit genetic information faithfully, meoisis must be a highly precise mechanism. Exact reproduction is essential to the functioning of the hereditary material. However, errors in the process of gametogenesis can occur, and the effects of such errors on the individual may sometimes be observed. Errors in mitosis also occur throughout an individual's life, but generally affect only one or two cells in the body. Errors in meiosis that involve entire chromosomes will be discussed first.

Polyploidy is the name given to the condition in which an individual has three or more complete sets of chromosomes, that is, sets in excess of the normal for a diploid species. If a zygote is produced from the union of a diploid gamete with a normal gamete, the result is a *triploid* individual, with three haploid sets of chromosomes. A *tetraploid* results when both paternal and maternal sets of chromosomes are duplicated. In plants, chemically induced polyploidy is a useful technique in the development of productive new varieties and hybrids. The tetraploid individual does not have the chromosome pairing problems found in, for example, triploids, as all the chromosomes in a tetraploid are able to pair. Gametogenesis can therefore proceed, and viable sperm or ova can be produced. Fertile hybrids between different species can be produced if they are tetraploids formed by the union of diploid gametes from both parent species. Normally, a hybrid between species is sterile because the disparity in chromosome number and conformation prevents pairing during gametogenesis. Polyploidy is, in general, lethal in animals, usually killing the embryo during development. Polyploidy therefore cannot be utilized in animals as it can in plants to produce viable interspecific hybrids.

In animals, it has sometimes been observed that complete chromosome sets are not duplicated, but that single chromosomes are overrepresented or underrepresented in the karyotype. The presence of extra chromosomes is called *polysomy*. If a chromosome lags behind during meiosis, it may be enclosed within the wrong daughter cell such that uneven products of meiosis occur. After fertilization, one daughter cell lacks this chromosome and is *monosomic,* the other has an extra copy of the chromosome and is *trisomic.* The deficient cell usually is not a viable zygote as the lack of so much genetic information is lethal. If the duplicated chromosome is a large one, its presence is also lethal. The sex chromosomes are exceptions to these rules. Trisomy of the smaller chromosomes is usually the only condition, other than polysomy of the sex chromosomes, observed in live progeny, but even this is usually associated with extensive phenotypic abnormalities. *Nullosomy* is the name of the lethal condition in which both members of a chromosome pair are lost.

3-9 Imbalance of the Sex Chromosomes

The existence of an extra X chromosome is an exception to the rule that an additional large chromosome is lethal. The X chromosome is a large chromosome, but individuals with such an imbalance are usually viable (although generally infertile). Figure 3-7 diagrams ways in which such a chromosomal imbalance can arise. An error during meiosis, giving rise to polysomy, is called *nondisjunction,* that is, the nonseparation of chromosomes. Nondisjunction was first discovered by C. B. Bridges in 1916 in his work with Drosophila eye colors and their relationship to sex inheritance. This occurrence provided convincing evidence that genes are indeed associated with the chromosomes.

The reason why an individual can tolerate the presence of an extra X chromosome, but not an additional large autosome, is directly related to the normal function of the sex chromosomes. The normal condition in the heterogametic sex is a gross imbalance in the genetic material on the sex chromosomes. The X chromosome is large and bears loci affecting many traits, while the Y chromosome is small, with few matching loci. The homogametic sex appears to have almost twice as many functional genes on its sex chromosomes as does the heterogametic sex.

Studies of the staining properties of male and female cell nuclei show that the homogametic sex normally has a densely staining section of material, called *sex chromatin,* which is lacking in the heterogametic sex. Certain techniques of cell preparation also reveal structures called *Barr bodies* within the nuclei. Their discoverer, M. Barr, working with cells from cats, was the first to note that their number is always one less than the number of X chromosomes. Hence, each cell nucleus in the heterogametic sex has no Barr body, and each nucleus of the homogametic sex has one Barr body. The presence or absence of Barr bodies can be used to determine the sex of embryos. An abnormal XXY

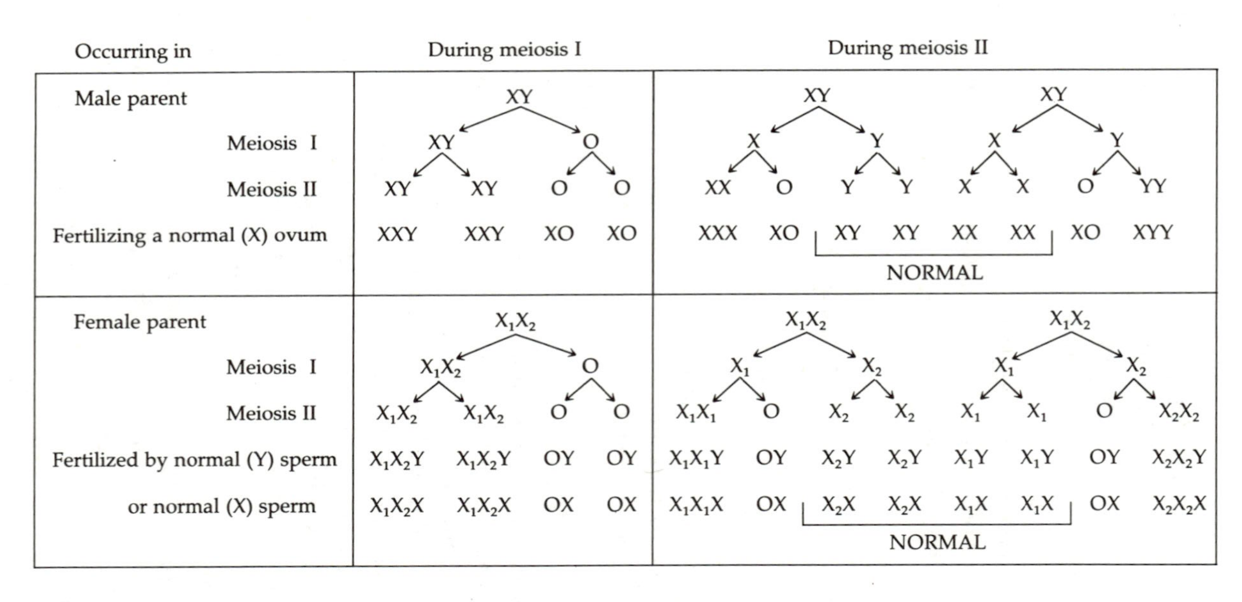

Figure 3-7 Ways in which nondisjunction of the sex chromosomes can occur. The abnormal products are: XYY is male, often subfertile or infertile; XXY is infertile male (Klinefelter syndrome in humans); XXX is female, often subfertile or infertile; XO is infertile female (Turner syndrome in humans); YO is lethal. These data are based on information, mostly from humans, regarding their frequencies of occurrence in live births. Frequencies range from $\frac{1}{250}$ for XYY to $\frac{1}{3000}$ for XO. [Note: If there are marker loci on the X-chromosome (for example, color genes such as black/orange in the cat) these may help to determine in which parent the nondisjunction occurred and whether it occurred at meiosis I or II.]

individual would have one Barr body in its cells. Accumulation of this type of evidence led to the formulation of an hypothesis by Mary Lyon in 1961, and independently by Liane Russell in the same year, which states that all X chromosomes in excess of one are switched off in each body cell. Thus, both the heterogametic and homogametic sexes have only one active X chromosome per cell. The phenomenon of the switching off of chromosomes is known as *dosage compensation,* which accounts for the tolerance for excess X chromosomes—all but one are switched off and only one per cell plays a part in cellular metabolism.

The switching off of X chromosomes occurs randomly at a certain point in development. One-half of the cells in the homogametic sex have one X active and one-half have the other X chromosome active, so the overall effect is similar to heterozygosity, although this form of heterozygosity differs from the intracellular heterozygosity of autosomal loci. An excellent example of the result of X chromosome inactivation is found in the color of the tortoiseshell cat. Black and orange are alleles at a locus on the X chromosome. The areas of black fur arise from cells in which the X chromosome carrying the orange allele was inactivated, and the orange areas from cells with inactivated black alleles. The calico cat, which is a white-spotted tortoiseshell, has a spotting gene that gives distinct areas of white and colored, either black or orange, fur. The spotting gene gives rise to areas of cell lines derived from the same origin, that is, from cells with one or other X chromosome inactivated early in development, as contrasted to the situation in nonspotted tortoiseshells where the switching off appears to occur later, creating a more diffuse pattern of black and orange.

The rare male tortoiseshell is an example of nondisjunction, as he has an XXY chromosome complement. He can survive as a result of the switching off of the additional X chromosome, but is infertile.

3-10 Gene Activity During Development

The switching off of an entire chromosome is an unusual event, serving the special purpose of balancing the male and female genotypes, that is, making them equivalent in terms of total active genes. At the gene level, different loci on the autosomes of each individual become inactive at different stages of development and in different cell lines. Development of the zygote into an embryo, through the fetal stage, and finally to a mature individual is a precisely orchestrated process. During development, highly specialized organs are produced which maintain their diverse functions throughout the life of the individual. Thus, the need for specific genes to be active is not constant through

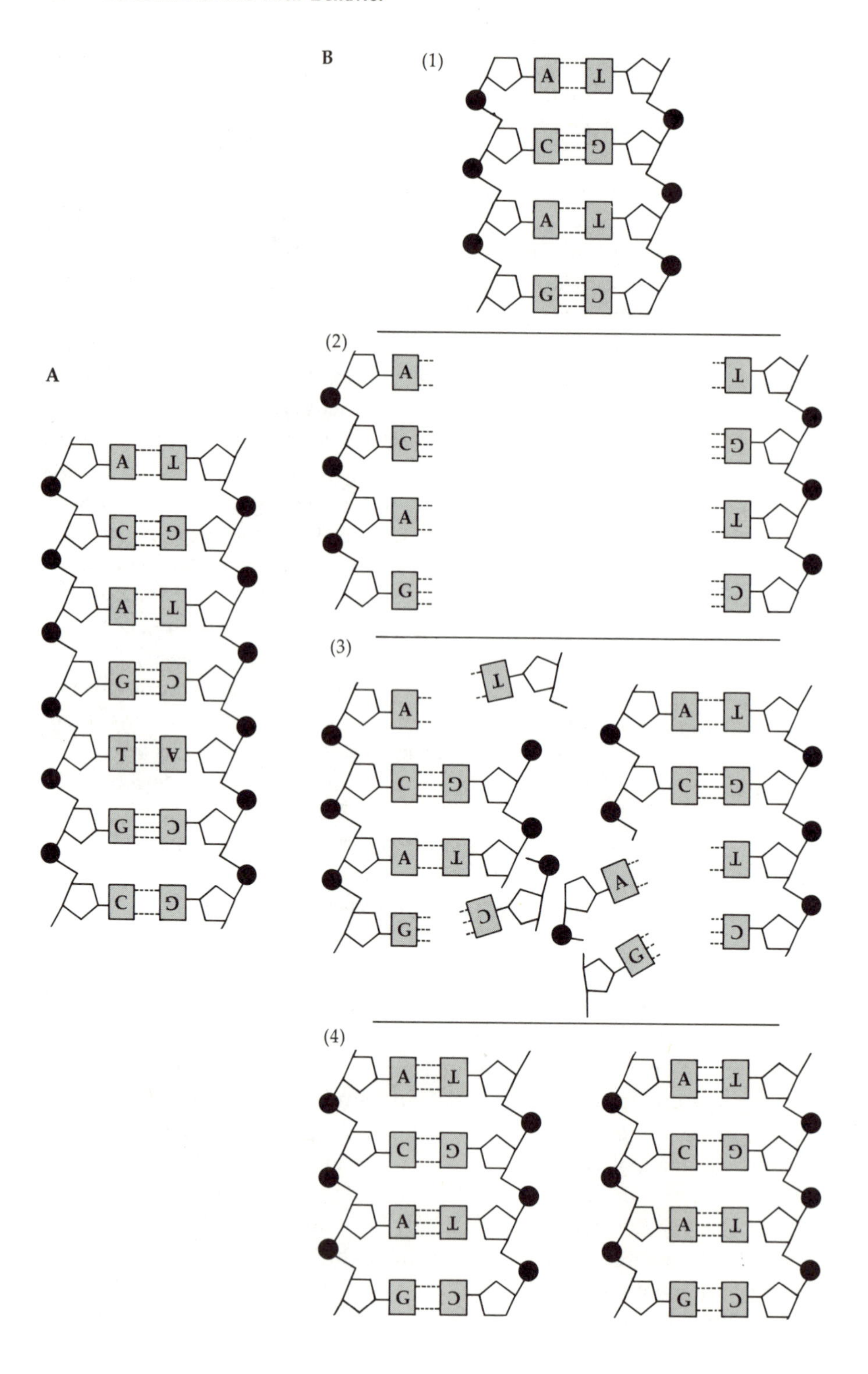
B
(1)
(2)
(3)
(4)
A

development and throughout life. For example, genes deciding how many digits an individual will have must act only at the stage of embryonic development when digit number is determined. At all other stages of the life cycle, these genes remain unused. The active genes in specialized cell populations are appropriate to the cellular function; for example, liver cells in the mature individual require a different set of functional genes than do kidney cells.

Much remains to be understood regarding the mechanisms that regulate genes during the development and life of an animal. Certainly much more genetic material than that on the additional X chromosome is quiescent during most of an animal's lifetime, but the details of the switching mechanism remain to be determined.

3-11 Chemical Structure of the Chromosomes and Mutation

Even for the purposes of animal breeding, some knowledge of chromosome structure and behavior is useful. Knowledge of the fine structure of chromosomes may prove important, although domestic animal breeding does not yet operate at this level.

The search for the chemical identity of the chromosomes culminated in 1953 when James Watson and Francis Crick determined that chromosomes were in fact a *double helical* structure of *deoxyribonucleic acid (DNA).* Figure 3-8 shows a diagrammatic representation of this molecule and its replication process. The bases in DNA, when read three at a time, form a code in which each unit specifies an amino acid or some kind of "punctuation," such as starting and end points. The DNA therefore acts as a template from which several kinds of *ribonucleic acid (RNA)* are synthesized. Messenger RNA (mRNA) leaves the nucleus and in turn serves as a template to put together, in concert

Figure 3-8 *(A)* DNA structure in diagrammatic form. G is guanine, C is cytosine, A is adenine, and T is thymine. The circles represent the phosphates; the pentagons, deoxyribose. The interrupted lines linking the bases (squares) indicate hydrogen bonds, three for GC pairs, two for AT pairs. *(B)* A schematic representation of DNA replication. The two strands of the DNA molecule (1) separate in the region undergoing replication (2). Free nucleotides pair with their appropriate partners and are linked together (3). The whole process proceeds in a zipper-like fashion until two complete DNA molecules are finally formed (4). The newly formed polynucleotide chains are indicated by shading. (After F. H. C. Crick, *Scientific American,* September 1957.)

with the ribosomal RNA (rRNA) and transfer RNA (tRNA) of the cell, amino acids in the sequence encoded. A series of bases therefore equates with a gene, and its product is a string of amino acids, that is, a polypeptide chain. Once released from the RNA template and the ribosome, the molecule takes up a three-dimensional structure, the configuration depending on attractions between its constituent amino acids, and may aggregate with other polypeptide chains. This three-dimensional form will decide the activity of the resultant protein in the metabolism of the cell.

The structure of DNA allows for precise replication, essential to inheritance of the genetic material, but also permits the occasional errors in duplication to be replicated. Replication of such errors is important in that it gives rise to genetic differences within a population. Without such variation, there is no raw material upon which selection can act.

Changes in the genetic information carried by a cell are called *mutations.* Mutations occur constantly in all cells of the body and are not, contrary to popular belief, rare events overall, although mutation rates for single loci are very low. Mutation is a point change in the genetic material which is then replicated during cell division and transmitted to all offspring which receive that particular chromosome. A mutation may result in a change in the phenotype, which is how mutations are eventually detected. Biochemical tests also reveal many mutations whose effects are not apparent in the appearance of the individual, but which act at the metabolic level.

Because the body is constantly exposed to environmental factors that cause mutation, such as ultraviolet light, chemicals, and radiation, evolution has provided a protective mechanism which repairs most mutations before they are replicated and transmitted to daughter cells. A system of enzymes makes up the repair mechanism, thus preventing the build-up of mutations in a cell which would eventually prove lethal.

An example of what happens when the repair mechanism fails to act is observed in the condition xeroderma pigmentosum in humans. This condition is controlled by an allele at a single autosomal locus. *DD* individuals are normal, *Dd* individuals have heavily freckled skin, while *dd* individuals have heavy freckling and skin that ulcerates where exposed to ultraviolet light, for example, on the face. The eventual result is cancer of these areas, which generally leads to the death of the affected individual. Xeroderma pigmentosum is thus a *semi-lethal* trait which is usually fatal before the individual reaches reproductive age. Studies have determined that the phenotypic effects of the *dd* genotype are caused by mutation build-up in the body cells, a result of the *dd* individual's lack of an enzyme that is essential for the mutation repair system.

3-12 Chiasmata and Crossing-over

During cell division, physical evidence for the cellular events that give rise to crossing-over between loci that are linked on the same chromosome (see Chapter 4) is observed. During the *diplotene* stage of meiosis I when the tetrad arrangement of chromatids occurs, cross-shaped figures called *chiasmata* (singular: chiasma) are seen (Figure 3-9). The sister chromatids are very closely

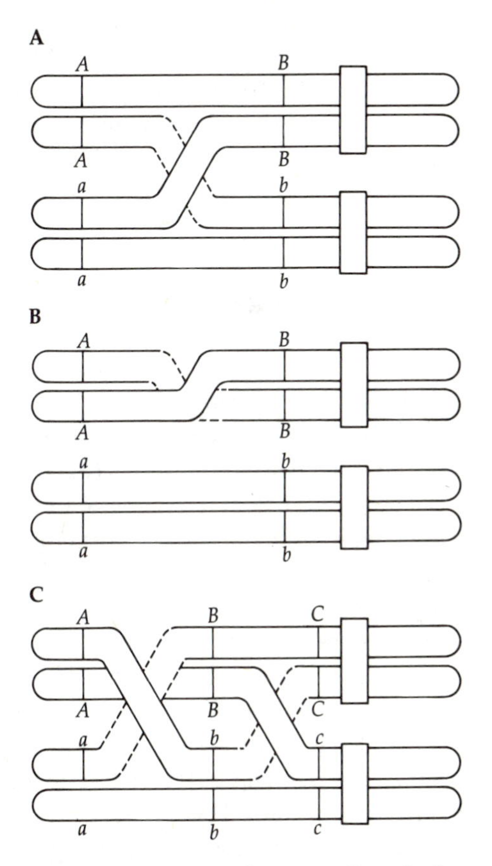

Figure 3-9 Diagram of a tetrad showing chiasmata. In a single crossover between chromosomes (panel *A*), *AB* and *ab* chromosomes have already duplicated. An exchange takes place at the chiasma between any two strands of the tetrad, resulting in four haploid cells of genetic constitution *AB, Ab, aB,* and *ab.* In a single crossover within a chromosome (panel *B*), the result is four haploid cells of genetic constitution *AB, AB, ab,* and *ab.* Thus, the crossing-over event is not detected because the products have the same genetic constitution as the original chromosome strands. In a double crossover between chromatids of two chromosomes (panel C) in which parentals are *ABC* and *abc,* the results are *AbC, ABc, aBC,* and *abc.*

associated and appear to have exchanged pairing partners along their length, giving rise to the observed chiasmata. In other words, each chromatid is paired with another, but this pairing partner is not the same member of the tetrad for all regions of each chromatid. Apparently breakage and reunion occur randomly at points along the chromatids, again mediated through the cellular enzyme system, such that new combinations of alleles result.

The precise mechanism of chiasmata formation may be a "copy choice" rather than a breakage process. In this theory, switching from one DNA strand as template to the other may occur as the DNA is replicated, resulting in a new DNA strand copied partly from one old strand and partly from the other at different points along its length. This theory does not account for the fact that all four tetrad members are intimately involved in the chiasmata. Both the copy choice and breakage theories have some inadequacies in explaining cellular events during crossing-over.

3-13 Chromosomal Rearrangements

Some chromosomal aberrations are caused by changes in the arrangements of the chromosome arms. If breakages occur in different chromosomes, and if the reunion ties together parts of *nonhomologous* or *heterologous* chromosomes (that is, not members of the same pair), the result is called a *translocation.* An equal transfer of material between such heterologous chromosomes is called a *reciprocal translocation.* Animals carrying the translocation chromosome may have lowered fertility because the translocation can interfere with the matching of pair members during meiosis I.

Inaccurate (nonreciprocal) repair of chromosome breakages may result in *duplication* or *deletion* of genetic material. The duplication may be a repeat of part of the chromosome's own genetic material if its source was a homologous chromosome, or may be a duplicate of the genetic material coming from a heterologous chromosome. The deletion chromosome is the matching chromosome produced during a nonreciprocal event that also gives rise to a duplication chromosome. One chromosome involved gains extra genetic material—the duplication—while the other chromosome in the event loses that portion of its chromosome arm—the deletion. Duplications and deletions vary in the severity of their effects on the phenotype, depending on the amount and content of the genetic material involved in the event.

An *inversion* may occur if there is more than one break in the same chromosome at the same time. The pieces may "flip over" before they are reattached such that the order of genes along part of the chromosome will

change. For example,

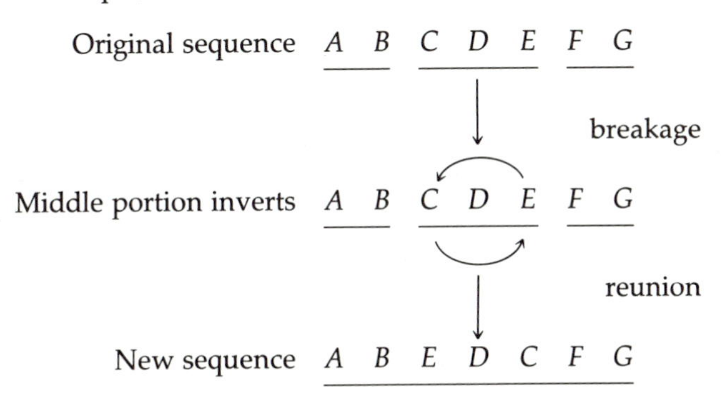

Inversions may interfere with pairing during meiosis, which may in turn reduce fertility in individuals carrying an inversion chromosome.

3-14 From Genotype to Phenotype

No gene affects the phenotype in an instantaneous fashion. Rather, a gene's effects are on aspects of the phenotype that are influenced by metabolic pathways in which the product of each particular allele is active. The base pairs in the DNA form the code for specific amino acids. Larger units of the chromosome, the series of base pairs, determine the order in which these amino acids will be linked together. The amino acid sequence in turn gives a product which assumes a particular three-dimensional configuration depending on the specific sequence of amino acids of which it is composed. The three-dimensional structure in turn determines the biochemical activity of that particular protein. More complex active proteins are produced when different protein chains link up together.

A single protein may be active in several metabolic pathways in the body. A change in the gene influencing the production of a protein will therefore affect all pathways in which that protein operates. The net result may be visible as a change in the phenotype. However, many mutations are not detected because they have only a minor effect on the protein structure. Therefore, knowledge of changes in the genotype depends on the observer's ability to detect their effect on the phenotype. This ability is augmented by chemical procedures such as electrophoresis, which gives a picture of the protein profile of the cell and can distinguish between different proteins in cells that appear alike. When an observable change occurs, it may be seen in a single trait or in several, that is, as a pleiotropic effect.

3-15 Practical Implications

The art of manipulating the organism at the chromosomal level is still in its infancy in animal breeding practice. Although major economic advantages have yet to be realized, some techniques have been developed that have enabled researchers to begin to explore this field. For example, the relatively new technique of embryo transfer does not manipulate the chromosomes per se, although it does implant entire genotypes into totally unrelated dams.

Manipulations performed on embryos can lead to a variety of products. With microsurgical techniques, complete nuclei can be transplanted into unrelated cytoplasm, resulting in a developing embryo whose intra- and extranuclear components are unrelated. Embryos early in development can be subdivided to give several embryos from one fertilized egg. In this way, identical offspring are produced: a clone of the original embryo. However, the cloning of a mature cow, for example, is a different and more formidable task. The differentiation and specialization of cells that occur during development make it exceedingly difficult to take any one cell from the adult animal and induce it to behave again as a recently fertilized egg. The total genetic information is present in the mature cell, but the mechanisms whereby genes are switched on and off in the sequence necessary to form a new individual must be understood.

Gene transplants have become feasible, although still difficult to achieve. Animal genes can be isolated and transplanted into bacteria in such a way that they will become active and produce their normal product. For example, when the mammalian gene that produces insulin is transplanted into a bacterium, the cells derived from that bacterium produce insulin in quantities that can be harvested and used to treat diabetes in humans. The logical ultimate step for gene transplants is the excision of genes that produce an inferior or faulty product, and their replacement with the desired gene. While not yet at this level, some transplants come close. A rat gene for growth hormone production has been transplanted into mice, which grow larger than their normal sibs. Some have transmitted their rat gene to their progeny. In humans, drug treatments are being developed to switch on a hemoglobin-producing gene that is normally switched off in adults to reduce the deleterious effects of the genetically transmitted conditions of sickle-cell anemia and thalassemia.

Selection programs for some species include chromosome studies. For example, routine cytological studies of Swedish Red and White cattle were instituted after a translocation associated with reduced fertility was found to be fairly widespread in the breed. Bulls for artificial insemination are routinely screened for this chromosomal abnormality, and are culled if they are found to carry the translocation.

The manipulations of genetic material in large animals are yet in their early stages, but hold great promise for the future.

3-16 Summary

Chromosomes were conclusively determined to be the carriers of Mendelian factors based on the facts that the nuclear material comprising the sperm contributes as much to the next generation as does the mass of the ovum; the chromosomes contained in the nucleus occur in pairs, as do Mendelian factors; and the chromosomes exhibit precise division behavior, segregating into the germ cells independently. Domestic species have characteristic chromosome numbers, which can be pictured in the karyotype. One pair of chromosomes is involved in sex determination, with one sex being homogametic, the other heterogametic.

The division of somatic cells is called mitosis and serves the function of cell duplication, with each cell inheriting a complete diploid set of chromosomes. Meiosis is the reduction division that takes place during gamete formation. The diploid chromosome complement becomes haploid during meiosis I, and meiosis II serves to separate duplicate chromosomes in the haploid cells, in the same manner as mitosis in somatic cells.

Errors may occur during cell division. Polyploidy is a duplication of entire chromosome sets over and above the normal diploid condition. Polysomy is an excess of just one chromosome of the set. Both conditions result in gross abnormalities, whereas the organism is generally able to survive an imbalance of the sex chromosomes. This is achieved through the process of inactivation of all X chromosomes except one in both normal cells and in cells with abnormal sex chromosome complements.

The process of mutation gives rise to genetic variation. Reassortment of genetic material takes place during crossing-over, which is related to the visible events of chiasmata during cell division. Larger reassortments result from chromosomal rearrangements such as translocations, duplications, deletions, and inversions. Their effects on the organism vary in severity.

References

Wilson, E. B. 1896. *The Cell in Development and Inheritance.* Macmillan, New York.

CHAPTER 4

Linkage and Sex-related Traits

4-1 Background

Enormous numbers of genes are contained within the genetic material of each individual. These genes are arranged on the physical units of inheritance, the chromosomes, as described in Chapter 3. The conclusion to be drawn from these facts is that each chromosome must carry genes at many loci, influencing a multitude of traits. Because such loci are physically linked together, they cannot be expected to obey Mendel's second law (the law of independent assortment) because the physical ligature between them would tend to cause the alleles to be inherited together rather than allowing them to reassort freely during gamete formation.

Chromosomes break and rejoin during the cycle of meiotic cell division, (Section 3-12). This breakage and reunion occur when chromosome pairs are lined up and before the pairs separate. The rejoining often occurs between lengths of separate chromatids, such that alleles at different loci may be reassorted during the tetrad phase of meiosis. Hence, when the chromosome pairs separate, they may carry with them new combinations of alleles along their arms. Logically, the new combinations must be represented in the progeny. Thus, linked traits do not segregate independently but neither are they tied together so tightly that new combinations are never seen.

The fraction of new combinations, called *recombinants,* appearing in each generation should indicate how closely the loci are linked together. The greater the distance between loci, the more chance there would be for a breakage and rejoining to occur and, therefore, the more recombinants that should be observed in the progeny. Conceivably, loci that are very far apart on the same chromosome would have so many breakages occurring between them that they would appear to be assorting independently. Alternatively, pairs of loci situated very closely together on the chromosome should have few intervening breakages, so new combinations would be hard to create and very few would be detected in the progeny.

The preceding paragraph describes the logical approach to linkage, which builds on what has been learned so far of Mendelism and the physical basis of heredity (Chapters 2 and 3). The question that must now be asked is, Do data on the inheritance of actual traits agree with these ideas?

4-2 Evidence of Linkage

The Tobiano pattern in horses is a white spotting pattern characterized by a predominantly colored head, white over the back, and white legs. The Tobiano pattern is determined by a single, autosomal dominant allele (T). As in other species, many blood proteins can be identified in horses and are often used to answer parentage questions. One of these blood constituents is serum protein albumin, which has variants controlled by alleles designated Alb^A, Alb^B, and Alb^I. These are three codominant alleles at one autosomal locus.

A Tobiano stallion ($T\_$) with albumin genotype $Alb^A Alb^B$ was mated to a group of nonspotted mares (tt) with a variety of albumin genotypes. He produced both spotted and nonspotted progeny, showing his genotype at this locus to be Tt. Some of his foals inherited the Alb^A allele from him, some the Alb^B allele.

The expected ratio of gametes from this stallion, using Mendel's second law, would be

$$1\ Alb^A T : 1\ Alb^B T : 1\ Alb^A t : 1\ Alb^B t$$

Of the 17 foals showing clearly which alleles came from the sire, 10 received the $Alb^B T$ combination, and 7 received the $Alb^A t$ combination. No progeny receiving $Alb^A T$ or $Alb^B t$ gametes appeared in this sample. So, rather than producing four kinds of sperm in equal proportions, the sire produced a majority of $Alb^B T$ and $Alb^A t$ sperm, with others possible, but not seen in this sample of foals. This evidence indicates that free assortment of alleles was not occurring and that the loci for Tobiano spotting and the serum protein albumin variants are linked on the same chromosome. The sire's chromosomes obviously are such that one carries the T allele linked with Alb^B, and its pair member carries t with the Alb^A allele. In genetic shorthand, this is commonly designated as

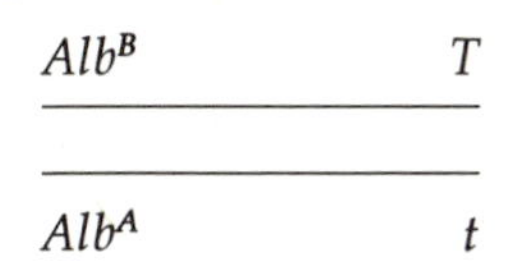

with the lines being used to represent the chromosome pair on which these loci are linked. The linkage appears to be close, as no recombinants were produced among the 17 foals in the sample.

The pleiotropic gene action in the example (Chapter 2) of the hornless condition in goats and its relationship to the development of the normal female might suggest that pleiotropy is the genetic cause of the joint occurrence of Tobiano and albumin *B*. Does in fact the Alb^B blood protein allele have an influence on color development such that it determines the Tobiano pattern, with the Alb^A allele influencing the development of solid color? This idea has proved incorrect, because many nonspotted (*tt*) horses have the Alb^B serum protein albumin variant. Evidently, all possible combinations do exist, with Alb^BT and Alb^At being the combinations found in this particular stallion. Pleiotropy, rather than close linkage, is indicated in the goat example because polled homozygosity and intersexuality generally appear in the same individual.

Established linkage between actual traits in domestic livestock, such as in the preceding example, is rare. The lack of examples does not, however, imply that linkage does not exist, but rather that the breeder is working with traits for which linkage is difficult to demonstrate. Commercially important production traits are not simple qualitative traits but are complex traits inherited in a quantitative manner. Researchers believe that genes at many loci influence these traits, with the effects of single genes being miniscule. Linkage between loci cannot be demonstrated until the influence of single alleles can be distinguished. *Linkage groups,* that is, collections of loci known to be on the same chromosome, must exist in these species, and much research is currently under way to establish gene maps for species such as cattle and horses, and to determine their degree of homology with the chromosome maps of humans and other species.

4-3 Calculation of Distances and Mapping

The domestic rabbit is an example of a species with some commercial importance as a meat producer and which also has a large number of qualitative loci determining important showring characteristics. This species presents a classic example of linkage involving three traits: self-colored versus Himalayan pattern, black color versus brown, and yellow fat versus white fat. Fat color is a qualitative trait influencing the acceptability of the meat to the consumer. The allele for white fat (Y) is dominant to that for yellow (y). Black color (B) is dominant to brown (b), and solid color (C) is dominant to Himalayan pattern (c^h). W. E. Castle's classic research in 1936 on the linkage relationships between these three traits serves as an excellent example for studying linkage.

The proper design of an experiment to study linkage and the distances between linked loci on a chromosome is important. A test cross is needed to an animal recessive at all loci. The other partner in the test mating must be an animal heterozygous at all three loci and hence capable of producing gametes that contain all possible combinations of alleles for the three traits being studied.

The test matings to study linkage are made in two stages. Initially, the heterozygote is produced from homozygous parents, with each parent homozygous for a different allele at each locus. The way in which alleles at the different loci are combined will depend on the mating made. When two dominant alleles are linked on the same chromosome, they are said to be in *coupling,* for example, B with Y. Linkage of a dominant with a recessive allele, such as B with y, is known as *repulsion.* Whether the genes of the heterozygote are in coupling or repulsion will affect the frequency of different gamete combinations produced by the heterozygote. For example, a mating of $bb\ CC\ yy$ and $BB\ c^hc^h\ YY$ homozygotes would produce a heterozygote whose gametes would be primarily bCy and Bc^hY since these are the parental linkage combinations. This mating of the homozygotes could be written in one of three ways:

$$bb\ CC\ yy \times BB\ c^hc^h\ YY \quad \text{gives} \quad Bb\ Cc^h\ Yy$$

or

$$\frac{bCy}{bCy} \times \frac{Bc^hY}{Bc^hY} \quad \text{gives} \quad \frac{bCy}{Bc^hY}$$

or

$$bCy/bCy \times Bc^hY/Bc^hY \quad \text{gives} \quad bCy/Bc^hY$$

All other gametes from the heterozygote, such as BCy, would be recombinants and would be found less frequently. In contrast, heterozygotes produced from an original mating of BCy and bc^hY homozygotes would produce BCy and bc^hY as the most frequent gamete combinations.

The most common progeny genotypes produced from the test matings with the homozygous triple recessive will indicate the combination of alleles at the three loci in the parents of the trihybrid. These combinations will be the key to begin a study of the order of the loci on the chromosome. As with test matings previously discussed, the homozygous recessive is used as a neutral background against which all allelic combinations in the gametes of the heterozygote will show up clearly.

Table 4-1 presents the data from Castle's cross of heterozygous rabbits with homozygous recessives. The immediate question concerns evidence of linkage. If there were no linkage, a ratio of equal numbers of each phenotype would be expected in the progeny listed. There are, however, 276 progeny of one kind and only 7 progeny of another combination, which is evidence that the traits are linked. A chi-square test (see Section 5-5.) would be used to establish that the observations deviate from the expectation of equal numbers by other than chance.

The next step is to determine which are the parental combinations of alleles at the different loci if this is not already known from the mating used to produce the multiple heterozygote. As already stated, the two most frequent progeny phenotypes should indicate the parental combinations, which are Bc^hY and bCy in this example. Obviously, these combinations must complement one another, with no allele represented in both.

Next, the gametes produced as a result of a double recombination event must be identified. Double recombination is the situation where breakage and reunion have occurred twice, once between each pair of loci. Double crossover types should be rarer than single recombinants. Again, complementarity is

Table 4-1 Progeny from Castle's test matings for linkage in rabbits

Progeny phenotype	Gamete from heterozygous parent		Number of progeny
black, colored, white fat	BCY		55
black, colored, yellow fat	BCy		108
black, Himalayan, white fat	Bc^hY		276
black, Himalayan, yellow fat	Bc^hy		7
brown, colored, white fat	bCY		16
brown, colored, yellow fat	bCy		275
brown, Himalayan, white fat	bc^hY		125
brown, Himalayan, yellow fat	bc^hy		46
		Total	908

* Data from W. E. Castle. Further data on linkage in rabbits, *Proc. Nat. Acad. Sci.* **22**:222–225, 1936.

anticipated, with no duplication of alleles. The 7 Bc^hy and 16 *bCY* progeny qualify for designation as double crossovers or double recombinants. Putting this evidence together with knowledge of the parental types allows determination of the order of the loci on the chromosome.

Consider the chromosomes diagrammed in Figure 4-1. If the configuration given as the original heterozygote were the true order of loci on the chromosome, the locus in the middle obviously would be identified as the one which is "flipped over" in the double crossover. That is, with a double crossover, *AbC* and *aBc* would be come *ABC* and *abc*: alleles at the outer loci are not affected by the double crossover, but the center locus would recombine with the other two. The real data show that the parental combinations of Bc^hY and *bCy* become Bc^hy and *bCY* in the rare double recombinants. The loci that remain together are the Bc^h and *bC* combinations. Hence, the *Y*/*y* locus is the one in the middle, and the correct order of alleles on the parental chromosomes is BYc^h and *byC*. Whether the *B*/*b* or C/c^h locus is written to the left or the right is not important, as there is no evidence to indicate the relationship of these loci to other structures (for example, the centromere or other loci) on the chromosome.

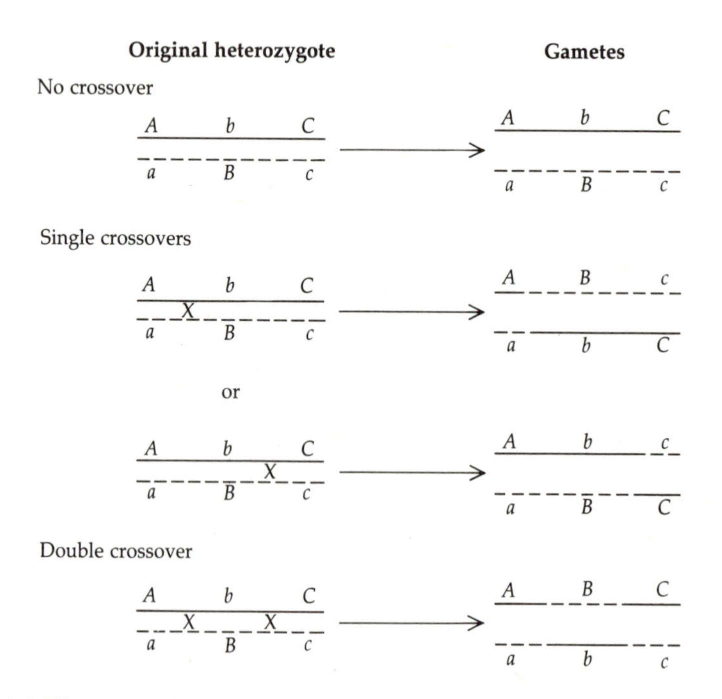

Figure 4-1 Diagrammatic representation of parental, single, and double crossover chromosome configurations. An X indicates the location of each crossover event.

The remaining part of the problem is the determination of how far apart the loci are on the chromosome. The entire data set contained in Table 4-1 is needed, and two calculations are made, one for the distance between the B and Y loci and one for Y and C. A third calculation of the distance between the outer loci, B and C, may be used to verify the other two calculations.

Distances are calculated using the number of recombinants between loci. Closely linked loci show fewer recombinants than loci far apart on the chromosome as there is less chance of breakage between loci that are physically close together.

Parental combinations for the B and Y loci are BY and by. Hence, any By and bY progeny are recombinants and must be counted in the calculation of distance. Note that double crossovers are automatically included in this calculation.

$$\text{number of } By \text{ and } bY \text{ progeny} = 108 + 7 + 16 + 125 = 256$$

which is $256/908 \times 100$ percent of all progeny = 28.19 percent. Thus, the estimated distance between these loci on the chromosome is 28.19 map units. The *map unit* or *morgan* (named in honor of geneticist T. H. Morgan) is the term used to indicate distance, and is equivalent to 1 percent crossing-over. The *centimorgan* refers to a distance of .01 morgan.

Similarly, for the second two loci: Yc^h and yC are parental types and

$$\text{number of } YC \text{ and } yc^h \text{ progeny} = 55 + 16 + 7 + 46 = 124$$

which is $124/908 \times 100$ percent of all progeny = 13.66 percent.

The map now appears as

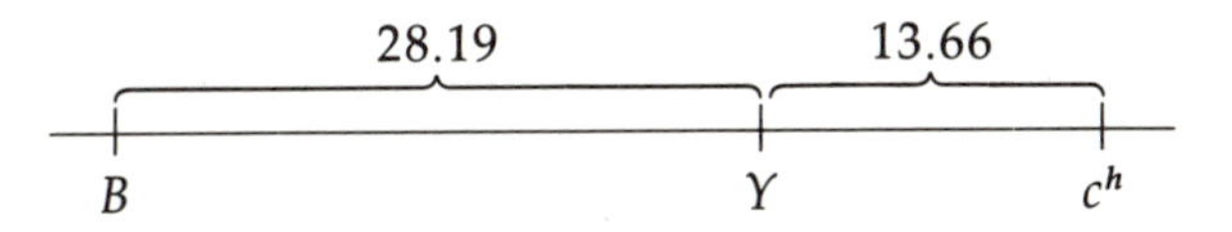

If the Y locus were ignored, the calculation of the distance between the B and C loci would be based on Bc^h and bC being parental types so that the number of recombinants would seem to be

$$\text{number of } BC \text{ and } bc^h = 55 + 108 + 125 + 46 = 334$$

This would give a distance of $334/908 \times 100$ percent = 36.78 map units, which does not equal the calculated sum of 28.19 + 13.66 = 41.85 map units. The discrepancy arises because the double crossovers have not been accounted for, as the second crossover restores the parental combination for the two outer loci (see Figure 4-1). Hence, the calculation must include the 23 double recombinant progeny *twice* since map distance is based on number of recombination events, and there are two such events in each double crossover progeny.

There are [233 + 101 + 2(16 + 7)] progeny that are crossovers between the outermost loci, which is 380/908 × 100 percent = 41.85 percent of all progeny. The previous calculations gave 28.19 + 13.66 = 41.85 map units for this distance, so the two calculations are in agreement.

Figure 4-2 summarizes the procedure to follow in solving a linkage prob-

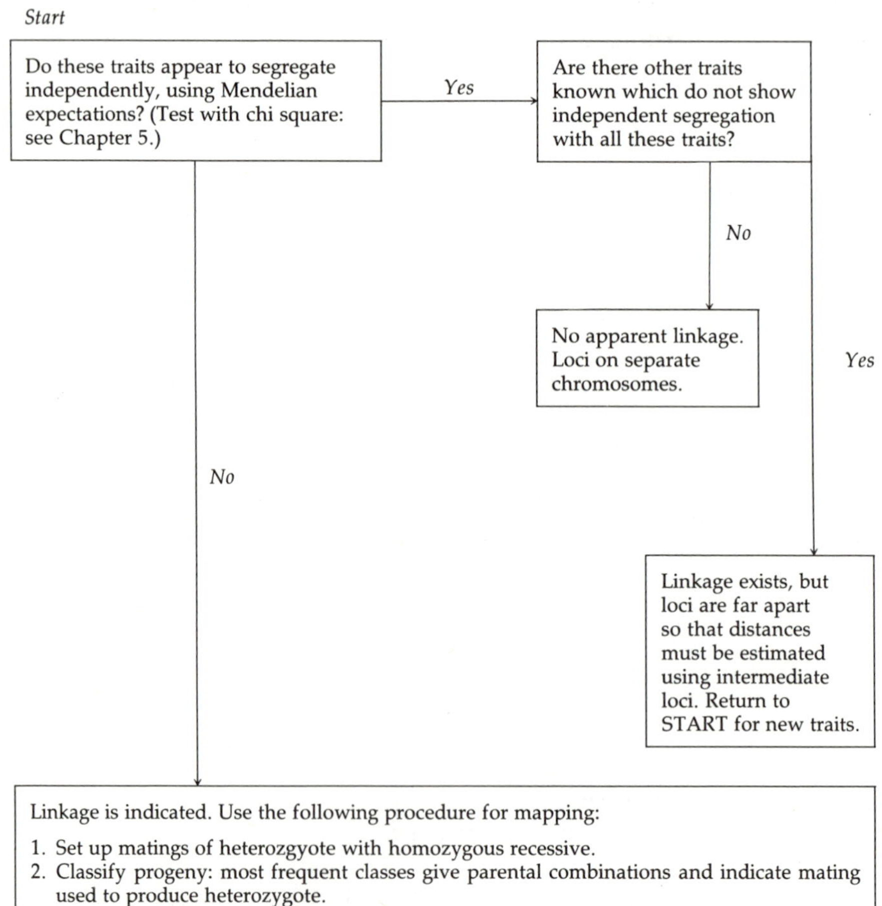

Figure 4-2 Approach to solving a linkage problem.

lem. As more linked loci are identified in a species, linkage groups are built up, one for each pair of chromosomes in that species. The process is slow. A researcher who studies two traits that happen to be qualitatively inherited and that happen to be controlled by loci on the same chromosome will find evidence of linkage. Most students of animal breeding study traits that do not

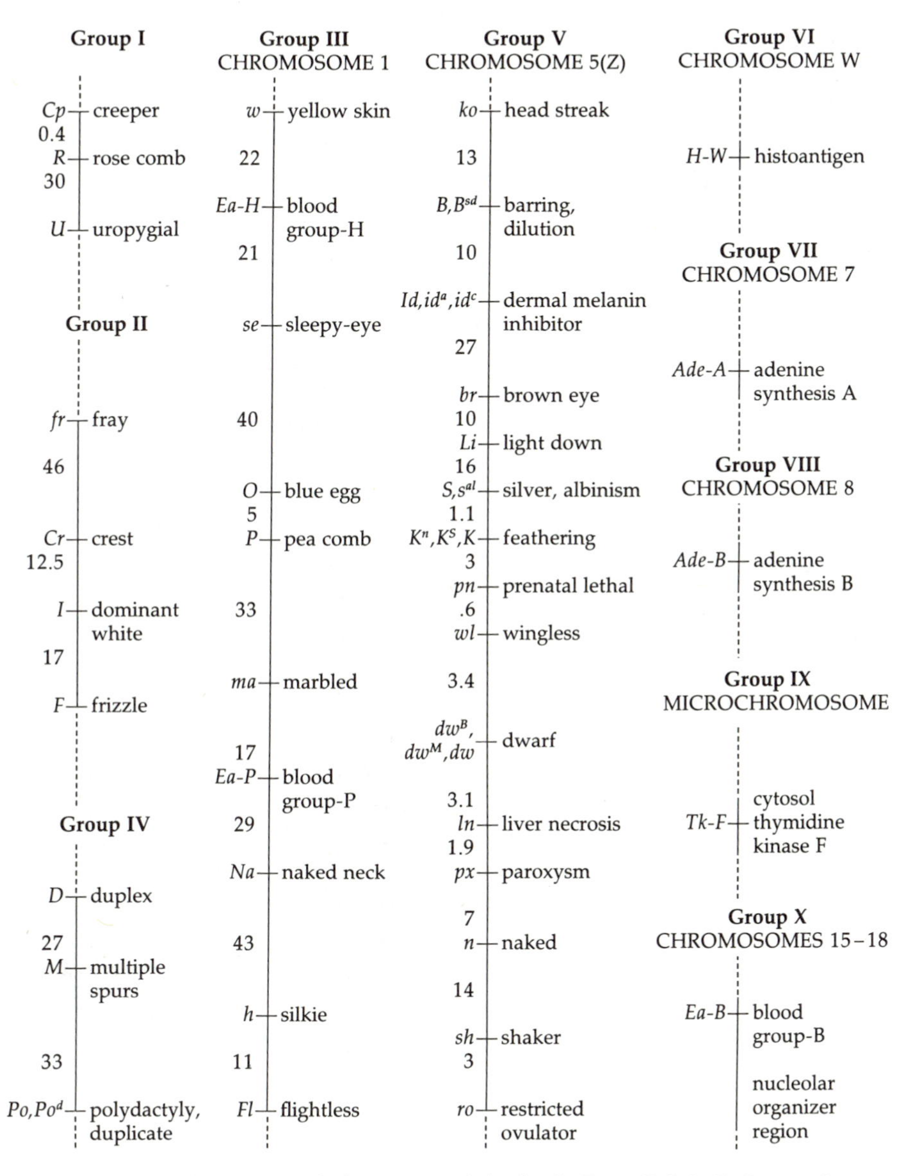

Figure 4-3 The chromosome linkage map of the fowl. (From Ralph G. Somes, Jr., 1978, *J. Hered.* **69**:402.)

meet all these requirements, so linkage groups are slow to be identified. Researchers have developed a fairly complex chromosome map for the popular laboratory mouse. The only domestic species which has been so widely studied is the chicken. The chromosome linkage map of the fowl is illustrated in Figure 4-3.

If two loci are 50 or more map units apart on the chromosome, the probability of multiple crossovers is so great that the genes at the two loci are essentially unlinked and therefore segregate independently. Therefore, maximum observed crossover frequency is 50 percent.

4-4 Sex Linkage

One class of exceptions to the rule that linkage is rarely noticed in domestic species is the relationship between certain characteristics and the sex of the individual. The mechanism of sex determination and the history of the realization that sex is indeed an inherited characteristic were discussed in Chapter 3. Although sex is determined by the inheritance of an entire chromosome rather than a single gene in domestic livestock, the techniques already discussed in relation to single traits can be used to study the inheritance of sex. Any trait controlled by a locus on the X chromosome will show a relationship with the inheritance of sex. The Y chromosome contains very few loci that have been identified as controlling traits unrelated to the sex of the individual, whereas the X chromosome carries genes for many traits. Sex linkage is therefore almost exclusively concerned with the inheritance of traits controlled by alleles on the X chromosome (Z chromosome in poultry).

A classical example of X linked inheritance in a domestic species is that of the inheritance of one of the many colors in cats. One of the color loci in this species has alleles for black (*B*) and orange (*b*) color (also called red, yellow, or "marmalade"). The X-linked black cat is purebreeding when mated to similar blacks. The orange is also purebreeding. However, when homozygous blacks and oranges are intermated, the result is a combination of black kittens, orange kittens, and tortoiseshell kittens. This latter color is a mixing of black and orange hair in the coat, indicating the codominance of the black and orange alleles. Why then is the entire litter not composed of tortoiseshell (*Bb*) kittens? The answer becomes evident when the kittens are classified both by color and by sex (summarized in Table 4-2). All females are tortoiseshell; males are the color of their mother, that is, either black or orange, depending on the specific mating. This is strong evidence that the trait is controlled by a locus on the X chromosome. An individual must have both a black and an orange allele in order to be tortoiseshell, which can, under normal conditions, occur only in

Table 4-2 Sex-linked inheritance in tortoiseshell cats*

Matings		Progeny	
Male	**Female**	**Male**	**Female**
X^BY black	× X^BX^B black	X^BY black	X^BX^B black
X^bY orange	× X^bX^b orange	X^bY orange	X^bX^b orange
X^BY black	× X^bX^b orange	X^bY orange	X^BX^b tortoiseshell
X^bY orange	× X^BX^B black	X^BY black	X^BX^b tortoiseshell
X^BY black	× X^BX^b tortoiseshell	X^BY black X^bY orange	X^BX^B black X^BX^b tortoiseshell
X^bY orange	× X^BX^b tortoiseshell	X^bY orange X^BY black	X^bX^b orange X^BX^b tortoiseshell

* X^B denotes the gene for X-linked black color and X^b denotes its allele for X-linked orange color.

females. Males have a single X chromosome, therefore only a single allele at this color locus.

Because the *B/b* color locus is situated on the X chromosome, the alleles are conventionally written X^B and X^b. A male may be black (X^B) or orange (X^b), depending on which allele he inherited from his dam. Here is an example of a mating where the outcome differs depending on which is the genotype of the sire and which is that of the dam. If the dam is black (X^BX^B) and the sire orange (X^bY), male kittens will be black (X^BY), since their X chromosome (with the color allele) comes from the dam, and their Y chromosome from the sire. If the sire is black (X^BY) and the dam orange (X^bX^b), then the male kittens will be orange (X^bY). For female kittens, the specific colors of the male and female parents is irrelevant because the female kittens will be heterozygous and tortoiseshell (X^BX^b) in either type of mating.

Tortoiseshell females will produce litters of differing constitutions, depending on the genotype of their mate. All their sons will be black or orange, in equal proportions, because half inherit the dam's gene for black color and half the allele for orange. Half the daughters will be tortoiseshell, the other half will resemble the sire. From a mating with a black male, 50 percent of the female offspring will be black; an orange sire will produce 50 percent orange daughters. The events (nondisjunction) giving rise to the rare tortoiseshell male were discussed in Chapter 3.

A sex-linked trait that is an abnormality rather than a variant of color is hemophilia. Blood clotting depends on the presence of certain factors in the blood, and the absence of any of these will lead to hemophilia or "bleeding disease". In this condition, the blood-clotting mechanism is faulty, so that even a minor injury can lead to severe loss of blood. The hemophiliac rarely survives to reproductive age; hence, this trait can be regarded as a semi-lethal in that it does not kill the individual in early development but prevents that animal from passing on genes to the next generation. Hemophilia occurs in both humans and dogs and has been reported in horses and swine.

Recessive sex-linked traits generally are identified by a characteristic pattern of inheritance. Typically, in mammals, unaffected parents will produce daughters that are all normal, but 50 percent of the sons of heterozygous females will show the recessive phenotype. Females can inherit sex-linked recessive traits, but, in order to show the trait, they must be homozygous, which means that the sire must have the allele to pass on to his daughters. He must show the appropriate phenotype for any recessive allele he carries because he has no corresponding dominant allele in the case of X-linked loci. As already indicated, for a trait like hemophilia this is unlikely to occur, as affected individuals rarely survive to reproduce. In general, therefore, the pattern of X-linked inheritance is such that only males show the trait if it is recessive and lethal or semi-lethal. For nonlethal traits, sex linkage is indicated by many more recessive phenotypes observed in male than in female progeny. The rate of occurrence depends on the frequency of the allele in the population (Chapter 6).

Breeders have made economic use of sex linkage in one domestic species, namely poultry. It is important to be able to sex chickens at hatching, but sexing on the basis of sex organs in the chick is notoriously difficult. However, sex-linked plumage color traits show up in chick down, and can be used for rapid sexing.

Barring is one trait that has been used for this purpose. Adults with the dominant barring allele show bars on their plumage. Chicks with the dominant allele have a pale head-spot which is lacking in recessive, nonbarred chicks. Recessive gold and dominant silver genes can also be used for this purpose, as the chick down has very different coloration. In both cases, the homogametic sex (which is the male in poultry) must contribute the recessive allele, otherwise all the progeny will be alike.

Another locus, that for dominant fast or recessive slow feathering, can be used in chicks that are all the same color (White Leghorns, the most common egg-laying breed). Fast featherers will have wing and tail feathers developing by the time they are a little more than a week old. Slow featherers show practically no feather development at this age. An experienced person can detect a difference in wing feathers at hatching time so that sexing need not be delayed until the chicks are older.

4-5 Sex-influenced Traits

One must exercise caution before assuming that sex linkage is in operation in any instance where inheritance of a trait appears to be different in the two sexes. Consider, for example, the inheritance of horns in sheep. Some breeds, such as the Suffolk, are naturally polled. Others, for example, the Dorset Horn, are horned in both sexes. A cross between two such breeds gives an F_1 in which males are horned, females polled. This result is unexpected since the F_1 from such a cross should be uniformly heterozygous and, therefore, all progeny should be of similar phenotype. From knowledge of these alleles in a variety of species, the allele for polledness would be expected to behave as a dominant, but in sheep it is dominant only in females. Males have an identical genotype but do show some horn growth, implying that the polled allele is not completely dominant in males. How does this situation differ from sex linkage? Figure 4-4 shows the expected results from this cross if indeed the trait were sex-linked. Suffolk rams mated to Dorset Horn ewes would give the observed results of polled females and horned males. The reciprocal mating would not: all progeny would be polled. The observed data always show a difference depending on the sex of the F_1 progeny, so sex-linkage is eliminated as an explanation.

If the trait is sex-influenced

	Suffolk female *HH*—polled
Dorset Horn male *hh*—horned	Males *Hh*—horned Females *Hh*—polled

	Dorset Horn female *hh*—horned
Suffolk male *HH*—polled	Males *Hh*—horned Females *Hh*—polled

The results are the same regardless of the breed of each parent in the cross. This outcome *agrees with the observations.*

If the trait were sex-linked, the results would have been

	Suffolk female X^HX^H—polled
Dorset Horn male X^hY—horned	Males X^HY—polled Females X^HX^h—polled

	Dorset Horn female X^hX^h—horned
Suffolk male X^HY—polled	Males X^hY—horned Females X^HX^h—polled

The results would have differed depending on the breed of each parent in the cross. This outcome was *not observed.*

Figure 4-4 Sex-influenced inheritance—horns in sheep.

In this example, the hormonal status of the individual has an influence on horn development, that is, on the pathway through which the genes express themselves in the phenotype. The presence of male hormones is conducive to horn growth, even in the heterozygote, whereas a female must be homozygous for the allele for horns before she grows horns. In purebred Dorset Horns, a female, even when homozygous, does not have the impressive horn growth found in males.

Another example of a *sex-influenced* trait is that of color in Ayrshire cattle. This dairy breed has a spotted pattern of red and white. The red varies in intensity, being bright in some animals and a darker mahogany color in others. The dark coloration is considered to be due to the presence of a dominant gene, but is also influenced by the sex of the animal. Males more often have the darker color than do females, because females must be homozygous to show the dark color, while heterozygous males will be dark. In both this example and that of horns in sheep, removal of the testes, the source of the male hormone testosterone, prevents development of the male expression of the genotype, and castrated males resemble females for this trait. Castration is a classical method used as a test in instances where the sex hormone is suspected of influencing a trait. Removal of the source of the hormone before the critical time period, and observation of the resulting effects on the phenotype, will demonstrate whether that hormone affects the phenotypic expression of the character. The hormonal status of the animal is part of the environment in which alleles express themselves. The traits discussed here, horns and color, are examples of traits for which the environment influences the phenotype.

4-6 Sex-limited Traits

The expression of another class of traits is related to the sex of the animal in yet a different way. Sex-linked traits are influenced by genes actually situated on the sex chromosomes; sex-influenced traits are controlled by autosomal loci, but their expression in the heterozygote depends on the sex of the individual. *Sex-limited* traits have a phenotypic expression that is limited to only one sex. These traits cannot be expressed in both sexes under normal conditions. Such traits as milk yield, egg-laying capacity and inherited testicular abnormalities are sex-limited characters.

Sex-limited traits are commercially the most important of these three types of sex-associated characteristics. The economically important traits in dairy animals and in egg-layers cannot be expressed in the male. At the same time, the male must be thoroughly evaluated genetically for these characters since he may be used to sire large numbers of offspring. Hence the need for complex sire evaluation programs based on pedigree and progeny evaluations (discussed in later chapters). The design and operation of such evaluation programs reflect the facts that these traits are quantitatively inherited and that their expression is limited to one sex.

The testicular abnormalities referred to may not be quantitatively inherited so it is appropriate to consider them here. Cryptorchidism is often considered to be an inherited condition in males and may be bilateral—neither testis descends into the scrotum—or unilateral—one testis fails to descend. Cryptorchidism is an economically important trait in that it can have a serious effect on the fertility of males being used in a breeding program. Some data have indicated that this condition is determined by an autosomal recessive allele, although other reports indicate that modifying loci and the environment may both play a part in the expression of this trait. One reason for the difficulty in determining the mode of inheritance is the fact that the trait is expressed in only one sex; genotypes of females must be inferred from breeding tests, which inevitably delays the determination of factors, genetic or otherwise, that control the trait.

4-7 Summary

Genes at different loci on the same chromosome are linked and may not segregate independently. Crossing-over and exchange of genetic material between paired chromosomes allow for new combinations of genes (recombinants). In test crosses between hybrids and homozygous recessives for linked genes, the parental combinations of gametes will be more frequent than recombinant gametes.

The percentage of crossovers among all offspring of a mating is defined as map distance:

$$\text{map distance} = \frac{\text{number of crossovers}}{\text{total number of offspring}} \times 100 \text{ percent}$$

Each map unit or morgan is equivalent to 1 percent crossing-over.

Loci may be more than 50 map units apart, but, because of multiple crossovers, the maximum number of recombinants is 50 percent, which is equivalent to independent segregation.

Linkage between loci less than 50 map units apart is indicated if traits do not segregate independently.

Steps to determine map distances are:

1. Mate known heterozygotes for two or more loci to animals which are homozygous recessive for those loci.

2. If three loci are involved, identify the center locus as the locus with the gene that is switched when the most frequent (parental) classes of progeny and the least frequent (double crossover) classes are compared.
3. Calculate map distances separately for each pair of loci as indicated.

Sex-linked traits are those traits associated with loci on the sex (X or Z) chromosome. Thus, the inheritance of such traits is related to the inheritance of sex. Recessive X-linked phenotypes are more common in males than females (the reverse is true for Z-linked traits) because the male will express whichever allele he receives from his mother. Reciprocal crosses between purebreeding males and purebreeding females give different results, with the male F_1 progeny resembling the female parent.

Sex-influenced traits are traits for which expression of the genotype depends on the sex of the animal. Reciprocal crosses between purebreeding males and females give the same result—F_1 males are always of one type and F_1 females of the other type.

Sex-limited traits can be expressed in only one sex; for example, milk production can be measured only for cows, not for bulls.

CHAPTER 5

Probability and Tests of Genetic Hypotheses

Mendelian genetics might be described as the study of probabilities. Probabilities involving the fraction one-half are most prevalent because a parent transmits a sample half of its genes, in fact a random allele at each locus, to its offspring. Discussions in previous chapters have indicated the need to use probability to explain genetic results. The purpose of this chapter is to describe those ideas of probability more precisely and to show that understanding some rather simple rules of probability can make the study of Mendelian and, in the next chapters, the study of population genetics easier. The most difficult aspect of putting genetic results into probability expressions is keeping in mind the terminology used in both genetics and probability.

5-1 Definitions

The two most critical steps in understanding probability are

1. carefully defining an event, and
2. knowing the difference between an event and an observation (see Figure 5-1).

An *event* is simply the occurrence of whatever it is stated to be. Consider, for example, horses which in an oversimplified way can be said to be either white or colored. Thus, with this definition of events, the two possible events for the phenotypes are white or colored (nonwhite). Other related events could be defined. For example, the events might have been defined as the presence or

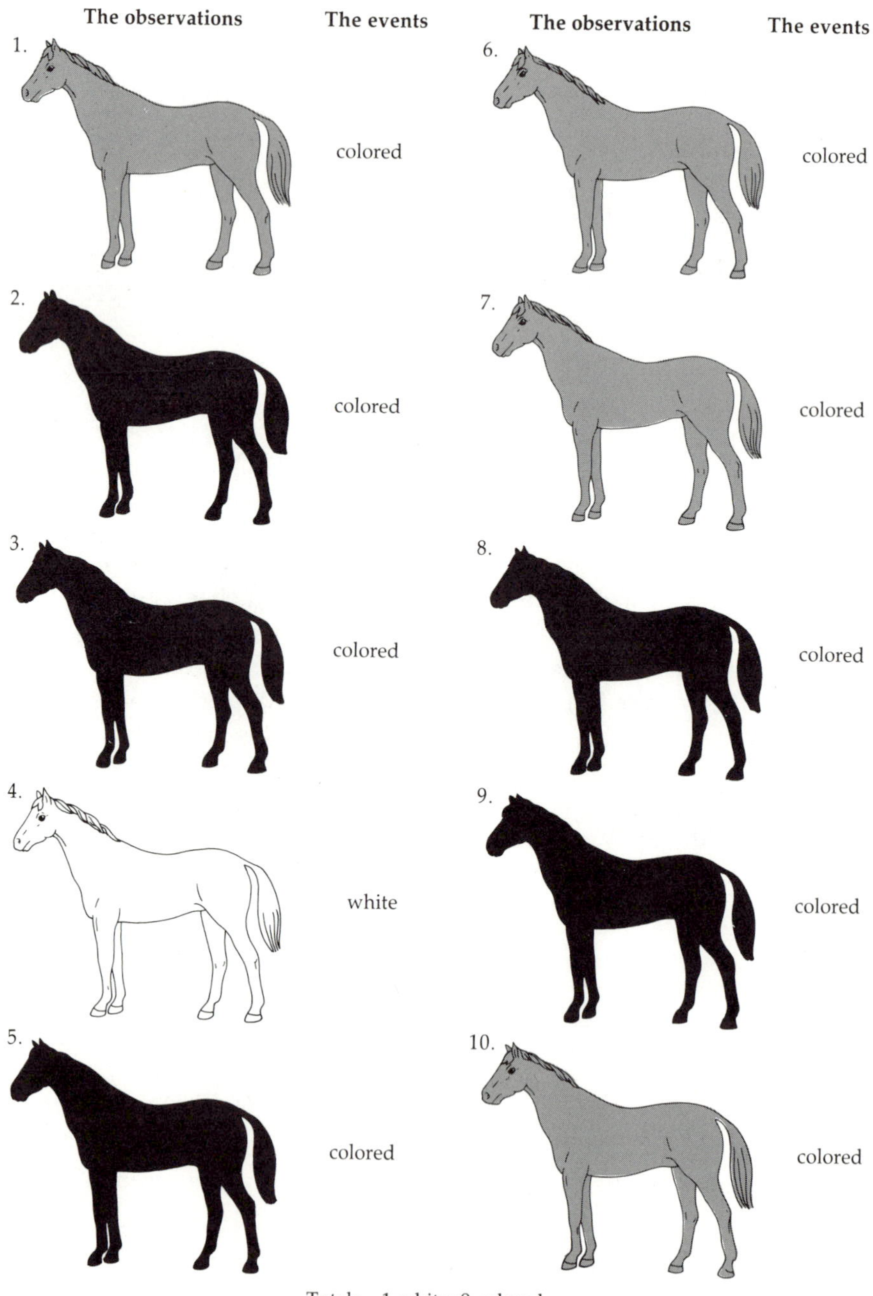

Figure 5-1 Events versus observations. Each observation will be of some event. In this figure the events are defined as white or colored (nonwhite).

absence of the allele, W, which causes the white phenotype, or as the presence or absence of the genotype, ww, which results in color of some kind. Care must be taken to distinguish what is defined to be an event from results or causes of that event. Often, in genetics, there is a strong and sometimes perfect relationship between genes, genotypes, and phenotypes, but whichever of these is the event must be defined clearly.

There may be some confusion between the words "event" and "observation." The example in Figures 5-1 and 5-2 demonstrates the difference. Colored (nonwhite) horses can be divided into genetically black and red horses. Horse breeders have given the name chestnut to red horses so the term chestnut will be used. Now the events can be defined differently, as in Figure 5-2. Event A will be the color white; event B will be the color black; and event C will be the color chestnut. No other events—colors—will be allowed for this example. Each animal provides one observation. Each observation will be event A, event B, or event C. The sum of the number of observations of events A, B, and C will be the total number of observations. For the example shown in Figure 5-2, of the ten horses, one is white (event A), five are black (event B), and four are chestnut (event C).

Sometimes, as with the white and nonwhite example, an event is the lack of another event. As another example, if event A is defined as a male birth then, if event A is not observed, the event observed is "not A" (in this case, not male is equal to female).

Often the occurrence of genes, or genotypes, cannot be determined exactly as can phenotypes, but their occurrence can be predicted with known probability. The *probability* P of an event is defined as

$$P(\text{event}) = \frac{\text{number of times the event occurs}}{\text{total number of observations}}$$

This definition applies to two usual cases:

1. when the probability is exact; for example, when a colored mare is known to be heterozygous, Bb, then under usual Mendelian assumptions the probability of B being passed to her offspring is one-half (the event is a B in the offspring received from the mother); and
2. when the probability is estimated from a sample of data; for example, suppose the event is the chestnut color in horses and that, of a sample of 1000 colored horses, 387 were chestnut so that the estimate for the population of colored horses is $P(\text{chestnut}) = \frac{387}{1000} =$ .387.

The study of Mendelian genetics usually deals with the case when exact probabilities are known and are used to explain what is expected to happen.

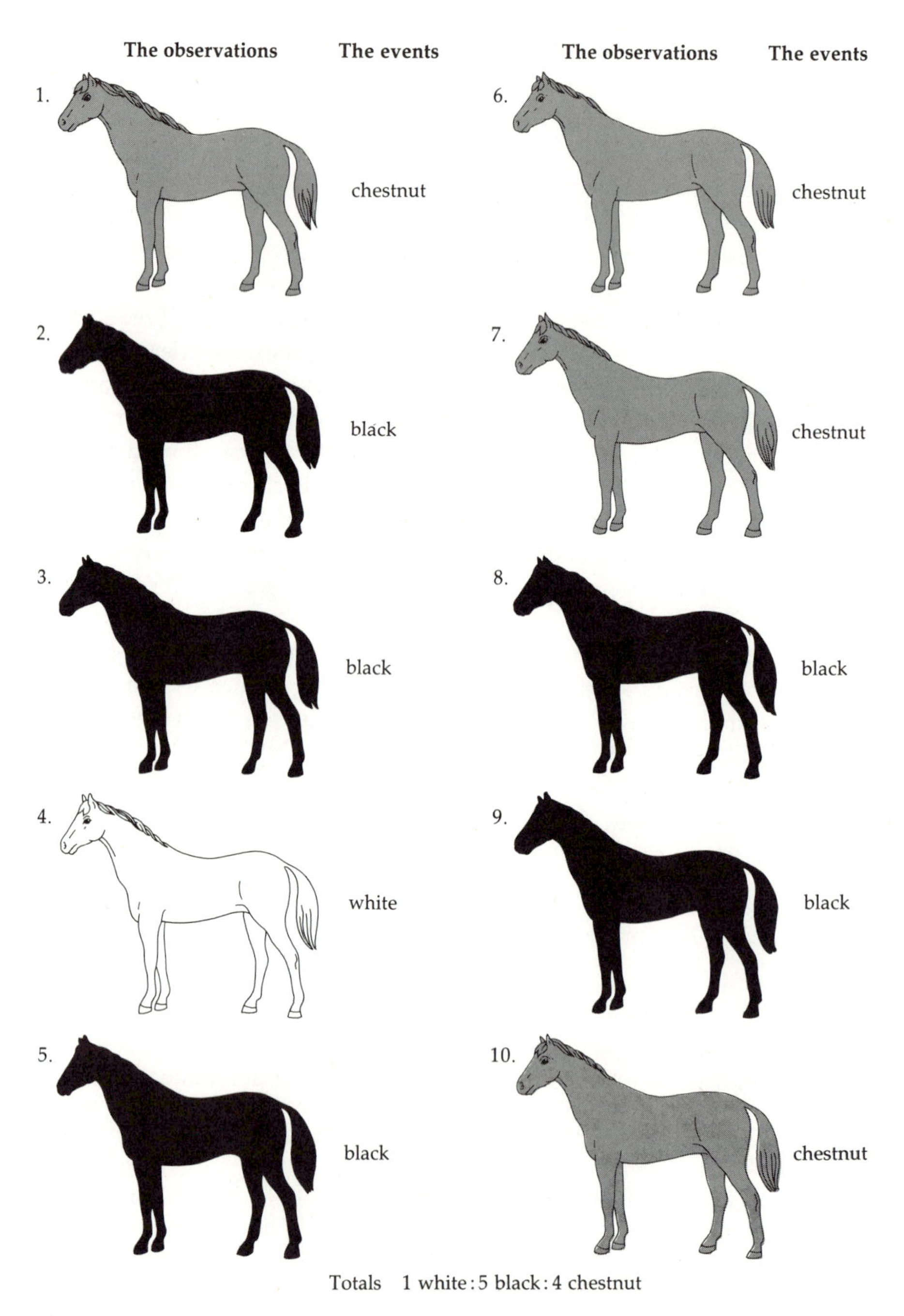

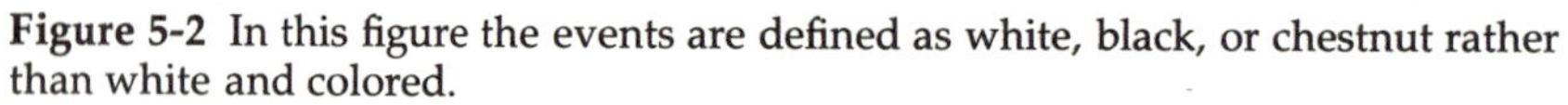

Figure 5-2 In this figure the events are defined as white, black, or chestnut rather than white and colored.

The second definition is especially important in population genetics. The empirical or estimated probability is a measurement of what has happened and can be used to predict what will happen. If conditions change, however, then the previously calculated empirical probability may no longer apply; for example, if the chestnut color becomes more popular, selection may result in a larger proportion of chestnuts. Another aspect of the empirical, or "sample," probability is that the sample size influences the accuracy of the estimate. A sample of 1000 horses would be expected to provide a more accurate estimate of the frequency of the chestnut color than a sample of 10 horses, as will be discussed later in this chapter.

The *frequency of occurrence* is numerically the same as probability of occurrence. The following example illustrates the difference in the normal use of the terms, frequency and probability. Suppose that the fraction of chestnut horses is exactly .387. Frequency refers to the fraction of chestnuts expected in a sample of horses, for example, 387 out of 1000, whereas probability refers to the chance a horse seen at random will be chestnut, that is, .387.

The *expected number* is the prediction of how many animals will exhibit a certain characteristic and is calculated as pN, where p is the probability of the characteristic and N is the total number of animals. Suppose a second sample of 1000 colored horses is examined. The expected number of chestnuts based on the first sample is $.387(1000) = 387$. The actual number of chestnuts turns out to be 400. If the total number of animals is small, the chance may not be great that the expected and actual number will agree closely. But if many samples of that size are measured, then the expected number will be close to the actual number.

The expected number is correct "on the average." For example, if the probability of a male birth is 50 percent, but the sample is a family of three, then a single family will never be $1\frac{1}{2}$ males and $1\frac{1}{2}$ females because the only possibilities are three and zero, two and one, one and two, or zero and three. But, when averaged over many families of three, the ratio of males to females will be about 50:50 or, equivalently, 1:1.

Two events are *mutually exclusive* if the occurrence of one means the other cannot occur. In the previous example, if a foal is white, then it cannot be colored. If the event is described as a combination of characteristics then, if the foal is male and chestnut, it cannot be male and black, female and chestnut, female and black, male and white, or female and colored.

Two events are *independent* if the occurrence of one does not affect the probability of occurrence of the other. An example, using the characteristics of sex and color, is that the chance the foal is male does not affect the chance it is black.

5-2 Basic Rules and Mathematical Definitions

Following are some important mathematical definitions and equivalences which will be used.

If n is a nonzero integer,

$$n \text{ factorial} = n! = (n)(n-1)(n-2) \cdots (1)$$

By definition,

$$\text{zero factorial} = 0! = 1$$

For example,

$$4! = (4)(3)(2)(1) = 24$$

If s is nonzero (for example, 1, 2, . . .) then

$$\text{zero to the power } s = 0^s = 0$$

and

$$s \text{ to the power zero} = s^0 = 1$$

By definition,

$$\text{zero to the power zero} = 0^0 = 1$$

If there are only two possible events, for example, allele B or allele b with probabilities p and q, then because the sum of the frequencies of both events must be 1, $p + q = 1$. That key equation establishes the two simple but important identities used often in population genetics:

$$p = 1 - q$$

and

$$q = 1 - p$$

Another useful identity also depends on the sum of the frequencies equalling 1.

Because

$$p + q = 1$$

then

$$(p + q)^2 = 1$$

and thus the expansion

$$(p^2 + 2pq + q^2) = 1$$

Rules for calculating probabilities

There are three basic rules for calculating probabilities of occurrence of events and combinations of events.

Rule 1 The probability that *one of a series of mutually exclusive events occurs* is the sum of the probabilities that the individual events will occur. If the mutually exclusive events are A, B, and C, then the probability of occurrence of event A or of event B or of event C is the probability of event A plus the probability of event B plus the probability of event C. In symbolic form:

$$P(\text{A or B or C}) = P(\text{A}) + P(\text{B}) + P(\text{C})$$

For example, if the events are white, black, and chestnut horses, then P(white or black) = P(white) + P(black).

Rule 2 The probability that *n independent events will all occur* is the product of their individual probabilities, that is,

$$P(\text{all } n \text{ occur}) = p_1 \times p_2 \times \cdots \times p_n$$

where p_i is the probability of event i. For example, if event A is a male foal, which is independent of event B, the color white,

$$P(\text{A and B both occur}) = P(\text{A}) \times \text{P(B)}$$

that is, P(foal is white and male) = P(foal is white) × P(foal is male).

As a numerical example, suppose two heterozygotes for the dominant phenotype black are mated, that is, *Bb* crossed with *Bb*. The probability of the dominant phenotype, black (*BB* or *Bb*), in the offspring is $\frac{3}{4}$ and of the recessive phenotype, red (*bb*), is $\frac{1}{4}$. What is the probability the first two offspring are black and the third is red? Because the color of each offspring is independent of the others, the answer is

$$\begin{aligned} P(\text{black, black, red in order}) &= P(\text{black}) \times P(\text{black}) \times P(\text{red}) \\ &= (3/4)(3/4)(1/4) \\ &= 9/64 \end{aligned}$$

This probability $\frac{9}{64}$ is not the same as the probability of obtaining two black and one red offspring out of three offspring. That probability also involves the number of mutually exclusive ways in which two black and one red offspring can occur, that is, black, black, and red; black, red, and black; and red, black, and black. Each of these three possible orders of occurrence is mutually exclusive of the others and is equally probable. Therefore,

$$\begin{aligned} P(\text{2 black and 1 red in any order}) &= P(\text{black, black, red in order}) \\ &\quad + P(\text{black, red, black in order}) \\ &\quad + P(\text{red, black, black in order}) \\ &= [P(\text{black}) \times P(\text{black}) \times P(\text{red})] \\ &\quad + [P(\text{black}) \times P(\text{red}) \times P(\text{black})] \\ &\quad + [P(\text{red}) \times P(\text{black}) \times P(\text{black})] \\ &= 3[P(\text{black}) \times P(\text{black}) \times P(\text{red})] \end{aligned}$$

Thus, for the previous example, $P(\text{2 black and 1 red in any order}) = 3(\frac{9}{64}) = \frac{27}{64}$.

Rule 3 provides a formula for combining probabilities for two mutually exclusive events.

Rule 3 If the probability of event A is p and the probability of event B (that is, "not A") is q, the probability that, in n observations, event A will occur exactly s times and event B exactly r times is

$$P(s \text{ of event A and } r \text{ of event B}) = (n!/s!r!)(p^s q^r)$$

Note that the second part of the expression, $p^s q^r$, gives the probability of s of event A and r of event B in a specified order, and the first part of the expression, $n!/s!r!$, gives the number of different mutually exclusive orders in which s of event A and r of event B can occur.

Continuing the previous example, if $P(\text{black}) = \frac{3}{4} = p$, and $P(\text{red}) = \frac{1}{4} = q$, then the probability of exactly two black and one red offspring in three offspring ($s = 2$, $r = 1$, and $n = 3$) is

$$P(\text{2 black and 1 red}) = (3!/2!1!)(3/4)^2(1/4)^1 = (3)(9/64) = 27/64$$

If there are m possible mutually exclusive events, rather than two, with probabilities $p_1, \ldots, p_m$, then the chance that in n observations the m events will occur exactly $s_1, \ldots, s_m$ times is

$$P(s_1 \text{ of event 1, } s_2 \text{ of event 2, } \ldots, s_m \text{ of event } m \text{ in any order})$$
$$= [n!/(s_1! \cdots s_m!)](p_1^{s_1} \times \cdots \times p_m^{s_m})$$

Rule 3 can be used to calculate the expected frequencies for the *binomial distribution,* that is, frequencies for combinations of two mutually exclusive events, or for the *multinomial distribution,* that is, frequencies for combinations of m mutually exclusive events.

The binomial and multinomial expansions

A simple computing procedure can be used to derive the formulas of Rule 3 for the frequencies involved with the binomial and multinomial distributions. The procedures are called the binomial and multinomial expansions. The procedure is

1. multiply symbolically the probability of each event by the symbol for the event,
2. sum these, and
3. expand that sum by multiplying it by itself n times, where n is the number of observations. (Multiplying a sum by itself n times is equivalent to finding the nth power of the sum.)

An example will make these steps more clear. Suppose that the probability of event A is p and of event B (that is, not A) is q. Then for n observations the steps are

1. multiply $p \times \text{A}$ and $q \times \text{B}$, which can be written $p\text{A}$ and $q\text{B}$;
2. add $p\text{A} + q\text{B}$; and
3. expand $(p\text{A} + q\text{B})^n$

For $n = 3$, the full expansion from step 3 in order of occurrence of A and B is

$$ppp\text{AAA} + ppq\text{AAB} + pqp\text{ABA} + qpp\text{BAA} + pqq\text{ABB} + qpq\text{BAB}$$
$$+ qqp\text{BBA} + qqq\text{BBB}$$

which can be written

$$p^3(\text{AAA}) + p^2q(\text{AAB}) + p^2q(\text{ABA}) + p^2q(\text{BAA}) + pq^2(\text{ABB}) + pq^2(\text{BAB}) + pq^2(\text{BBA}) + q^3(\text{BBB})$$

The different orders of the A's and B's have been kept separate to this point even though the individual probabilities are the same for all orders of, for example, 2 A's and 1 B. When the orders with the same number of A's and B's are combined the expression becomes

$$p^3(\text{3A's}) + 3p^2q(\text{2A's and 1B}) + 3pq^2(\text{1A and 2B's}) + q^3(\text{3B's})$$

The coefficients are the same as those given by rule 3.

The approach is the same for the multinomial expansion, but just takes longer to write out. For example, with the three events A, B, and C, with corresponding probabilities p, q, and r, the steps for the trinomial expansion are

1. $p \times \text{A}$, $q \times \text{B}$; and $r \times \text{C}$,
2. $p\text{A} + q\text{B} + r\text{C}$; and
3. $(p\text{A} + q\text{B} + r\text{C})^n$, which can be expanded.

5-3 The *If* Principle and Conditional Probabilities

Two important and equivalent phrases dealing with probabilities are *if* or *given that.* The phrases simply mean that some event has happened or is certain to happen. As a simple example, suppose the event is defined as a male birth. If a male birth has occurred, the probability that that birth was male is 1, or 100 percent. But if a female birth has occurred, the probability that that birth was a male birth is 0. A slightly more difficult question to answer is, if the first foal of a mare is a male, what is the probability the first two foals are both male? The answer is not $\frac{1}{2} \times \frac{1}{2} = \frac{1}{4}$, as might be wrongly suspected. Because the first foal is known to be a male, P(first is a male) $= 1$, not $\frac{1}{2}$. Because the sex of the second foal is unknown, P(second will be a male) $= \frac{1}{2}$. Thus, the probability the first two are male given that the first is a male $= 1 \times \frac{1}{2} = \frac{1}{2}$. If the question had been, what is the probability the first two foals are male given that the first was a female, then the answer would have been $0 \times \frac{1}{2} = 0$ as expected. A less trivial example was presented in Chapter 2 with the brief discussion of testing for carriers of recessive alleles. Although not emphasized, probabilities of detection were computed given that the animal being tested was, in fact, a carrier of the recessive allele (a heterozygote). Other applications of the *if* principle will be discussed next.

Probabilities calculated based on the *if* or *given that* principle are often called *conditional probabilities;* for example, the probability some event A occurs depends on the condition that event B has occurred or will occur.

The usual mathematical expression for a conditional probability is

$$P(\text{event A occurs given that event B has occurred or will occur})$$
$$= P(\text{A}|\text{B}) \qquad \text{where ``}|\text{'' indicates ``given that''}$$

The conditional probability is

$$P(\text{A}|\text{B}) = P(\text{A, B})/P(\text{B})$$

where $P(\text{A, B}) = P(\text{event A occurs and event B occurs})$ is the probability of the joint occurrence of events A and B, and $P(\text{B})$ is the probability of the overall occurrence of event B. Note that $P(\text{A}|\text{B})$ is read "the probability that event A will occur given that event B has occurred or will occur."

If events A and B are independent so that $P(\text{A and B}) = P(\text{A}) \times P(\text{B})$, as most genetic examples thus far have been, then

$$\begin{aligned} P(\text{A}|\text{B}) &= [P(\text{A}) \times P(\text{B})]/P(\text{B}) \\ &= P(\text{A}) \end{aligned}$$

which shows that the chance of event A occurring does not depend on (that is, is independent of) whether event B occurs.

If events A and B are not independent (the occurrence of either A or B is dependent on the occurrence of the other) so that

$$P(\text{A,B}) \neq P(\text{A}) \times P(\text{B})$$

then

$$P(\text{A}|\text{B}) \neq P(\text{A})$$

which shows that the chance of event A does depend on event B.

The answer to the simple question, given that the first foal is male, what is the probability the first two are male, can be written in terms of conditional probabilities.

First, assume

$$P(\text{male}) = 1/2$$

and

$$P(\text{female}) = 1/2$$

Event A is that the first two foals are male, and event B is that the first foal is male. Thus, the table of joint probabilities for two births is

Foals		
First	**Second**	**Joint probability**
Male_1	Male_2	1/4
Male_1	Female_2	1/4
Female_1	Male_2	1/4
Female_1	Female_2	1/4

Then

$$\begin{aligned} P(\text{male}_1 \text{ and } \text{male}_2|\text{male}_1) &= P(\text{male}_1, \text{male}_2)/P(\text{male}_1) \\ &= \frac{(1/4)}{(1/2)} \\ &= 1/2 \end{aligned}$$

This example is somewhat difficult to interpret because the probability of event A is the product of two independent probabilities, $P(\text{male}_1)P(\text{male}_2) = \frac{1}{4}$. Then, because the first foal is known to be a male,

$$P(\text{male}_1, \text{male}_2|\text{male}_1) = [P(\text{male}_1)P(\text{male}_2)]/P(\text{male}_1) = P(\text{male}_2)$$

Nevertheless, the key to calculating conditional probabilities is knowing the joint probabilities.

The example of the sex-linked color characteristics in cats which results in tortoiseshell females will be used to illustrate the more usual calculations involving conditional probabilities.

Assume for the example that male and female kittens are equally likely and that a black tom cat (genotype X^BY) mates with a tortoiseshell female (genotype X^BX^b). The joint probabilities are the expected frequencies of the resulting kittens:

$$X^BY \quad \times \quad X^BX^b$$
$$\downarrow$$
$$(1/4)X^BY + (1/4)X^bY + (1/4)X^BX^B + (1/4)X^BX^b$$

That is, $\frac{1}{4}$ black, male + $\frac{1}{4}$ orange, male + $\frac{1}{4}$ black, female + $\frac{1}{4}$ tortoiseshell, female. In table form, the phenotypic frequencies that correspond to the joint probabilities, P(color, sex), are

	Black	Orange	Tortoiseshell	Sex
Male	1/4	1/4	0	1/2
Female	1/4	0	1/4	1/2
Color	1/2	1/4	1/4	

For example, P(black, female) = $\frac{1}{4}$, and P(orange, female) = 0.

The overall sex and color frequencies which are given in the margins are sometimes called marginal probabilities. For example, the marginal probability of a male is the sum of all the joint probabilities of the mutually exclusive events which include a male; for example, P(male of any color) = P(black, male) + P(orange, male) + P(tortoiseshell, male) = $\frac{1}{4} + \frac{1}{4} + 0 = \frac{1}{2}$.

Some examples of the conditional probabilities of various colors given the sex of the kitten are of the form

$$P(\text{color}|\text{sex}) = P(\text{color, sex})/P(\text{sex})$$

Thus,

$$\begin{aligned} P(\text{black}|\text{male}) &= P(\text{black, male})/P(\text{male}) \\ &= \frac{1/4}{1/2} = 1/2 \end{aligned}$$

$$\begin{aligned} P(\text{tortoiseshell}|\text{male}) &= P(\text{tortoiseshell, male})/P(\text{male}) \\ &= 0/(1/2) = 0 \end{aligned}$$

and

$$\begin{aligned} P(\text{tortoiseshell}|\text{female}) &= P(\text{tortoiseshell, female})/P(\text{female}) \\ &= \frac{1/4}{1/2} = 1/2 \end{aligned}$$

The formula for a conditional probability was given as

$$P(\text{A}|\text{B}) = P(\text{A, B})/P(\text{B})$$

The conditional probability of event B given that event A has occurred can be written similarly:

$$P(\mathrm{B}|\mathrm{A}) = P(\mathrm{A, B})/P(\mathrm{A})$$

In the same example,

$$P(\text{sex}|\text{color}) = P(\text{sex, color})/P(\text{color})$$

where $P(\text{sex, color}) = P(\text{color, sex})$, because it makes no difference which is written first for joint probabilities.

These conditional probabilities apply only to the specifications of the situation: equal chance of male and female kittens, and the mating of a black tom cat to a tortoiseshell female. Thus,

$$P(\text{male}|\text{orange}) = P(\text{male, orange})/P(\text{orange})$$
$$= \frac{1/4}{1/4} = 1$$

and

$$P(\text{female}|\text{black}) = P(\text{female, black})/P(\text{black})$$
$$= \frac{1/4}{1/2} = 1/2$$

Suppose, as another example, that two loci are 20 map units apart, and that heterozygotes in the repulsion phase are mated to a homozygous recessive; that is,

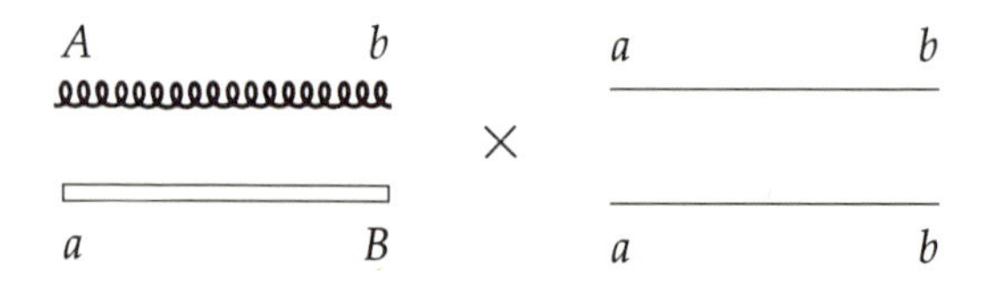

This would allow measurement of the frequencies of parental type gametes (*Ab* and *aB*) and of recombinant type gametes (*AB* and *ab*) produced by the heterozygote.

The chromosomes are represented differently to illustrate the recombinant gametes and to distinguish the chromosomes of the heterozygous parent from the homozygous parent used as a test cross.

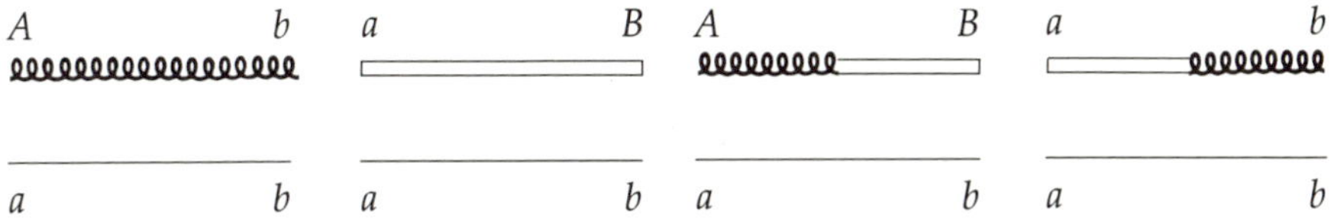

A b	a B	A B	a b
a b	a b	a b	a b
$P(A, b) = .4$	$P(a, B) = .4$	$P(A, B) = .1$	$P(a, b) = .1$

Because this is a test cross, genes at the A and B loci contributed by the heterozygous parent can be identified. Thus, $P(A, b)$ can be used to describe $P(Aa, bb)$.

The table of joint probabilities is

		First locus phenotype		Marginal probability, second locus
		A	a	
Second locus phenotype	B	.1	.4	.5
	b	.4	.1	.5
Marginal probability, first locus		.5	.5	

The frequencies of the parent type gametes are the sum of the joint probabilities of the mutually exclusive phenotypes:

$$P(A, b) + P(a, B) = .4 + .4 = 80 \text{ percent parental type gametes}$$

The frequencies of the recombinant type gametes are the sum of the joint probabilities:

$$P(ab) + P(AB) = .1 + .1 = 20 \text{ percent recombinant type gametes}$$

The four possible conditional probabilities for P(first locus phenotype|second locus phenotype) are

$$P(A|B) = P(A, B)/P(B) = .1/.5 = .2$$
$$P(a|B) = P(a, B)/P(B) = .4/.5 = .8$$
$$P(A|b) = P(A, b)/P(b) = .4/.5 = .8$$
$$P(a|b) = P(a, b)/P(b) = .1/.5 = .2$$

These calculations show that 20 percent of progeny with characteristic *B* are expected to have characteristic *A* and, similarly, 80 percent of progeny with characteristic *B* are expected to have characteristic *a*.

5-4 The Perception of Probability

Although the true probabilities of inheritance do not change, the perception of those probabilities may change depending on the evidence available. An example that occurs often is loci having alleles with low frequency, as will become apparent in the chapter on population genetics. For instance, a bull was mated by artificial insemination to 30 daughters of known mulefoot carrier bulls. Mulefoot is a recessive characteristic in cattle, so half of the 30 mates would be expected to carry the mulefoot gene. Because all the test progeny were normal, the bull was assumed to be homozygous normal, and the probability that his sperm carried only the normal gene was thought to be 100 percent with a high degree of confidence. After 60,000 additional matings by artificial insemination, a mulefoot calf was born, and immediately the perception changed. Half his sperm were then known to carry the mulefoot gene. The perception of the probability that any one of his sperm carried the normal gene changed from 1.00 to .5, even though the true probability was always .5.

5-5 Testing of Genetic Hypotheses

Probabilities are critical in testing whether the results of test matings are consistent with a stated genetic hypothesis. The probabilities are calculated as if the hypothesis is true. For example, do the results correspond to those expected if the trait is autosomal recessive? An important principle of science is that a hypothesis cannot be proved—there is always a chance, although perhaps only a very small chance, that more data will show the hypothesis to be incorrect. Rather than stating that results prove a hypothesis, the correct statement is that the results agree with the hypothesis, and the hypothesis is not rejected. Two kinds of possible errors can be made in testing a hypothesis, as shown in Figure 5-3. If the hypothesis is true but the sample of observations leads to rejection of the hypothesis (saying it is not true), statisticians call the mistake a type I error. The type II error occurs when the hypothesis is not true but the sample of observations agrees closely enough with the hypothesis that the hypothesis is not rejected. The probability of these errors can be calculated only under the assumption that the hypothesis is true, and the true situation is

Decision is	Hypothesis is True	Hypothesis is False
Accept, do not reject the hypothesis.	Decision is correct.	Decision is wrong, a type II error.
Reject the hypothesis.	Decision is wrong, a type I error.	Decision is correct.

Figure 5-3 Types of errors in testing genetic hypotheses.

known when neither can be known for sure. Thus, acceptance or rejection of the hypothesis is right or is wrong.

Traditional statistical testing involves a probability of type I error of 5 percent or of 1 percent because the first statistical tables, having been laborious to compute, were prepared for only 5 percent or 1 percent. For example, the hypothesis is assumed true and is rejected only if the chance that the observed results differ from the expected results is less than 5 percent. The probability of a type II error is not commonly used in hypothesis testing.

Testing with the binomial expansion

One method of testing a genetic hypothesis is by calculating the exact probability that the observed results would occur if the hypothesis were true. For example, assume the hypothesis is that black color in horses is dominant to chestnut color. Suppose a black stallion known to have sired chestnut foals, and thus under the hypothesis assumed to be heterozygous, is mated to 10 chestnut mares, which under the hypothesis are assumed to be homozygous. The resulting progeny are seven black and three chestnut foals. What is the chance of obtaining this result based on the hypothesis? The test procedure is to assume the expected probabilities of black and chestnut foals of .5 and .5 are true and calculate the probabilities of all possible combinations of black and chestnut foals totaling 10. The combinations are 10 black and 0 chestnut, 9 black and 1 chestnut, . . . , 0 black and 10 chestnut, and their expected frequencies can be calculated using either rule 3 of Section 5.2 or the binomial expansion. The resulting calculations listed in Table 5-1 show that there is only about a 25 percent chance ($\frac{252}{1024}$) of having exactly the expected numbers of 5 and 5. If the hypothesis is true, there is just as great a chance that the sample ratio would be 3 : 7 and 7 : 3. The probability of a ratio of 7 : 3 or greater and, as well, the probability of a ratio of 3 : 7 or less is

$$176/1024 + 176/1024 = .34$$

Table 5-1 Probabilities of combinations of ten offspring when 50 percent of each are expected (test cross of heterozygote for a dominant trait)

Number of black	Number of chestnut	Probability of such a result
10	0	1/1024 = .001
9	1	10/1024 = .010
8	2	45/1024 = .044
7	3	120/1024 = .117
6	4	210/1024 = .205
5	5	252/1024 = .246
4	6	210/1024 = .205
3	7	120/1024 = .117
2	8	45/1024 = .044
1	9	10/1024 = .010
0	10	1/1024 = .001
		Total = 1.000

Thus, a ratio as different from 5 : 5 as 7 : 3 is likely—in fact, should occur over one-third of the time. Such a result would not suggest rejecting the hypothesis of equal frequencies of black and chestnut.

The binomial probabilities are not symmetrical when the probability of an event is not one-half. The interpretation of results different from the expected ratio also is somewhat more difficult. In dogs, for example the Collie, the allele for sable is dominant to the allele for black and tan. The allele for black and tan, when homozygous, produces black and tan with white markings, called tricolor. Suppose two sable Collies, both of which had a tricolor parent, are mated. Because both would have to be heterozygous if the hypothesis were true, the expected phenotypes of the pups would be 75 percent sable and 25 percent the recessive black and tan. Suppose, out of a litter of 10 pups, that one-half were sable and one-half were black and tan. Table 5-2 gives the probabilities of the possible combinations of sable and black and tan pups in litters of size 10, based on the binomial expansion of $(.75 \text{ sable} + .25 \text{ black and tan})^{10}$.

The table shows that a litter of 5 and 5 is not very likely (.058), but would be expected only about once in every 20 litters of size 10. Litters of 10 pups with fewer than 5 sables are quite unlikely, and litters of all 10 black and tan would be expected only once out of every million litters.

Table 5-2 Probabilities of combinations of 10 offspring when .75 sable and .25 black and tan are expected (mating of heterozygotes for a dominant trait)

Number of sable	Number of black and tan	Probability of such a litter
10	0	.056
9	1	.188
8	2	.282
7	3	.250
6	4	.146
5	5	.058
4	6	.016
3	7	.003
2	8	.0004
1	9	.00003
0	10	.000001

The lack of symmetry of the probabilities is evident. The expected numbers of 7.5 sable and 2.5 black and tan cannot occur, but the combinations with highest probability are 8 : 2 and 7 : 3, as might be predicted. Lack of symmetry makes difficult the determination of the probability of a ratio as much different from the expected ratio as that observed. Even though the probabilities are difficult to determine, the important point is that differences from the hypothesized ratio can occur by chance in either direction.

Testing with the chi-square approximation

The binomial expansion can be used to determine the chance of various combinations for sample sizes smaller or larger than 10. The computations for larger samples, however, can become overwhelming unless a computer is used. Therefore, a chi-square approximation, which is based on the normal or so-called bell-shaped distribution, is often used to test whether the observed results agree or disagree with the hypothesis.

A histogram of the expected progeny numbers when the probabilities are .50 for each of 2 events appears surprisingly similar to the normal or bell-shaped distribution, even with a sample size as small as 10 [Figure 5-4(*A*)]. The approximation to the normal distribution is not so close when the probabilities are .75 and .25 [Figure 5-4(*B*)].

With larger and larger sample sizes a plot of the binomial outcomes be-

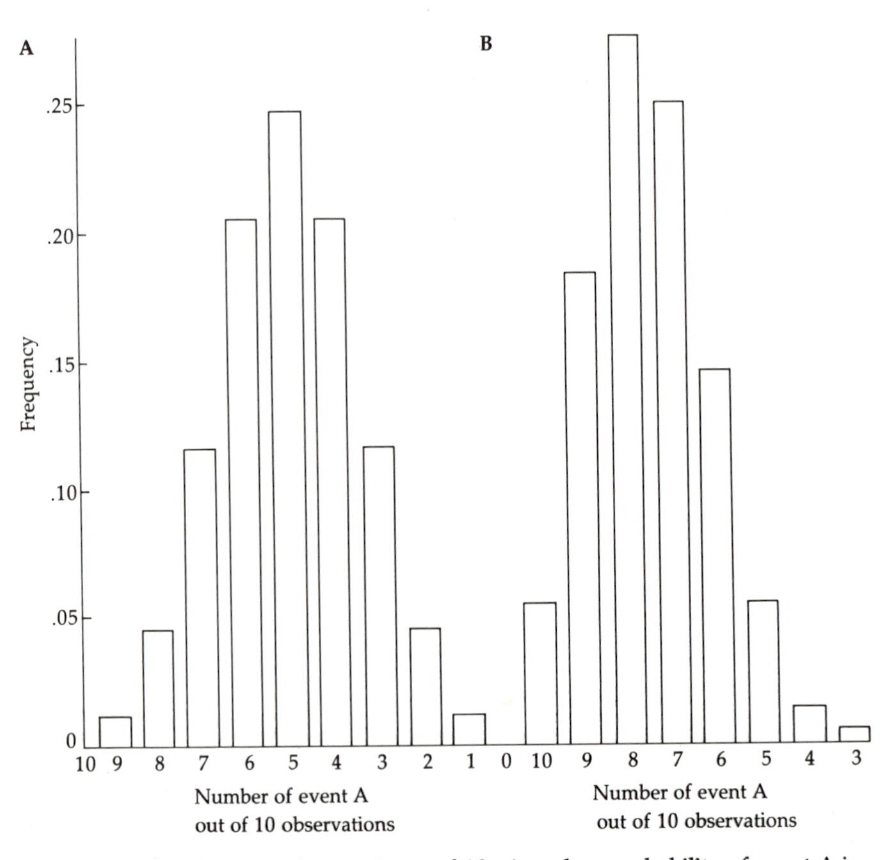

Figure 5-4 *(A)* Frequency of event A out of 10 tries when probability of event A is .50. *(B)* Frequency of event A out of 10 tries when probability of event A is .75.

comes more and more similar to the normal distribution even when the expected numbers are not .50 of each. The histograms for samples of 100 are shown in Figure 5-5(*A*) and (*B*) for probabilities of .50 and .75. If the dots indicating tops of the bars of the histograms are connected close approximations to the normal distribution are obtained. The approximation to the normal distribution can be used to test whether the results of matings agree with the genetic expectations. For example, what is the chance a ratio of 60 black to 40 chestnut foals, or a ratio even more different from expectation, would result from mating heterozygous black stallions to chestnut mares? The calculations could be done with the binomial expansion, as with sample size of 10, but the computations without a computer are exceedingly tedious. Therefore, the statistical procedure called the *chi-square test* is applied to approximate the probability that a difference as great or greater than expectation would be observed.

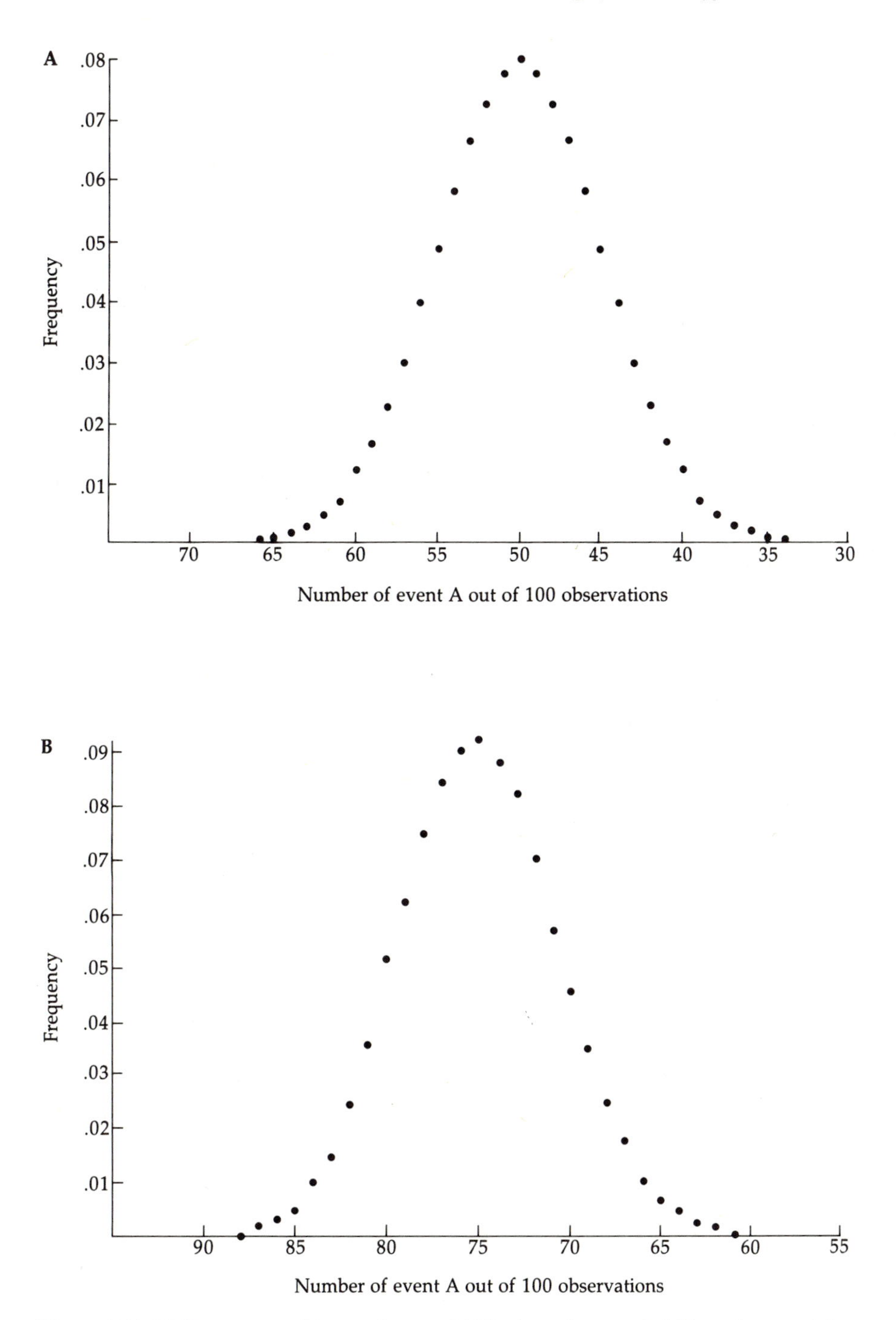

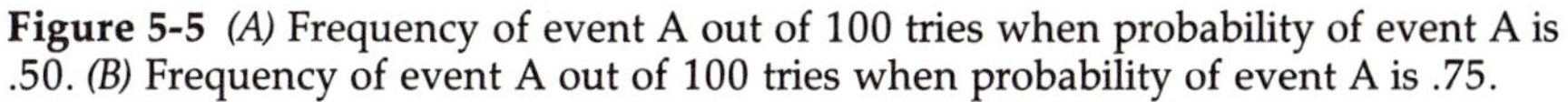

Figure 5-5 *(A)* Frequency of event A out of 100 tries when probability of event A is .50. *(B)* Frequency of event A out of 100 tries when probability of event A is .75.

Table 5-3 Critical values of the chi-square distribution

Degrees of freedom	Probability of such a value of the chi-square statistic by chance								
	0.995	**0.975**	**0.9**	**0.5**	**0.1**	**0.05**	**0.025**	**0.01**	**0.005**
1	.000	.000	0.016	0.455	2.706	3.841	5.024	6.635	7.879
2	0.010	0.051	0.211	1.386	4.605	5.991	7.378	9.210	10.597
3	0.072	0.216	0.584	2.366	6.251	7.815	9.348	11.345	12.838
4	0.207	0.484	1.064	3.357	7.779	9.488	11.143	13.277	14.860
5	0.412	0.831	1.610	4.351	9.236	11.070	12.832	15.086	16.750
6	0.676	1.237	2.204	5.348	10.645	12.592	14.449	16.812	18.548
7	0.989	1.690	2.833	6.346	12.017	14.067	16.013	18.475	20.278
8	1.344	2.180	3.490	7.344	13.362	15.507	17.535	20.090	21.955
9	1.735	2.700	4.168	8.343	14.684	16.919	19.023	21.666	23.589
10	2.156	3.247	4.865	9.342	15.987	18.307	20.483	23.209	25.188
11	2.603	3.816	5.578	10.341	17.275	19.675	21.920	24.725	26.757
12	3.074	4.404	6.304	11.340	18.549	21.026	23.337	26.217	28.300
13	3.565	5.009	7.042	12.340	19.812	22.362	24.736	27.688	29.819
14	4.075	5.629	7.790	13.339	21.064	23.685	26.119	29.141	31.319
15	4.601	6.262	8.547	14.339	22.307	24.996	27.488	30.578	32.801
16	5.142	6.908	9.312	15.338	23.542	26.296	28.845	32.000	34.267
17	5.697	7.564	10.085	16.338	24.769	27.587	30.191	33.409	35.718
18	6.265	8.231	10.865	17.338	25.989	28.869	31.526	34.805	37.156
19	6.844	8.907	11.651	18.338	27.204	30.144	32.852	36.191	38.582
20	7.434	9.591	12.443	19.337	28.412	31.410	34.170	37.566	39.997
21	8.034	10.283	13.240	20.337	29.615	32.670	35.479	38.932	41.401
22	8.643	10.982	14.042	21.337	30.813	33.924	36.781	40.289	42.796
23	9.260	11.688	14.848	22.337	32.007	35.172	38.076	41.638	44.181
24	9.886	12.401	15.659	23.337	33.196	36.415	39.364	42.980	45.558
25	10.520	13.120	16.473	24.337	34.382	37.652	40.646	44.314	46.928

26	11.160	13.844	17.292	25.336	35.563	38.885	41.923	45.642	48.290
27	11.808	14.573	18.114	26.336	36.741	40.113	43.194	46.963	49.645
28	12.461	15.308	18.939	27.336	37.916	41.337	44.461	48.278	50.993
29	13.121	16.047	19.768	28.336	39.088	42.557	45.722	49.588	52.336
30	13.787	16.791	20.599	29.336	40.256	43.773	46.979	50.892	53.672
31	14.458	17.539	21.434	30.336	41.422	44.985	48.232	52.192	55.003
32	15.135	18.291	22.271	31.336	42.585	46.194	49.481	53.486	56.329
33	15.816	19.047	23.110	32.336	43.745	47.400	50.725	54.776	57.649
34	16.502	19.806	23.952	33.336	44.903	48.602	51.966	56.061	58.964
35	17.192	20.570	24.797	34.336	46.059	49.802	53.203	57.342	60.275
36	17.887	21.336	25.643	35.336	47.212	50.998	54.437	58.619	61.582
37	18.586	22.106	26.492	36.335	48.363	52.192	55.668	59.893	62.884
38	19.289	22.879	27.343	37.335	49.513	53.384	56.896	61.162	64.182
39	19.996	23.654	28.196	38.335	50.660	54.572	58.120	62.428	65.476
40	20.707	24.433	29.051	39.335	51.805	55.758	59.342	63.691	66.766
41	21.421	25.215	29.907	40.335	52.949	56.942	60.561	64.950	68.053
42	22.139	25.999	30.765	41.335	54.090	58.124	61.777	66.206	69.336
43	22.860	26.786	31.625	42.335	55.230	59.304	62.990	67.460	70.616
44	23.584	27.575	32.487	43.335	56.369	60.481	64.202	68.710	71.893
45	24.311	28.366	33.350	44.335	57.505	61.656	65.410	69.957	73.166
46	25.042	29.160	34.215	45.335	58.641	62.830	66.617	71.202	74.437
47	25.775	29.956	35.081	46.335	59.774	64.001	67.821	72.443	75.704
48	26.511	30.755	35.949	47.335	60.907	65.171	69.023	73.683	76.969
49	27.250	31.555	36.818	48.335	62.038	66.339	70.222	74.920	78.231
50	27.991	32.357	37.689	49.335	63.167	67.505	71.420	76.154	79.490

Source: Values from 1 to 30 degrees of freedom from C. M. Thompson, *Biometrika* **32:**188–189, 1941, with permission of the publisher. Values from 31 to 50 degrees of freedom from F. J. Rohlf and R. R. Sokal, *Statistical Tables,* 2nd ed., W. H. Freeman and Company, New York, 1981.

The chi-square test also can be used for multinomial cases when there are more than two possible events.

When using the chi-square test, the number of *degrees of freedom* must also be determined to be used with the calculated chi-square statistic to estimate the probability that the observed results could have happened by chance if the hypothesis were true. The number of degrees of freedom is usually one less than the number of classes into which events can be classified. In the binomial case with two classes, for example black and chestnut horses, the degrees of freedom are $2 - 1 = 1$. Thus if the sample size is n, and s of the observations are black, then the number of chestnuts is known to be $t = n - s$. For example, if 6 are black, then $10 - 6 = 4$ must be chestnut. The degrees of freedom is the number of independent classes ($2 - 1 = 1$ in this example).

The symbol for the chi-square statistic is the Greek letter chi, χ, with an exponent of 2, that is, χ^2. When testing a hypothesis, the calculated chi-square statistic is compared to critical values of the χ^2 distribution associated with the proper degrees of freedom (Table 5-3). If the chi-square statistic is greater than the critical value for the specified probability level of a type I error (usually .05), then the observations are said not to agree with the hypothesis. A chi-square statistic less than the critical value leads to the statement that the hypothesis is not rejected.

The calculation of the chi-square statistic is easy. If there are N classes so that degrees of freedom $= N - 1$,

$$\begin{aligned}\chi^2_{N-1} = & \frac{(\text{number observed in class 1} - \text{number expected in class 1})^2}{\text{number expected in class 1}} \\ & + \frac{(\text{number observed in class 2} - \text{number expected in class 2})^2}{\text{number expected in class 2}} \\ & + \cdots \\ & + \frac{(\text{number observed in class } N - \text{number expected in class } N)^2}{\text{number expected in class } N}\end{aligned}$$

where the subscript, $N - 1$, on χ^2 refers to the number of degrees of freedom. In statistical shorthand this is

$$\chi^2_{N-1} = \sum_{i=1}^{N} \left(\frac{(O_i - E_i)^2}{E_i} \right)$$

where $(O_i - E_i)^2/E_i$ is the squared difference between the observed and expected number in class i divided by the expected number in class i and $\Sigma_{i=1}^{N}$ means that the expression in parentheses is to be summed for all values of

Table 5-4 The relationship between size of χ^2 statistic and the genetic hypothesis

Calculated χ^2	Agreement between observed and expected	Probability of difference due to chance	Status of hypothesis that gave expected numbers
Large	Poor	Small	May reject
Small	Good	Large	Do not reject

the subscript i (class $= 1, 2, \cdots$, up to N). The calculated χ^2 statistic corresponds to a probability that such a deviation from expected could occur by chance. Some of the probabilities are found in Table 5-3. As illustrated in Table 5-4, if the χ^2 statistic is large, which happens when the observed and expected numbers do not agree well, then there is only a small chance the observed numbers could have happened if the hypothesis was true. Then, depending on the degrees of freedom and the level selected for the type I error, the hypothesis that gave the expected numbers will likely be rejected.

The chi-square approximation becomes more accurate as sample size becomes large and when each class has roughly the same expected number. The chi-square test gives a poor approximation with small sample sizes and with a small expected number (5 or less) in any of the classes.

As an example, suppose the chi-square approximation is used to test the hypothesis that black is dominant to chestnut in horses. The progeny resulting from mating the black heterozygous stallion to 10 chestnut mares were 7 black and 3 chestnut, when the expected numbers were 5 black and 5 chestnut. For this example, $N = 2$ is the number of classes, and the degrees of freedom are $N - 1 = 1$. The formula for the chi-square statistic ($N = 2$) is

$$\chi_1^2 = \frac{(O_1 - E_1)^2}{E_1} + \frac{(O_2 - E_2)^2}{E_2}$$

When observed and expected numbers are substituted for their symbols:

$$\chi_1^2 = \frac{(7 - 5)^2}{5} + \frac{(3 - 5)^2}{5} = 1.6$$

FromTable 5-3, the probability of obtaining such a chi-square statistic by chance, for degrees of freedom of 1, is between .5 and .1 (the actual probability of .22 is not shown in the table). The probability is large enough to suggest that such a result could well have happened, that is, 22 percent of the time, if the hypothesis is true.

If the χ^2 statistic had been 3.841 or larger, Table 5-3 indicates such a result would be expected 5 percent of the time or less, and the hypothesis would be rejected if a type I error of 5 percent is specified.

If 100 such matings resulted in 60 black and 40 chestnut foals when 50 of each were expected, the chi-square statistic would be

$$\chi_1^2 = \frac{(60-50)^2}{50} + \frac{(40-50)^2}{50} = 4.0$$

with probability of such a result due to chance of less than .05 (in fact, .045). Had the binomial expansion been used, the probability of such a result or worse by chance would be calculated as .057. Therefore, from a practical standpoint, the approximate probability from the chi-square approximation is close to the true probability obtained from the binomial expansion.

This last example, however, illustrates the arbitrary nature of testing hypotheses. Because of computational limitations when the theory of hypothesis testing was developed, most tests of significance of differences between observed and expected results have been made at the 5 percent (or 1 percent) level of probability of a type I error. In this example, the probability from the chi-square statistic, .045, is less than .05; under the 5 percent rule, the hypothesis is rejected. The exact binomial probability, however, is .057, which would indicate acceptance (that is, not rejection) of the hypothesis.

The determination of the proper hypothesis to test is not always easy. As an example, geneticists had long suspected that the gene for roan in horses is lethal in the fetal stage when homozygous. Table 5-5 gives the results of the study of the Belgian registry by Hintz and Van Vleck (1979).

If the gene for roan is lethal when homozygous, then the expected ratio of roans to nonroans in roan by roan matings is 2 : 1 because the other one-quarter die. All roans would be heterozygous, and the expected ratio of offspring

Table 5-5 Roan and nonroan foals registered in 1937 from roan by roan and roan by nonroan matings

Type of mating	Male foals		Female foals		Combined	
	Roan	Nonroan	Roan	Nonroan	Roan	Nonroan
roan × roan	56	32	74	35	130	67
roan × nonroan	119	117	165	181	284	298

Source: H. F. Hintz and L. D. Van Vleck, *J. Heredity* **70**:145–146, 1979.

would be 2 roan : 1 nonroan : 1 died before observation is possible. Thus, Hintz and Van Vleck tested the expected ratio of 2 : 1. They found the observed ratio to be 1.94 : 1 with a chi-square statistic of .0406 corresponding to a probability of such a result by chance of .90. They concluded the results were in agreement with the hypothesis that the homozygous condition is lethal.

Another approach would have been to test the hypothesis that roan is not lethal. Then the expected ratio would be at least the classical 3 : 1 for a dominant trait. (Actually, more than 75 percent would be expected because if the gene is not lethal when homozygous some of the roan parents might be homozygous rather than heterozygous.) The chi-square statistic when the observations are compared to an expected 3 : 1 ratio is 8.543 with probability of .005, which would lead to the rejection of the hypothesis of a simple dominant mode of inheritance.

Although in this case testing two separate hypotheses poses no problem, the guideline followed by statisticians is not to test sequential hypotheses with the same data; that is, do not keep changing the hypothesis until a good agreement is obtained between the observations and the expected numbers. Should a geneticist reject one hypothesis then formulate an alternative hypothesis, the new hypothesis should be tested with new observations. Another obvious guideline is that a hypothesis should have a reasonable biological basis.

Testing multinomial hypotheses with chi-square

The multinomial expansion can be used to calculate probabilities for observed results given assumed probabilities of occurrence of individual events based on hypotheses involving more than two classes. Deciding which combinations are more extreme than those observed is difficult. As a consequence, the chi-square test is usually used. An example is testing for linkage between two loci. The hypothesis to be tested usually is stated as the two loci are independent. If the loci are independent a test cross should give equal frequencies of the four types of progeny. For example, Castle (1936) reported the following results in rabbits from a test cross (dihybrid mated with double homozygous recessive):

Class	Number
Himalayan spotting, black	283
Himalayan spotting, brown	171
solid color, black	163
solid color, brown	291
	908

In fact there are three hypotheses. The fetal viability of each phenotype is also involved. If the locus for spotting were independent of the locus for hair color, if blacks and browns were equally viable fetuses, and if solids and spotteds were equally viable, the expected numbers for the four classes would be equal; that is, $\frac{908}{4} = 227$ of each. Degrees of freedom are for four classes: $4 - 1 = 3$. The chi-square statistic is

$$\chi_3^2 = \frac{(283 - 227)^2}{227} + \frac{(171 - 227)^2}{227} + \frac{(163 - 227)^2}{227} + \frac{(291 - 227)^2}{227} = 63.72$$

with probability of such a result by chance, $P < .005$. Thus, the joint hypotheses of independence and equal viabilities are rejected. The test does not indicate which, if any, of the individual hypotheses should be rejected.

A useful property of the chi-square statistic is that it can be separated into statistically independent parts corresponding to the individual hypotheses with corresponding degrees of freedom. In the rabbit example, $\chi_3^2 = 63.72$ can be broken into three χ_1^2 statistics, each with a single degree of freedom. One test of viability would be whether Himalayans and solids (two classes and one degree of freedom) occur equally:

$$\chi_1^2 = \frac{(283 + 171 - 454)^2}{454} + \frac{(163 + 291 - 454)^2}{454} = 0$$

with probability 1.0 so that there is no evidence for different fetal viability between Himalayans and solids. These results from an actual experiment show that, even though with large samples the exact expected numbers rarely occur, these numbers can occur. A similar test can be made for viability of blacks and browns:

$$\chi_1^2 = \frac{(283 + 163 - 454)^2}{454} + \frac{(171 + 291 - 454)^2}{454} = .28$$

with the probability between .90 and .50 of such a result by chance. Again, there is no evidence for differences in fetal viability between blacks and browns.

The third degree of freedom is for crossover versus noncrossover types. The most numerous types, Himalayan black and solid brown, are parent types, so the crossover types are Himalayan brown and solid black. The chi-square statistic is

$$\chi_1^2 = \frac{(283 + 291 - 454)^2}{454} + \frac{(171 + 163 - 454)^2}{454} = 63.44$$

with the probability $<.005$ of such a result by chance. This indicates that essentially all of the difference of the observed numbers from the expected numbers in the four classes can be attributed to linkage. The hypothesis of independence is rejected. The three chi-square statistics with one degree of freedom sum to the overall chi-square statistic with three degrees of freedom: $.00 + .28 + 63.44 = 63.72$.

It is not correct, however, to estimate the crossover fraction from these records, recalculate the expected numbers in the four classes based on the estimated crossover frequency, and then do a chi-square test. Almost certainly the hypothesis would not be rejected. The proper procedure would be to estimate the crossover frequencies and use those frequencies to compute expected numbers to test against numbers obtained from another set of matings.

Testing using confidence ranges

A third method of testing, for binomial events, whether the observed results agree with the expected results is available although seldom used. The method also depends on the approximation of the binomial frequencies by the normal distribution. A standard statistical rule for the normal distribution is that the range,

$$(\text{average} - \text{standard deviation}) \text{ to } (\text{average} + \text{standard deviation})$$

will include 68 percent of the observations of a normally distributed variable (Figure 5-6). The standard deviation is the square root of the variance. Both are measures of variability and will be discussed more fully in later chapters when concepts of quantitative genetics are introduced.

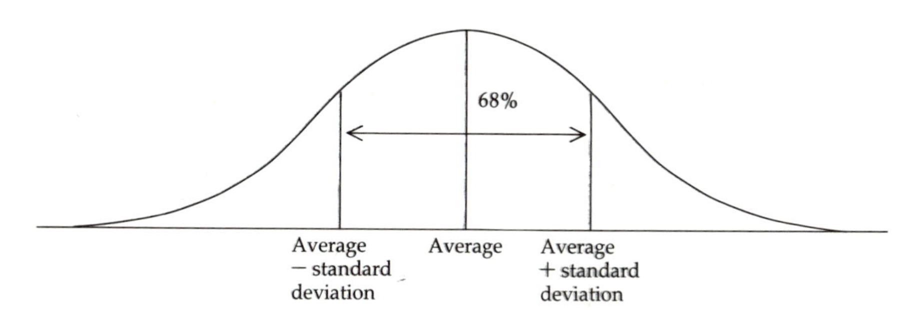

Figure 5-6 The range from one standard deviation less than the average to one standard deviation greater than the average includes 68 percent of observations for a normally distributed trait.

Frequency, p, assumed known The variance of the average of a binomial variable (for example, black color) is $[p(1-p)]/n$, and thus depends on the frequency of the event, p, and the sample size, n. The range can be adjusted to any confidence level desired. Thus, the general formula for the confidence range on the binomial probability is

$$p - t\sqrt{p(1-p)/n} \quad \text{to} \quad p + t\sqrt{p(1-p)/n}$$

where t determines the confidence level used to test the hypothesis, for example, 68 percent, 95 percent, 99 percent. Table 5-6 gives the correspondence between t and the confidence level.

The usual rule is that if the range includes the observed frequency then the hypothesis is not rejected, but if the observed frequency is outside the range, the hypothesis is rejected. The 95 percent confidence range corresponds to a type I error of 5 percent, and the 99 percent confidence range to a type I error of 1 percent.

In the previous example of mating a heterozygous black stallion to 10 chestnut mares, the expected frequency, p, of a black foal was .5. Thus, under the *if* principle, if $p = .5$ and $n = 10$, does the confidence range include the observed case of 7 black and 3 chestnut foals? The hypothesis is $p = .5$ so that $\sqrt{p(1-p)/n} = .16$. The 95 percent range is $.5 - 1.96(.16)$ to $.5 + 1.96(.16)$ or .19 to .81. Because the range includes the frequency of .70 which was observed, the hypothesis is not rejected. In fact, had 8 black out of 10 been observed, the hypothesis also would not have been rejected. A similar decision would have been made after examining the exact probabilities calculated from the binomial

Table 5-6 Values to multiply by the standard deviation to obtain different confidence ranges

Percentage included in range (confidence level)	**Multiplier** t
50	.67
60	.84
68	1.00
70	1.04
80	1.28
90	1.65
95	1.96
99	2.58

expansion and after using the chi-square approximation. The confidence range method, as with the chi-square test, works well for large samples and expected frequencies near .5, and less well for small samples and expected frequencies approaching 1.0 or 0.0.

Sample estimate of p The confidence range is more commonly used in population genetics, which will be introduced in the next chapter. The frequency of an allele in the population is often estimated and never known exactly. The confidence range in this case depends on the estimate of p, denoted $\hat{p}$ and read "p hat." The hat, ˆ, is a common notation to distinguish an estimate from the true frequency. The *if* principle cannot be applied because the true frequency is unknown. The estimate of the standard deviation also depends on $\hat{p}$ and sample size n. The confidence range is $\hat{p} \pm t\sqrt{[\hat{p}(1-\hat{p})]/n}$, which is the same as before except that $\hat{p}$ is substituted for p. Because the true p is not known or assumed, the interpretation of the range is subtly different: the range is 95 percent (or whatever level is chosen) certain to include the true p. When p is assumed known, the interpretation is that the sample average of 95 percent of all possible samples is within the range.

5-6 Summary

Important terms in discussing probability include event, observation, probability, and mutually exclusive, independent, and correlated events.

The basic rules for calculating probabilities are:

1. If events A and B are mutually exclusive, then

$$P(\text{A or B occurs}) = P(\text{A occurs}) + P(\text{B occurs})$$

2. If two events, A and B, are independent, then

$$P(\text{A and B both occur}) = P(\text{A occurs}) \times P(\text{B occurs})$$

3. If two events, A and B, have probabilities of occurrence of p and $q = 1 - p$, then with n observations the probability that event A will occur exactly s times and event B will occur exactly r times ($s + r = n$) is

$$P(s \text{ of A and } r \text{ of B}) = (n!/s!r!)(p^s q^r)$$

This formula is the outcome of the binomial expansion

$$(p\text{A} + q\text{B})^n$$

These three rules can be expanded to multinomial cases where more than two different events can occur.

4. If two events, A and B, are not independent, then

   P(A occurs given that B occurs) = P(A and B both occur)/P(B occurs)

   that is,

$$P(\text{A}|\text{B}) = P(\text{A, B})/P(\text{B})$$

There are several possible tests of genetic hypotheses. All are based on probabilities that the observed results will occur if the hypothesis is true:

1. From the binomial expansion, calculate the exact probability the observed or a worse result could have occurred by chance.
2. For large sample size, calculate a chi-square value with $N-1$ degrees of freedom,

$$\chi^2_{N-1} = \sum_{i=1}^{N} \left(\frac{(O_i - E_i)^2}{E_i} \right),$$

   which corresponds to an approximate probability the difference between observed (O_i) and expected (E_i) numbers could have occurred by chance.
3. Calculate the confidence range based on the hypothesis that p is the expected frequency of event A, and n is the sample size

$$p \pm t \sqrt{[p(1-p)]/n}$$

   where t is a number corresponding to a probability the result could occur by chance. If the range includes the proportion found in the sample, there is no reason to reject the hypothesis.
4. If p is estimated from a set of data (for example, the frequency of an allele) as $\hat{p}$, then the confidence range

$$\hat{p} \pm t \sqrt{[\hat{p}(1-\hat{p})]/n}$$

   would be expected to contain the true frequency, p, a certain fraction of the time (confidence) corresponding to the number used for t. ($t = 1.96$ corresponds to a 95 percent confidence level.)

References

Castle, W. E. 1936. Further data on linkage in rabbits. *Proc. Nat. Acad. Sci.* **22:**222–225.

Hintz, H. F., and L. D. Van Vleck. 1979. Lethal dominant roan in horses. *J. Heredity* **70:**145–146.

CHAPTER 6

Basic Principles of Population Genetics

An understanding of Mendel's fundamental laws of inheritance allows the prediction of the genotypic and phenotypic distributions of the progeny that result from matings between parents of known genotypes. For example, the mating of heterozygous polled cattle ($Pp \times Pp$) is expected to produce the genotypic ratio of 1 PP : 2 Pp : 1 pp in the progeny. Further, since P is dominant to its allele p, the phenotypic distribution is expected to be three polled progeny to one horned. Chance deviations from the expected genotypic and phenotypic ratios can occur, the probabilities of which can be calculated as shown in Chapter 5.

To predict the phenotypic and genotypic frequencies of progeny resulting from matings among the total population the probabilities of matings between individuals of various genotypes are needed. That is, for a population of individuals having genotypes PP, Pp, and pp, the frequency of matings between, for example, PP males and pp females must be known to predict the frequencies in their progeny.

Population genetics is the study of gene and genotypic frequencies within a population and the prediction of these frequencies in subsequent generations. This chapter describes concepts of population genetics as they apply to qualitative traits.

6-1 Gene and Genotypic Frequencies

For the polled or horned condition in cattle there are two alleles involved at an autosomal locus. The frequency of each allele and of each genotype must be known to describe a population relative to this locus. A frequency is denoted as

$f(\ )$; hence, $f(P)$ represents the frequency of the P allele and $f(Pp)$ the frequency of the heterozygote in the population. Because each individual of a diploid species carries two genes at a particular autosomal locus, in a population of n individuals there are $2n$ total genes at that locus.

Genotypic frequency

The frequency of a genotype is defined as the proportion of the n individuals in the population with a particular genotype. If the number of individuals in a population of polled and horned cattle with each genotype is represented by n subscripted with the genotype, then

$$n_{PP} = \text{ the number of homozygous polled, } PP$$
$$n_{Pp} = \text{ the number of heterozygous polled, } Pp\text{, and}$$
$$n_{pp} = \text{ the number of horned, } pp$$

The genotypic frequencies can then be represented

$$f(PP) = n_{PP}/n$$
$$f(Pp) = n_{Pp}/n,$$

and

$$f(pp) = n_{pp}/n$$

The sum of frequencies of all possible events must equal 1:

$$f(PP) + f(Pp) + f(pp) = 1$$

Gene frequencies

The frequency of a gene is defined as the proportion of the $2n$ genes represented by a particular allele. If n_P is the number of P alleles in the population, then

$$f(P) = n_P/2n$$

The number of P alleles is twice the number of homozygous polled animals (they each carry two P alleles) plus the number of heterozygotes, (each of which carries one P allele); that is,

$$n_P = 2n_{PP} + n_{Pp}$$

Similarly, n_p is the number of heterozygotes plus twice the number of horned animals:

$$n_p = n_{Pp} + 2n_{pp}$$

Gene frequencies may also be calculated from genotypic frequencies. If $f(PP)$, $f(Pp)$, and $f(pp)$ are the frequencies of the three genotypes, then

$$f(P) = f(PP) + (1/2)f(Pp)$$

and

$$f(p) = (1/2)f(Pp) + f(pp)$$

Example 6-1 illustrates the calculation of gene and genotypic frequencies of coat color in a sample of Shorthorn cattle.

Example 6-1

In Shorthorn cattle, three coat colors are red, roan, and white. In a sample of 1000 Shorthorns assume the number of animals with each coat color is

Color	Genotype	Number
Red	RR	$n_{RR} = 360$
Roan	Rr	$n_{Rr} = 480$
White	rr	$n_{rr} = 160$
		$n = 1000$

The genotypic frequencies are

$$\text{f(red)} = f(RR) = 360/1000 = .36$$
$$\text{f(roan)} = f(Rr) = 480/1000 = .48$$
$$\text{f(white)} = f(rr) = 160/1000 = .16$$

Because of codominance these are also the phenotypic frequencies. In this sample, there are a total of 2000 genes:

$$\begin{aligned} n_R &= 2n_{RR} + n_{Rr} \\ &= 720 + 480 = 1200 \end{aligned}$$

$$\begin{aligned} n_r &= n_{Rr} + 2n_{rr} \\ &= 480 + 320 = 800 \end{aligned}$$

Therefore,

$$f(R) = n_R/2n = 1200/2000 = .6$$

and

$$f(r) = n_r/2n = 800/2000 = .4$$

Note also that the same results are obtained using genotypic frequencies to calculate gene frequencies:

$$\begin{aligned} f(R) &= f(RR) + (1/2)f(Rr) \\ &= .36 + (1/2)(.48) = .6 \end{aligned}$$

and

$$\begin{aligned} f(r) &= (1/2)f(Rr) + f(rr) \\ &= (1/2)(.48) + .16 = .4 \end{aligned}$$

Frequencies are equivalent to probabilities. For example, the probability of randomly drawing a *PP* individual from a population is equal to $f(PP)$. Likewise, the probability of randomly drawing a gamete carrying the *P* gene from the pool of gametes is $f(P)$. Equating frequencies and probabilities becomes important when determining the chance of a mating between two individuals of given genotypes.

Random mating

Random mating, also called *panmixia,* is a mating system in which each individual has an equal opportunity to mate with any individual of the opposite sex. When the impact of alternative mating strategies on gene and genotypic frequencies is considered, random mating is often used as the basis of comparison. That is, how do the gene and genotypic frequencies obtained from a particular mating strategy differ from those that would have resulted had the population mated at random?

Probability of mating

The probability of random mating between two animals of given genotypes is the product of the frequencies of the two genotypes in the population. For a particular autosomal locus with two alleles, for example, *B* and *b*, and three genotypes, *BB, Bb,* and bb, there are nine possible mating combinations with the following frequencies:

Genotype of parent		
Male	**Female**	**Frequency of mating**
BB	*BB*	$f(BB) \times f(BB)$
BB	*Bb*	$f(BB) \times f(Bb)$
BB	*bb*	$f(BB) \times f(bb)$
Bb	*BB*	$f(Bb) \times f(BB)$
Bb	*Bb*	$f(Bb) \times f(Bb)$
Bb	*bb*	$f(Bb) \times f(bb)$
bb	*BB*	$f(bb) \times f(BB)$
bb	*Bb*	$f(bb) \times f(Bb)$
bb	*bb*	$f(bb) \times f(bb)$

An assumption in determining frequencies of these matings is that genotypic frequencies are the same in males as in females. In some cases, for example, with sex-linked alleles, this assumption may not be valid and the frequencies of genotypes must be established for each sex. Nevertheless, with random mating, the probability of a particular mating is always the product of the genotypic frequencies of the mates.

6-2 The Hardy-Weinberg Law

With the rediscovery in 1900 of Mendel's work, scientists began to reconcile evolutionary theory with the particulate theory. One question was, what happens to gene and genotypic frequencies from one generation to the next? In 1908, G. H. Hardy and W. Weinberg, working independently, developed the fundamental relationship between gene and genotypic frequencies over many generations under certain assumptions. The Hardy-Weinberg law may be stated as follows:

> For a large random mating population, in the absence of forces that change gene frequencies (mutation, migration, and selection), the gene and genotypic frequencies remain constant from one generation to the next.

When gene and genotypic frequencies remain constant a population is said to be *in equilibrium.* The necessary conditions for equilibrium are

1. a large population,
2. random mating, and
3. the absence of forces acting to change gene frequency.

Chapter 7 discusses the consequences when these conditions are not met.

Beginning with the following discussion of gene and genotypic frequencies in populations in Hardy-Weinberg equilibrium, a shorthand notation for gene frequencies will be used. For an autosomal locus with two alleles, a common notation is to represent the frequency of one allele, say *B*, with the letter p and the frequency of the other allele, *b*, with the letter q. That is,

$$f(B) = p$$
$$f(b) = q$$

Gene and genotypic frequencies at Hardy-Weinberg equilibrium

A population in Hardy-Weinberg equilibrium for a single autosomal locus with two alleles exhibits the existence of a distinct relationship between gene and genotypic frequencies. Assume a population at equilibrium with $f(B) = p$ and $f(b) = q$. The genotypic frequencies are

Genotype	Frequency
BB	p^2
Bb	$2pq$
bb	q^2

This means the frequency of genotype *BB* is equal to the square of the frequency of the *B* allele, the frequency of the heterozygote *Bb* is two times the product of the frequency of *B* and *b*, and the frequency of genotype *bb* is the square of the frequency of the *b* allele.

Why does the relationship between gene and genotypic frequencies exist? Under the conditions of the Hardy-Weinberg law, genes pair at random. The probability of drawing a *B* allele at random from the total gene pool of the male or female parents is p, and the probability of drawing a *b* allele is q. Because each genotype has two genes, the probability of having two *B* alleles is p times p, or p^2. These results for genotypic frequencies are expressed by the binomial expansion

$$(pB + qb)^2 = p^2BB + 2pqBb + q^2bb$$

The sum of the probabilities of drawing B or b is 1, or

$$p + q = 1$$

Similarly, the sum of the frequencies of all possible events must equal 1; thus

$$p^2 + 2pq + q^2 = 1$$

That the gene and genotypic frequencies remain constant from one generation to the next can be easily demonstrated. The frequency of the B allele, in both the male and female pools of gametes, is p and of the b allele, q. The union of these gametes is random. The frequency of zygotes from P (the union of two gametes) for four types of unions is as follows:

Gamete from		Frequency of gamete from		
Male	**Female**	**Male**	**Female**	***P*(union of gametes)**
B	B	p	p	p^2
B	b	p	q	pq
b	B	q	p	pq
b	b	q	q	q^2

Therefore, in the progeny, $f(BB) = p^2$, $f(Bb) = pq + pq = 2\text{pq}$, and $f(bb) = q^2$, which are the same as in the parent population. The gene frequencies are also the same:

$$\begin{aligned} f(B) &= f(BB) + (1/2)f(Bb) \\ &= p^2 + pq \end{aligned}$$

which algebraically reduces to

$$\begin{aligned} &= p(p + q) \\ &= p \qquad \text{(recall that } p + q = 1\text{)} \end{aligned}$$

and

$$\begin{aligned} f(b) &= (1/2)f(Bb) + f(bb) \\ &= pq + q^2 \\ &= q(p + q) \\ &= q \end{aligned}$$

For a population in Hardy-Weinberg equilibrium, the relationship between gene and genotypic frequencies can be used to estimate gene frequencies. The frequency of b can be estimated as $\sqrt{f(bb)}$. The estimated frequency of B is then $1 - f(b)$. This approach is useful for traits with a dominant allele because the phenotype for BB is the same as for Bb; "gene counting," as was done with the red, roan, and white Shorthorns in Example 6-1, is not possible. Estimating q from $f(bb)$ is not valid if the population is *not* in equilibrium. To distinguish an estimated gene frequency from the true gene frequency the hat, ˆ (introduced in Chapter 5), is placed over the notation for frequency. Thus, $\hat{q}$ represents an estimate of the frequency of that allele. Example 6-2 illustrates how to estimate gene and genotypic frequencies when one allele is dominant.

Example 6-2

The distribution of coat color in a sample of 1000 Angus cattle was determined to be

Phenotype	Genotype	Number
Black	BB or Bb	640
Red	bb	360
		1000

If the population is in equilibrium, then the estimate of q is

$$\hat{q} = \sqrt{f(bb)}$$
$$= \sqrt{360/1000} = .6$$

Then $\hat{p}$ is $(1 - \hat{q})$ or .4; $\hat{p}$ and $\hat{q}$ can be used to estimate $f(BB)$ and $f(Bb)$:

$$f(BB) = \hat{p}^2 = .16$$

and

$$f(Bb) = 2\hat{p}\hat{q} = .48$$

In Example 6-1 exact gene frequencies were obtained for the sample by gene counting, which was possible since the heterozygote had a distinct phenotype. When one allele is dominant the gene frequencies can only be estimated because the true distribution of homozygotes and heterozygotes cannot be known.

When will a population reach equilibrium such that $f(B) = p$, $f(b) = q$ and $f(BB) = p^2$, $f(Bb) = 2pq$, and $f(bb) = q^2$? In 1918 E. N. Wentworth and B. L. Remick showed that, for one autosomal locus with two alleles, a population not in equilibrium reaches equilibrium in the first generation after random mating *if* the gene frequencies are the same in male and female parents. Hence, in progeny generated by random mating, $f(b) = \sqrt{f(bb)}$; that is, $q = \sqrt{f(bb)}$ under the assumptions of the Hardy-Weinberg law. If, however, the initial gene frequencies in males and females are different, two generations of random mating are required to reach equilibrium. In the first generation the gene frequencies become equal in both sexes and are the average of the frequencies in the parents. In the second generation the genotypic frequencies reach equilibrium. Example 6-3 demonstrates a population that reaches equilibrium in one generation of random mating.

Example 6-3

Assume that the following frequencies of phenotypes were observed in a population of Shorthorn cattle for both males and females:

Phenotype	Genotype	Frequency of genotype
Red	*RR*	.6
Roan	*Rr*	.4
White	*rr*	.0

For this population

$$f(R) = f(RR) + (1/2)f(Rr)$$
$$= .6 + .2 = .8$$

and

$$f(r) = .2$$

The population is obviously not in equilibrium because no white cattle were observed despite the fact that $f(r) = .2$.

This population is now randomly mated. The probabilities of unions between particular gametes are

Gametes		
Male	Female	Probability of union
R	R	$(.8)(.8) = .64$
R	r	$(.8)(.2) = .16$
r	R	$(.2)(.8) = .16$
r	r	$(.2)(.2) = .04$

Hence, the expected genotypic frequencies in the progeny are

$$f(RR) = .64$$
$$f(Rr) = .32$$
$$f(rr) = .04$$

The progeny population is at equilibrium; the gene and genotypic frequencies will remain constant for all future generations under the assumptions of the Hardy-Weinberg law.

6-3 Sex-linked Loci

The calculation of genotypic frequencies for a sex-linked locus with two alleles is similar to that for autosomal loci with the obvious difference of the heterogametic state of one of the sexes (males in mammals, females in birds). Assume that $f(B) = p$ and $f(b) = q$ for two alleles at a sex-linked locus. At equilibrium, the genotypic frequencies within each sex are

	Sex		
Genotype	Mammals	Birds	Frequency within sex
X^BY	Male	Female	p
X^bY	Male	Female	q
X^BX^B	Female	Male	p^2
X^BX^b	Female	Male	$2pq$
X^bX^b	Female	Male	q^2

The sum of genotypic frequencies is $p + q = 1$ in the heterogametic sex and $p^2 + 2pq + q^2 = 1$ in the homogametic sex. In mammals the probability of an X^BY male in the *population of males* is p. However, the probability of an X^BY

male in the *entire population* is $\frac{1}{2}p$ because only one-half of the total population is expected to be male.

In the homogametic sex, the frequency of gametes carrying the B allele is p and of those carrying the b allele is q. The heterogametic sex, however, produces three types of gametes — one-half carrying either the B or b allele and the other half carrying the Y chromosome. The frequency of genotypes in the progeny is the probability of union of male and female gametes of each type under random mating. For mammals the genotypes and frequencies are

Gamete from		Frequency of gametes from			
Male	**Female**	**Male**	**Female**	**P(union)**	**Progeny sex**
X^B	X^B	$(1/2)p$	p	$(1/2)p^2$	Female
X^B	X^b	$(1/2)p$	q	$(1/2)pq$	Female
X^b	X^B	$(1/2)q$	p	$(1/2)pq$	Female
X^b	X^b	$(1/2)q$	q	$(1/2)q^2$	Female
Y	X^B	1/2	p	$(1/2)p$	Male
Y	X^b	1/2	q	$(1/2)q$	Male

The frequency of males in the progeny is

$$\begin{aligned} f(\text{males}) &= (1/2)p + (1/2)q \\ &= (1/2)(p+q) \\ &= 1/2 \end{aligned}$$

and of females is

$$\begin{aligned} f(\text{females}) &= (1/2)p^2 + pq + (1/2)q^2 \\ &= (1/2)(p^2 + 2pq + q^2) \\ &= 1/2 \end{aligned}$$

The expected frequencies of genotypes within sex are obtained as conditional probabilities (Section 5.3). For example, the frequency of X^BX^B genotypes if the progeny is female is expected to be

$$P(X^BX^B|\text{female progeny}) = \frac{P(X^BX^B \text{ and female})}{P(\text{female})} = \frac{(1/2)p^2}{1/2} = p^2$$

This is the same frequency as in females of the parent population. Similar calculations provide the expected frequencies of the other genotypes in the progeny. Therefore, under the conditions of the Hardy-Weinberg law, the gene and genotypic frequencies for sex-linked loci also are constant from one generation to the next. If a population is in equilibrium, estimates of gene frequencies can be obtained directly from the phenotypic frequencies *within* the heterogametic sex:

$$f(B) = f(X^BY) = p$$

and

$$f(b) = f(X^bY) = q$$

For autosomal loci, a population not in Hardy-Weinberg equilibrium reaches equilibrium after one generation of random mating if the gene frequencies are the same in males and females. The same is true for sex-linked loci. If the initial gene frequencies are not equal in both sexes for autosomal loci, two generations of random mating are required to bring the population to equilibrium. For sex-linked loci, however, more than two generations of random mating are required to reach equilibrium. Assume a mammalian population with the following genotypic frequencies:

Genotype	f(Genotype within sex)
X^BY	.5
X^bY	.5
X^BX^B	.36
X^BX^b	.48
X^bX^b	.16

In males $f(B)$ is .5, denoted p_M, and in females .6, denoted p_F. The product of the frequencies of gametes produced by both sexes under random mating,

$$[(1/2)p_M + (1/2)q_M + (1/2)Y)] \times (p_F + q_F)$$

gives the expected frequencies of genotypes in the progeny:

Genotype	f(Genotype in progeny)	f(Within sex)
X^BY	$(1/2)p_F = .30$	.6
X^bY	$(1/2)q_F = .20$	.4
X^BX^B	$(1/2)p_Mp_F = .15$	.3
X^BX^b	$(1/2)p_Mq_F + (1/2)q_Mp_F = .25$	.5
X^bX^b	$(1/2)q_Mq_F = .10$	.2

The frequencies of the alleles in the male progeny are exactly the same as the frequencies of the alleles in the female parents because males receive the *B* or *b* allele only from their mothers. The frequencies of alleles in the female progeny are the averages of the gene frequencies of the parents because females receive half their genes from each parent:

$$f(B \text{ in females}) = (1/2)p_M + (1/2)p_F$$
$$= .55$$

which is the same as

$$f(B \text{ in females}) = f(X^B X^B \text{ female}) + (1/2)f(X^B X^b \text{ female})$$
$$= .3 + .25 = .55$$

In the second generation of progeny from random mating, $f(B)$ in males will be .55, and $f(B)$ in females the average of frequencies of their parents, or $\frac{1}{2}(.6 + .55) = .575$. The frequencies for three generations are

Generations of random mating	*f*(*B*) Males	*f*(*B*) Females	Absolute difference
Initial	.5	.6	.1
1	.6	.55	.05
2	.55	.575	.025
3	.575	.5625	.0125

The absolute difference in gene frequency between sexes is halved from one generation to the next. The frequencies approach equality in both sexes when

$$f(B \text{ in equilibrium}) = (1/3)f(B \text{ in the initial population of males})$$
$$+ (2/3)f(B \text{ in the initial population of females})$$

For the given example,

$$f(B \text{ at equlibrium}) = (1/3)(.5) + (2/3)(.6)$$
$$= .5667$$

Intuitively, this equilibrium frequency makes sense. For sex-linked loci, three *X* chromosomes are involved: in mammals the male has one and the female has two. The contribution of genes from the male parents is one of every three and from the female parents two of every three. The frequencies of the alleles at equilibrium are expected to be a weighted average of the frequencies within sex for the initial generation, the weights being $\frac{1}{3}$ for males and $\frac{2}{3}$ for females.

6-4 Multiple Alleles at a Single Locus

At loci with more than two alleles, the consequences of random mating are the same as for two alleles. Using a three-allele system as an example, assume that

$$f(A_1) = p$$
$$f(A_2) = q$$
$$f(A_3) = r$$

with the subscripts designating alleles at the A locus. At equilibrium the genotypic frequencies can be derived from the trinomial expansion

$$(pA_1 + qA_2 + rA_3)^2$$

which is

$$f(A_1A_1) = p^2$$
$$f(A_1A_2) = 2pq$$
$$f(A_1A_3) = 2pr$$
$$f(A_2A_2) = q^2$$
$$f(A_2A_3) = 2qr$$
$$f(A_3A_3) = r^2$$

As in the case of two alleles, the sum of gene frequencies must be 1:

$$p + q + r = 1$$

Therefore, the sum of all genotypic frequencies is 1 and $(p + q + r)^2$ is

$$p^2 + 2pq + 2pr + q^2 + 2qr + r^2 = 1$$

The probability of drawing, for example, an A_1A_2 animal at random from the population is $f(A_1A_2) = 2pq$.

Under the assumptions of the Hardy-Weinberg law, the gene and genotypic frequencies for multiple alleles remain constant over generations. If the gene frequencies are the same for both sexes, a population not in equilibrium will reach equilibrium after one generation of random mating.

For sex-linked loci, the genotypic frequencies at equilibrium for males are

Genotype	Frequency
$X^{A_1}Y$	p
$X^{A_2}Y$	q
$X^{A_3}Y$	r

The frequencies for females are the same as for autosomal loci.

Calculating the gene frequencies for multiple alleles for situations where each genotype has a distinct phenotype is the same as for the two-allele system; that is, count genes and divide by the total number of genes. If dominance exists between pairs of alleles, the approach becomes more complex, as shown in Example 6-4.

Example 6-4

The following coat color phenotypes were observed in 1000 mink for genotypes at a locus having three possible alleles, A_1, A_2, and A_3:

Genotype	Genotypic frequency	Phenotype	Number observed
A_1A_1	p^2		
A_1A_2	$2pq$	Natural dark	910
A_1A_3	$2pr$		
A_2A_2	q^2	Steelblu	80
A_2A_3	$2qr$		
A_3A_3	r^2	Platinum	10
			1000

Note that A_1 is dominant to A_2 and A_3, while A_2 is dominant to A_3. Knowing that platinum represents the recessive genotype, the estimate of $f(A_3)$ is

$$\hat{r} = \sqrt{f(A_3A_3)} = \sqrt{10/1000} = .1$$

However, because of dominance, one cannot use the square roots of the other phenotypic frequencies as estimates of gene frequencies; $\hat{r}$ and the expected genotypic frequencies can be used to estimate q. The frequency of steelblu is .08, being the sum of $q^2 + 2qr$; hence,

$$q^2 + 2qr = .08$$

and with $\hat{r} = 0.1$,

$$\begin{aligned} q^2 + 2q(.1) &= .08 \\ q^2 + .2q &= .08 \\ q^2 + .2q - .08 &= 0 \\ (q + .4)(q - .2) &= 0 \end{aligned}$$

so that

$$q = -.4 \text{ or } .2$$

Because frequencies must be $\geqq 0$, $\hat{q} = .2$, and because $p + q + r = 1$, $\hat{p}$ must be .7.

6-5 Multiple Loci

A population with equal gene frequencies in males and females, which is not in equilibrium for autosomal loci, requires only one generation of random mating to reach equilibrium. Intuitively it would seem that, if two different loci independently reach equilibrium, they should also be jointly at equilibrium. This assumption may not be true.

Consider two loci with two alleles each. Let $f(A) = p_1$ and $f(a) = q_1$ at the first locus, and $f(B) = p_2$ and $f(b) = q_2$ at the second locus. To be jointly in equilibrium, the frequencies of AB, Ab, aB, and ab gametes must equal the products of the respective gene frequencies p_1p_2, p_1q_2, q_1p_2, and q_1q_2. If the gametic frequencies are equal to these products, then it is easy to show by substitution that

$$f(AB)f(ab) = f(Ab)f(aB)$$

and that

$$f(AB)f(ab) - f(Ab)f(aB) = 0$$

If the loci are not in joint equilibrium then

$$f(AB)f(ab) - f(Ab)f(aB) \neq 0$$

and the difference will be denoted as d. Therefore, d represents the difference in production of gametes in the coupling state versus the repulsion state. Two different situations must be considered to demonstrate disequilibrium:

1. the loci are on separate chromosomes, or are far enough apart on one chromosome so that they appear to segregate independently; and
2. the loci are linked.

Independent loci

For independent loci (as defined above) the disequilibrium value, d, is halved with each generation of mating. In the strictest sense of the definition, the population would never reach equilibrium in any finite number of generations. For practical purposes, however, equilibrium is reached in a relatively few generations because d becomes small rapidly. The disequilibrium value in the nth generation, d_n, is equal to $(\frac{1}{2})^n d_0$, where d_0 is the initial value (see Example 6-5).

Example 6-5

Assume the following frequencies of joint genotypes:

				Totals (*B* locus)
	$f(AA\ BB) = .39$	$f(Aa\ BB) = .06$	$f(aa\ BB) = .05$	$f(BB) = .50$
	$f(AA\ Bb) = .08$	$f(Aa\ Bb) = .08$	$f(aa\ Bb) = .04$	$f(Bb) = .20$
	$f(AA\ bb) = .13$	$f(Aa\ bb) = .06$	$f(aa\ bb) = .11$	$f(bb) = .30$
Totals (*A* locus)	$f(AA) = .60$	$f(Aa) = .20$	$f(aa) = .20$	

For this population

$$f(A) = .7$$
$$f(a) = .3$$
$$f(B) = .6$$
$$f(b) = .4$$

The population obviously is not in equilibrium at either locus because $f(AA) \neq f(A) \times f(A)$, and $f(BB) \neq f(B) \times f(B)$. The gametes produced by the genotypes in this population are

		Gametic frequency			
Genotype	**Genotypic frequency**	***AB***	***Ab***	***aB***	***ab***
AA BB	.39	.39			
AA Bb	.08	.04	.04		
AA bb	.13		.13		
Aa BB	.06	.03		.03	
Aa Bb	.08	.02	.02	.02	.02
Aa bb	.06		.03		.03
aa BB	.05			.05	
aa Bb	.04			.02	.02
aa bb	.11				.11
	Totals	.48	.22	.12	.18

Now

$$\begin{aligned} d_0 &= f(AB)f(ab) - f(Ab)f(aB) \\ &= (.48)(.18) - (.22)(.12) \\ &= .06 \end{aligned}$$

If gametes combine randomly, the genotypic frequencies are

			Sperm			
			AB	***Ab***	***aB***	***ab***
		Gametic frequencies	**.48**	**.22**	**.12**	**.18**
Ova	***AB***	**.48**	*AA BB* (.2304)	*AA Bb* (.1056)	*Aa BB* (.0576)	*Aa Bb* (.0864)
	Ab	**.22**	*AA Bb* (.1056)	*AA bb* (.0484)	*Aa Bb* (.0264)	*Aa bb* (.0396)
	aB	**.12**	*Aa BB* (.0576)	*Aa Bb* (.0264)	*aa BB* (.0144)	*aa Bb* (.0216)
	ab	**.18**	*Aa Bb* (.0864)	*Aa bb* (.0396)	*aa Bb* (.0216)	*aa bb* (.0324)

The new genotypic frequencies become

				Totals (*B* locus)
	$f(AA\ BB) = .2304$	$f(Aa\ BB) = .1152$	$f(aa\ BB) = .0144$	$f(BB) = .36$
	$f(AA\ Bb) = .2112$	$f(Aa\ Bb) = .2256$	$f(aa\ Bb) = .0432$	$f(Bb) = .48$
	$f(AA\ bb) = .0484$	$f(Aa\ bb) = .0792$	$f(aa\ bb) = .0324$	$f(bb) = .16$
Totals (*A* locus)	$f(AA) = .49$	$f(Aa) = .42$	$f(aa) = .09$	

While each locus is itself in equilibrium, the two loci are not jointly in equilibrium. The gametes produced by this generation have the following frequencies:

$$f(AB) = .45$$
$$f(Ab) = .25$$
$$f(aB) = .15$$
$$f(ab) = .15$$

and $d_1 = .03$, which is equal to $(\frac{1}{2})^1 d_0$. After five more generations of random mating, $d_6 = .0009$, which is equal to $(\frac{1}{2})^6 d_0$.

Linked loci

Loci that are linked approach equilibrium more slowly than do loci segregating independently, and the closer the linkage, the slower the approach to equilibrium.

Linkage between loci influences the proportion of gametic types produced, favoring the parental type gametes over the recombinant type gametes (see Section 4.2). The recombinant gametes reflect crossing-over between the homologous pair of chromosomes. Frequencies of gametes produced from a particular population are calculated just as for independent loci in Example 6-5, except for the treatment of contributions from the dihybrid *Aa Bb*. From Example 6-5 it is evident that the dihybrid is the only genotype in which crossing-over rates affect the frequency of gametes produced. That is, crossing-over does not influence gamete frequencies produced by genotypes homozygous for at least one locus. When obtaining the frequencies of gametes produced by the dihybrid, the two possible types of dihybrids must be considered: *AB/ab* and *Ab/aB*. Example 6-6 shows how these two different dihybrids contribute to the gametic pool.

Example 6-6

Assume in Example 6-5 that the original dihybrids were equally represented in the population such that the frequencies of joint genotypes are

				Totals (B locus)
	$f(AA\ BB) = .39$	$f(Aa\ BB) = .06$	$f(aa\ BB) = .05$	$f(BB) = .50$
	$f(AA\ Bb) = .08$	$f(AB/ab) = .04$ $f(Ab/aB) = .04$	$f(aa\ Bb) = .04$	$f(Bb) = .20$
	$f(AA\ bb) = .13$	$f(Aa\ bb) = .06$	$f(aa\ bb) = .11$	$f(bb) = .30$
Totals (A locus)	$f(AA) = .60$	$f(Aa) = .20$	$f(aa) = .20$	

As in Example 6-5,

$$f(A) = .7$$
$$f(a) = .3$$
$$f(B) = .6$$
$$f(b) = .4$$

and the population is not in equilibrium at either locus or jointly. The gametes produced by the genotypes in this population are the same for all genotypes except the dihybrids (*Aa Bb*) as those in Example 6-5:

		Gametic frequency			
Genotype	**Frequency**	***AB***	***Ab***	***aB***	***ab***
AA BB	.39	.39			
AA Bb	.08	.04	.04		
AA bb	.13		.13		
Aa BB	.06	.03		.03	
Aa Bb	.08	— (see page 137) —			
Aa bb	.06		.03		.03
aa BB	.05			.05	
aa Bb	.04			.02	.02
aa bb	.11				.11

The contributions from the dihybrids, which must take into account crossing-over, are

		Gametic frequency			
Genotype	Frequency	AB	Ab	aB	ab
AB/ab	.04	$(1/2)(1-r)(.04)$	$(1/2)r(.04)$	$(1/2)r(.04)$	$(1/2)(1-r)(.04)$
Ab/aB	.04	$(1/2)r(.04)$	$(1/2)(1-r)(.04)$	$(1/2)(1-r)(.04)$	$(1/2)r(.04)$

where r is the proportion of recombinants and cannot exceed .5. Gametic frequencies in progeny are calculated exactly as in Example 6-5 except that the dihybrids are identified.

6-6 Summary

Population genetics deals with gene and genotypic frequencies in populations and the prediction of these frequencies in subsequent generations. Gene frequency is the proportion of the total number of genes represented by a particular allele. If no allele is dominant, gene frequencies may be determined by simply counting alleles. Genotypic frequency is the proportion of total animals with a particular genotype.

The Hardy-Weinberg law states that gene and genotypic frequencies remain constant from one generation to the next if the following conditions are met:

1. large population,
2. random mating, and
3. the absence of any force (mutation, migration, or selection) acting to change gene frequency.

A population that meets these assumptions is in equilibrium. If $f(B) = p$ and $f(b) = q$, then the genotypic frequencies can be determined from the binomial expansion

$$(pB + qb)^2 = p^2BB + 2pqBb + q^2bb$$

For multiple alleles at a single locus, the multinomial expansion is used to obtain genotypic frequencies at equilibrium.

Random mating is a mating system in which each individual has an equal opportunity to mate with any individual of the opposite sex.

For populations in equilibrium, the frequency of the recessive allele, f(b), may be estimated as

$$\hat{q} = \sqrt{f(bb)}$$

A population not in equilibrium for an autosomal locus will reach equilibrium in one generation of random mating if the gene frequencies are equal in both sexes; if the gene frequencies are not equal, two generations of random mating are required.

At equilibrium, gene frequencies at sex-linked loci can be estimated directly from phenotypes of the heterogametic sex. If the gene frequencies are the same in both sexes, populations not in equilibrium will reach equilibrium in one generation of random mating. If the gene frequencies are different between sexes, the population needs more generations of random mating to reach equilibrium than with autosomal loci. With sex linkage in mammals the frequency of genes in male progeny is equal to the frequency in their mothers, and the frequency in female progeny is the average of frequencies in both parents.

For two loci considered jointly, populations not in equilibrium do not reach joint equilibrium in one generation of random mating. A measure of disequilibrium when considering two loci jointly is

$$d = f(AB)f(ab) - f(Ab)f(aB)$$

At equilibrium, $d = 0$. For independent loci, d in the nth generation is

$$d_n = (1/2)^n d_0$$

where d_0 is the measure of disequilibrium in the initial generation. For linked loci

$$d_n = (1 - r)^n d_0$$

where r is the measure of linkage, measured in map units.

CHAPTER 7

Forces That Change Gene and Genotypic Frequencies

Chapter 6 presented the Hardy-Weinberg law, which states that gene and genotypic frequencies remain constant in a population which meets the conditions necessary for equilibrium. In this chapter changes in gene and genotypic frequencies are examined when any of these conditions are not met.

7-1 Nonrandom Mating

Random mating is a mating system in which each breeding animal has an equal opportunity to mate with any animal of the opposite sex. Suppose, however, a set of rules is imposed which dictates the mating strategy. As examples, (1) only individuals of the same phenotype are allowed to mate, and (2) only individuals of different phenotypes are mated. These two strategies are called assortative mating, *positive assortative mating* when individuals of the same phenotype are mated and *negative assortative mating* when individuals with different phenotypes are mated. Assortative mating will be discussed further in Chapter 15. The present concern is the impact of nonrandom mating on gene and genotypic frequencies of the progeny. Nonrandom mating in the absence of selection alters genotypic frequencies but does not always change gene frequencies from one generation to the next.

The expected frequency of progeny genotypes with random mating is the product of male and female gametic frequencies. With nonrandom mating, however, this approach is not valid. The rules governing the matings influence the frequencies of matings in the population, and hence the probabilities of union between gametes from each sex. For example, assume a population of polled and horned cattle and the strategy of mating like phenotypes. The frequency of horned males among all males is q^2 and of horned females is also q^2. The probability of mating between a horned male and horned females when the population is randomly mated is the product of these genotypic frequencies; that is

$$\begin{aligned} P(\text{horned male} \times \text{horned female}) &= P(\text{horned male}) \times P(\text{horned female}) \\ &= q^2 \times q^2 \\ &= q^4 \end{aligned}$$

With positive assortative mating the probability of this mating is different. The probability of a horned female is still q^2; however, the probability that she is mated to a horned male is 1 rather than q^2 because the mating rule dictates that she be mated to horned males. Hence, the probability of the mating of two horned cattle is now q^2, $(q^2 \times 1)$, instead of q^4. This strategy increases the frequency of matings between horned cattle. The influence of nonrandom mating on genotypic frequencies in the progeny will be demonstrated by example for two situations: (1) for each genotype having a distinct phenotype (Example 7-1), and (2) for complete dominance (Example 7-2).

Example 7-1

Assume a mating strategy of positive assortative mating for coat color in Shorthorn cattle where initially

Genotype	Frequency	Phenotype
RR	p^2	Red
Rr	$2pq$	Roan
rr	q^2	White

Under the mating strategy, three types of matings can occur: red by red, roan by roan, and white by white. The frequency of *RR* females is p^2; however, once an *RR* female is chosen, the genotype of her mate is fixed (there is no random element associated with choosing her mate). Figure 7-1 is a visual representation of the mating scheme for this example. Females to be bred are held in one large pen. The frequency with which females of a particular genotype enter the breeding chute is equal to the genotypic frequency in the female population. Which pen the female is placed in to be bred depends on her phenotype; that is, red females are placed in the pen of red bulls. It is easy to

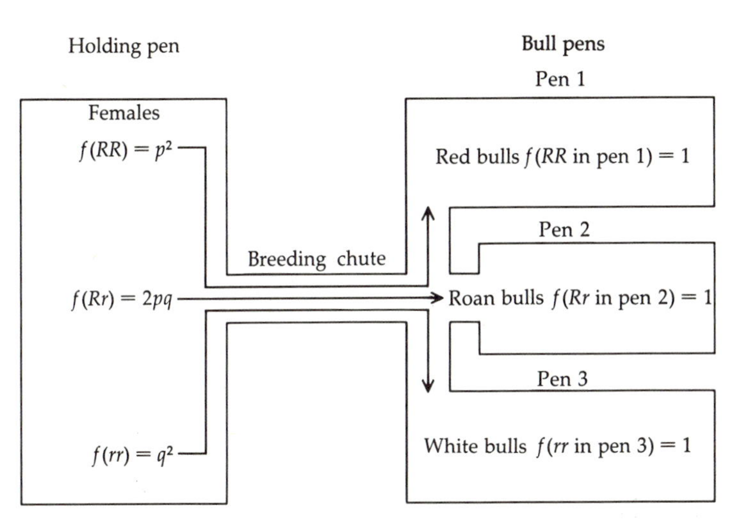

Figure 7-1 Breeding pen representation of the mating strategy used in Example 7-1.

see that the probability a red female is mated to a red bull is 1 because the pen into which she is placed contains only red bulls.
Therefore, the frequency of RR by RR matings is p^2. Likewise, $f(Rr \times Rr) = 2pq$ and $f(rr \times rr) = q^2$. The contributions of each mating to the expected frequency of all progeny genotypes can be summarized as

		Expected contributions to genotypic frequencies of all progeny		
Mating	***f*(mating)**	***RR***	***Rr***	***rr***
$RR \times RR$	p^2	p^2	—	—
$Rr \times Rr$	$2pq$	$(1/2)pq$	pq	$(1/2)pq$
$rr \times rr$	q^2	—	—	q^2

The expected contributions to progeny genotypic frequencies are obtained by the product of f(mating) and P(progeny genotype|the mating). For example, $f(Rr \times Rr) = 2pq$, and $P(RR|Rr \times Rr)$ is $\frac{1}{4}$. The contribution to the frequency of RR progeny from this mating is $\frac{1}{4}(2pq) = \frac{1}{2}pq$. The expected progeny genotypic frequencies are obtained by summing each column:

$$f(RR \text{ in progeny}) = p^2 + (1/2)pq$$
$$f(Rr \text{ in progeny}) = pq$$
$$f(rr \text{ in progeny}) = q^2 + (1/2)pq$$

Note, (1) the frequency of heterozygotes is halved, and the frequency of each homozygote increases, that is, genotypic frequencies are changed, and (2) the gene frequencies are the same in parents and progeny in this example.

$$\begin{aligned} f(R \text{ in progeny}) &= f(RR) + (1/2)f(Rr) \\ &= [p^2 + (1/2)pq] + (1/2)pq \\ &= p^2 + pq \\ &= p(p + q) \\ &= p \end{aligned}$$

Example 7-2

Assume positive assortative mating for coat color in Angus cattle, where in the population of parents at equilibrium $f(B) = \frac{2}{3}$ and $f(b) = \frac{1}{3}$ such that

Genotype	Frequency	Phenotype
BB	$p^2 = 4/9$	Black
Bb	$2pq = 4/9$	Black
bb	$q^2 = 1/9$	Red

The mating rule is the same as in Example 7-1; however, the example is more complex because of dominance. A black female may be *BB* or *Bb*, and although a male's phenotype must be black, his genotype also can be *BB* or *Bb*. Because the parent population is in equilibrium, $f(BB) = p^2$ in females, but what proportions of the black mates are *BB* and *Bb*? The answer involves conditional probabilities (see Chapter 5). The probability of a *BB* male given that his phenotype is black is

$$\begin{aligned} P(BB|\text{male with a black coat}) &= P(BB \text{ and black})/P(\text{black coat}) \\ &= (4/9)/(8/9) = 1/2 \end{aligned}$$

Likewise for this example,

$$\begin{aligned} P(Bb|\text{male with a black coat}) &= P(Bb \text{ and black})/P(\text{black coat}) \\ &= (4/9)/(8/9) = 1/2 \end{aligned}$$

Therefore,

$$\begin{aligned} f(BB \text{ female} \times BB \text{ male}) &= f(BB \text{ in females}) \times f(BB|\text{male with a black coat}) \\ &= (4/9) \times (1/2) = 2/9 \end{aligned}$$

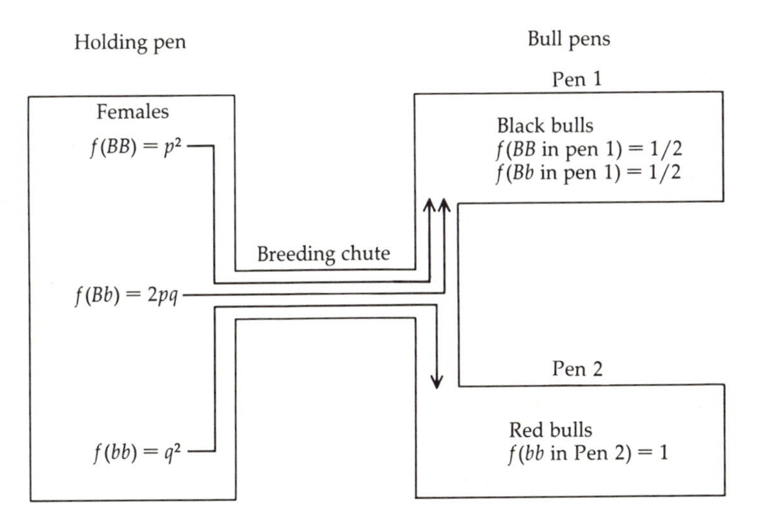

Figure 7-2 Breeding pen representation of the mating strategy used in Example 7-2.

Figure 7-2 shows the breeding pen diagram appropriate for this example. Since the males are placed in pens based on their phenotype, pen 1 contains males with two different genotypes. Which genotype of black male actually mates with a female is random within that pen.

The table for expected contributions to progeny genotypes is

Mating genotypes			**Expected contribution to progeny genotypic frequencies**		
Male	**Female**	***f*(mating)**	***BB***	***Bb***	***bb***
BB	*BB*	(1/2)(4/9) = 2/9	2/9	—	—
BB	*Bb*	(1/2)(4/9) = 2/9	1/9	1/9	—
Bb	*BB*	(1/2)(4/9) = 2/9	1/9	1/9	—
Bb	*Bb*	(1/2)(4/9) = 2/9	1/18	1/9	1/18
bb	*bb*	(1)(1/9) = 1/9	—	—	1/9
Overall expected progeny frequencies			1/2	1/3	1/6

As with Example 7-1, the frequency of heterozygotes in the progeny decreases and the frequencies of homozygotes increase with positive assortative mating. The gene frequencies also do not change.

$$\begin{aligned} f(B) &= f(BB) + (1/2)f(Bb) \\ &= 1/2 + (1/2)(1/3) \\ &= 2/3 \end{aligned}$$

In both examples, the frequencies of matings are different from those expected with random mating. With positive assortative mating, the genotypic frequencies in the progeny are different from their parents but gene frequencies are the same. In both examples, the frequency of heterozygotes decreased and each type of homozygote increased. If positive assortative mating is continued for many generations, ultimately all animals will be homozygous; that is, all heterozygotes will be lost in the population.

Although positive assortative mating in the absence of selection does not change gene frequency, negative assortative mating generally does change gene frequency, as shown in Example 7-3.

Example 7-3

Assume the same population as in Example 7-2 for coat color in Angus, with $f(B) = \frac{2}{3}$ and $f(b) = \frac{1}{3}$ and

Genotype	Frequency	Phenotype
BB	$p^2 = 4/9$	Black
Bb	$2pq = 4/9$	Black
bb	$q^2 = 1/9$	Red

The mating system is negative assortative mating; that is, black females are mated to red males and red females to black males. Figure 7-2 still applies with respect to how the animals are penned; however, the arrows showing where females are placed change; for example, red females now move into pen 1 with the black bulls. The table of expected contributions to progeny genotypes is

Mating genotypes			Expected contribution to genotypic frequencies of progeny		
Male	Female	f(mating)	BB	Bb	bb
BB	bb	$(1/2)(1/9) = 1/18$	—	1/18	—
Bb	bb	$(1/2)(1/9) = 1/18$	—	1/36	1/36
bb	BB	$(1)(4/9) = 4/9$	—	4/9	—
bb	Bb	$(1)(4/9) = 4/9$	—	2/9	2/9
	Overall expected progeny frequencies		0	3/4	1/4

Note that there is no mating that results in *BB* progeny in this example, so that the genotypic frequencies in progeny are obviously different from the frequencies in the parents. What about gene frequencies? In the progeny, the gene frequencies are

$$\begin{aligned} f(B) &= f(BB) + (1/2)f(Bb) \\ &= 0 + (1/2)(3/4) \\ &= 3/8 \end{aligned}$$

and

$$\begin{aligned} f(b) &= 1 - f(B) \\ &= 5/8 \end{aligned}$$

Hence the gene frequencies have changed from the $f(B) = 2/3$ and $f(b) = 1/3$ in the population of parents.

Example 7-3 demonstrates the potential of mating systems to alter gene frequencies. With negative assortative mating the reason for the change in gene frequency is the disproportionate use of males as compared to random mating. In Example 7-3, the red males (*bb*) represent $\frac{1}{9}$ of the total bull population but mate with $\frac{8}{9}$ of the female population and produce $\frac{8}{9}$ of the progeny. Conversely, the black males represent $\frac{8}{9}$ of the male population but mate with only $\frac{1}{9}$ of the females. With random mating, *bb* males are involved in only $\frac{1}{9}$ of the matings and $B\_$ males in $\frac{8}{9}$ of the matings. The disproportionately high use of *bb* males accounts for the increase in the frequency of the *b* allele. With positive assortative mating the proportional use of males is exactly the same as with random mating if the parent population is in equilibrium.

In summary, nonrandom mating changes genotypic frequencies from one generation to the next; however, the gene frequencies may or may not remain constant. The key to determining how nonrandom mating influences genotypic and gene frequencies lies in properly defining the frequency of matings for the particular mating strategy. The breeding pen diagrams are useful in visualizing how the frequencies of matings are obtained. With the breeding pen approach, the females are always brought to the males. For many situations, the approach produces incorrect results if males are brought to female pens. The frequency of males in the population does not always equal the frequency with which they mate; that is, males may mate with more than one female.

7-2 Migration

Migration is the movement of individuals from one breeding population to another. Assume there are two populations and that, for a particular autosomal locus, $f(a) = q_1$ in population 1 and q_2 in population 2. Now suppose animals from population 2 migrate to population 1 as depicted in Figure 7-3. The assumption is that migration is random, with the gene frequencies in the migratory animals equal to the gene frequencies in the population from which they emigrated; that is, $f(a)$ in the migrants is q_2. After migration, population 1 increases in size with a proportion of the animals, m, having come from population 2:

$$m = n_2/(n_1 + n_2)$$

where n_1 is the number of native animals in population 1 and n_2 the number of migrant animals from population 2. The remaining proportion, $1 - m$, are the native animals of population 1. What are the gene frequencies in the expanded population 1? Let q_1' represent $f(a)$ after migration, such that

$$q_1' = (1 - m)q_1 + mq_2$$

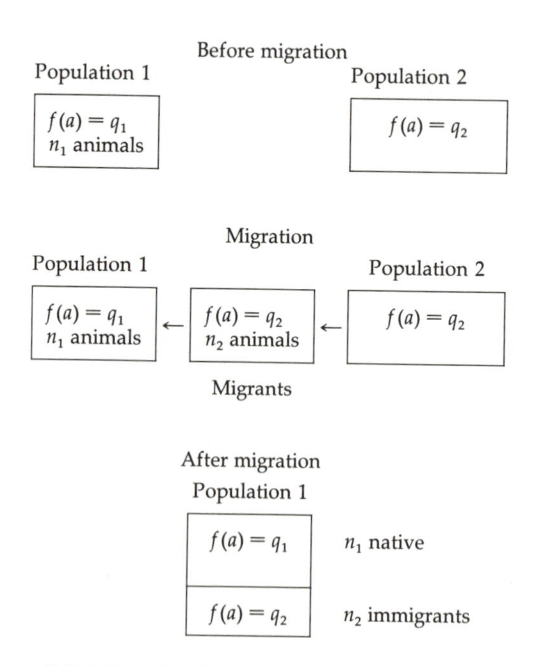

Figure 7-3 Migration from population 2 to population 1.

which can be written equivalently as

$$q_1' = q_1 + m(q_2 - q_1)$$

The change in gene frequency caused by migration, $m(q_2 - q_1)$, depends on the initial difference in gene frequencies between the two populations $(q_2 - q_1)$ and on the proportion of migratory animals, m. If the gene frequency in both populations is equal $(q_1 = q_2)$, then random migration does not alter frequency.

The difference in gene frequencies between the two populations is smaller after migration:

$$\begin{aligned}(q_1' - q_2) &= q_1 + m(q_2 - q_1) - q_2 \\ &= (q_1 - q_2) - m(q_1 - q_2) \\ &= (q_1 - q_2)(1 - m)\end{aligned}$$

where $q_1 - q_2$ is the initial difference in gene frequency between the populations. The difference decreases depending on m. Further migration would continue to reduce this difference. Example 7-4 demonstrates the influence of migration on gene frequencies.

Example 7-4

Assume two populations where $f(a) = q_1 = .2$ in the first population, and $f(a) = q_2 = .6$ in the second population. Assume the number of animals in population 1 was 8000 and that 2000 emigrate from population 2 to population 1. Thus $m = 2000/(8000 + 2000) = .2$. The frequency of a after migration, q_1', is

$$\begin{aligned}q_1' &= q_1 + m(q_2 - q_1) \\ q_1' &= .2 + .2(.6 - .2) \\ &= .2 + .08 = .28\end{aligned}$$

The difference in gene frequencies was initially

$$\begin{aligned}q_1 - q_2 &= .2 - .6 \\ &= -.4\end{aligned}$$

and after migration

$$\begin{aligned}q_1' - q_2 &= .28 - .6 \\ &= -.32\end{aligned}$$

7-3 Mutation

A *mutation* is a sudden heritable change in genetic material. The phenomenon may be the creation of a new allele or the change of one allele to an already existing allele. Mutation is the ultimate source of new genetic material for evolutionary processes and selection. In this section the influence on gene frequency of one-time mutations, or *nonrecurrent mutations,* will be ignored. This type of mutation is of little consequence unless it provides a selective advantage for the individual with the "new" allele. *Recurrent mutations,* defined as mutational events with characteristic rates, are of greater importance. Simply stated, recurrent mutation of one allelic form to another is assumed to occur at a constant rate. The mutation rate is usually small, ranging from 10^{-4} to 10^{-8} for most reported mutations. In any given generation, mutations do not measurably affect gene frequency. However, over many generations, the rate of mutation does influence gene frequency.

The mutation rate, u, is the probability that a particular allele B mutates to allele b, $P(B \rightarrow b) = u$. For the population, the frequency of mutations is the product of the mutation rate and the gene frequency, p_0:

$$f(B \rightarrow b) = up_0$$

The product, up_0, is the fraction of B genes lost from the population. In the next generation, if the frequency of b alleles mutating to B alleles is ignored, the new gene frequency is

$$\begin{aligned} f(B) = p_1 &= p_0 - up_0 \\ &= p_0(1 - u) \end{aligned}$$

and

$$f(b) = q_1 = q_0 + up_0$$

Due to mutational loss $f(B)$ decreases. The frequency of $B \rightarrow b$ for the next generation is up_1 so that, although the mutation rate, u, is constant, the frequency of actual mutations in the population decreases as p decreases. For the second generation,

$$p_2 = p_1 - up_1 = p_1(1 - u)$$

Because $p_1 = p_0(1 - u)$, then

$$p_2 = p_0(1 - u)^2$$

Thus after n generations,

$$p_n = p_0(1 - u)^n$$

One-way mutations are considered in Example 7-5.

Example 7-5

Assume $f(B) = .8$ at locus 1, and $f(C) = .8$ at locus 2. Assume also the mutation rate is 10^{-4} at locus 1 for $B \rightarrow b$, and 10^{-8} at locus 2 for $C \rightarrow c$. The following frequencies are expected for B and C after many generations:

Generation	$f(B)$	$f(C)$
0	.8	.8
10	.7992	.7999
100	.7920	.7999
1,000	.7239	.7999
10,000	.2942	.7999
100,000	.0000	.7992

Both genes, B and C, will ultimately be lost in the population, although the process is slow even for a "high" mutation rate. The number of generations required for $f(B)$ to become less than some frequency can be predicted. For example, how many generations (t) are required before $p_t \leqq .7$, if p_0 is .8, given the value of u? The equation to solve is

$$p_t = .7 = .8(1 - u)^t$$

and

$$(1 - u)^t = .7/.8 = .875$$

By use of logarithms

$$t \log(1 - u) = \log(.875)$$

and

$$t = \log(.875)/\log(1 - u)$$

For the two genes, *B* and *C*,

$$t_B = \log(.875)/\log(1 - 10^{-4}) \cong 1.3 \times 10^3 = 1300 \text{ generations}$$
$$t_C = \log(.875)/\log(1 - 10^{-8}) \cong 1.3 \times 10^7 = 13{,}000{,}000 \text{ generations}$$

Recurrent mutations can occur in both directions, from *B* to *b* and from *b* to *B*. Let *u* be the mutation rate $B \rightarrow b$, and *v* be the rate for $b \rightarrow B$. These two rates need not be equal. If $f(B) = p_0$ in the initial population, then

$$p_1 = p_0 - up_0 + vq_0$$

where up_0 is the frequency of the mutation *B* to b which results in a loss of *B* alleles, and vq_0 is the frequency of the mutation of *b* to *B* which results in a gain of *B* alleles. The three possible outcomes are

1. $up_0 > vq_0$ and *f*(*B*) decreases,
2. $up_0 < vq_0$ and *f*(*B*) increases, and
3. $up_0 = vq_0$ and *f*(*B*) is stable, or at equilibrium

Whether *u* is greater than or less than *v* does not determine whether *f*(*B*) decreases or increases. The change in *f*(*B*) depends on the frequencies of mutations, up_0 and vq_0. The third possible outcome indicates that equilibrium can occur. In fact, in the absence of other forces which influence gene frequencies, *f*(*B*) will increase or decrease until the population reaches equilibrium. Therefore, the value of *f*(*B*) at equilibrium is of interest. Let $p_E = f(B)$ and $q_E = f(b)$ at equilibrium. Given that *u* and *v* are constants, and that equilibrium occurs when the number of alleles changing from $B \rightarrow b$ equals the number changing from $b \rightarrow B$, that is, when $vq = up$, then p_E or q_E can be found as follows:

$$vq_E = up_E$$

and

$$vq_E = u(1 - q_E)$$

so that

$$\mathrm{v}q_E + uq_E = u$$

Hence,

$$q_E = u/(u + v)$$

and

$$p_E = 1 - q_E = v/(u + v)$$

Recurrent mutations occurring in both directions are considered in Example 7-6.

Example 7-6

Assume $p_0 = .8$, $u = 4.2 \times 10^{-5}$ for $B \rightarrow b$, and $v = 2.1 \times 10^{-5}$ for $b \rightarrow B$. What are p_1 and p_E?

$$\begin{aligned} p_1 &= p_0 - up_0 + vq_0 \\ &= .8 - (.336 \times 10^{-4}) + (.042 \times 10^{-4}) \\ &= .8 - (.294 \times 10^{-4}) \end{aligned}$$

The change in gene frequency is essentially undetectable, although $f(B)$ is decreasing. At equilibrium

$$q_E = \frac{4.2 \times 10^{-5}}{(4.2 \times 10^{-5}) + (2.1 \times 10^{-5})} = 2/3$$

and

$$p_E = 1 - q_E = 1/3$$

The frequency of the B allele would continue to decrease to $\frac{1}{3}$ and then stabilize.

7-4 Selection

In previous sections the discussion was based on the assumptions that all individuals of a population have an equal opportunity to breed regardless of their genotype, and that each genotype has the same reproductive capability. If genotypes have different probabilities of survival or potential to reproduce, gene frequencies will change from one generation to the next. An obvious example is a recessive gene that causes a lethal condition in the recessive homozygote. Since the individual with this genotype dies, its genes are lost to the population and the frequency of that gene decreases.

Selection is a major force that causes change in gene frequencies. *Natural selection* refers to the influence of the environment on the probability that a particular phenotype survives and reproduces. Not all phenotypes are equally fit to compete in a particular environment. *Fitness* is the capability of a phenotype and the corresponding genotype to survive and reproduce in a given environment. *Artificial selection* refers to a set of rules designed by humans to govern the probability that an individual survives and reproduces. Individuals capable of surviving may not be allowed to survive under an artificial selection program because their appearance or performance does not meet some standard. This section will consider natural selection; artificial selection is discussed in Chapter 9.

The influence of natural selection on gene frequencies will be examined using a single locus with two alleles, *A* and *a*. The examples considered in this section can be classified according to the relative fitness of the heterozygous genotype. First, the heterozygote may be identical to one of the homozygotes, as in the case of complete dominance. That is, *Aa* is equally as fit as *AA*. Selection may be for or against the *A*_ genotype. Second, the heterozygote may be intermediate in fitness to the two homozygotes, as with incomplete dominance and no dominance. Finally, the heterozygote may be the most favored genotype, as is the case with overdominance. The following sections present equations which permit prediction of the change in gene frequencies for each case.

Complete dominance, selection favoring the A_ genotype

Assume a population in which $f(A) = p_0$, and $f(a) = q_0$ for two alleles at an autosomal locus, and that initial genotype frequencies are $f(AA) = p_0^2$, $f(Aa) = 2p_0q_0$, and $f(aa) = q_0^2$. Individuals with at least one *A* gene have an advantage in survival, and, in terms of fitness, the genotype *Aa* is equally as fit as the *AA* genotype. Animals with the *aa* genotype are relatively less fit by the fraction *s*. This situation can be represented as follows:

Genotype	Initial frequency	Relative fitness
AA	p_0^2	1
Aa	$2p_0q_0$	1
aa	q_0^2	$1 - s$

A fitness value of 1 does not mean that all animals of the genotype live and reproduce. Accidental death, predation, or disease may cause losses of that genotype. A fitness value of 1 is assigned to the genotype which has the highest proportion of animals likely to survive, and then other genotypic fitness values are defined relative to that genotype. For the case of complete dominance, the fitness values of *AA* and *Aa* are both 1. These two genotypes are equally fit. The fitness value for the *aa* genotype is $1 - s$ where s can have any value from 0 to 1. When s is close to zero, selection against the *aa* genotype is small; however, for situations where *aa* is lethal, selection against *aa* is complete and $s = 1$.

Determining the influence of natural selection on gene frequency requires the calculation of the gene frequencies in the gametes produced by surviving animals (that is, parents of the next generation). At birth, the frequency of the *AA* genotype is p_0^2. However, the frequency of *AA* animals is higher in the population of those that survive to be parents simply because a higher proportion of *aa* animals is lost. The proportion of the initial population that produces gametes is the sum of products of genotypic frequencies and fitness values:

$$\text{Proportion producing gametes} = (1)p_0^2 + (1)2p_0q_0 + (1-s)q_0^2$$

Because $p_0^2 + 2p_0q_0 + q_0^2 = 1$, the proportion of animals successfully producing gametes can also be written as

$$\text{Proportion producing gametes} = 1 - sq_0^2$$

Of the surviving animals, then,

$$f(AA \text{ in survivors}) = \frac{p_0^2}{1 - sq_0^2}$$

$$f(Aa \text{ in survivors}) = \frac{2p_0q_0}{1 - sq_0^2}$$

$$f(aa \text{ in survivors}) = \frac{(1-s)q_0^2}{1 - sq_0^2}$$

The frequency of the *A* gene in the survivors, p_1, is

$$\begin{aligned} p_1 &= f(AA \text{ in survivors}) + (1/2)f(Aa \text{ in survivors}) \\ &= \frac{p_0^2}{1 - sq_0^2} + \frac{p_0q_0}{1 - sq_0^2} \\ &= \frac{p_0}{1 - sq_0^2} \end{aligned}$$

and q_1, the frequency of the *a* gene in the survivors is $1 - p_1$. Note that $p_1 > p_0$ because $1 - sq_0^2 < 1$. This increase in the frequency of the *A* allele is expected because the only individuals not producing gametes have the *aa* genotype.

If the survivors are mated at random to produce the next generation, then *f*(*AA* in progeny) will be p_1^2, *f*(*Aa* in progeny) will be $2p_1q_1$, and *f*(*aa* in progeny) will be q_1^2. The process is repeated until eventually the *A* allele becomes fixed. *Fixation* is defined as the frequency of one allele being 1, in this case when $f(A) = 1$. The next two examples illustrate fitness and the association of fitness to gene and genotypic frequencies. Example 7-7 demonstrates that equal fitness of the three genotypes has no influence on gene or genotypic frequencies. Example 7-8 shows the expected change in gene frequencies when selection is against the *aa* genotype.

Example 7-7 Equal fitness of genotypes

Rodent species are heavily subjected to predation. For a particular locus, whether or not an individual falls prey to a predator is independent of its genotype under the assumptions of the Hardy-Weinberg law. An equal proportion of individuals from each genotype is lost. To demonstrate this concept, assume 10 percent of the population is killed by predators, and that initially $p = .6$ and $q = .4$. If predation is random with respect to this locus,

Genotype	Initial frequency	Proportion lost	Proportion remaining
AA	.36	.036	.324
Aa	.48	.048	.432
aa	.16	.016	.144
Total	1.0	.1	.9

The *f*(A) after predation is

$$p = \frac{.324 + (1/2)(.432)}{.90} = .6$$

and $f(AA) = .324/.90 = .36$. The gene frequency is still in equilibrium for this locus because the relative fitness of each genotype is the same.

Example 7-8 Selection favoring *AA* and *Aa*

Assume in mice that the *aa* genotype produces a coat color different from the wild type (*AA* or *Aa*), and mice of this coat color are more easily spotted by predators. Although mice of all genotypes are killed by predators, a smaller frequency of the wild type is killed; relative fitness favors the $A\_$ genotypes. Assume $s = .5$ describes the relative fitness of the *aa* genotype; then with initial frequencies as in Example 7-7,

Genotype	Initial frequency	Relative fitness
AA	.36	1.0
Aa	.48	1.0
aa	.16	.5

Using the formula for $f(A$ in survivors)

$$p_1 = \frac{p_0}{1 - sq_0^2} = \frac{.6}{1 - (.5)(.16)}$$
$$= .652$$

and

$$q_1 = 1 - p_1 = .348$$

Random mating of these survivors gives

Progeny genotype	Frequency in first generation
AA	$p_1^2 = .425$
Aa	$2p_1q_1 = .454$
aa	$q_1^2 = .121$

Note the decrease in the expected number of *aa* homozygotes in the progeny.

As mentioned earlier, a special case of selection against the *aa* genotype is when $s = 1$, which occurs when all *aa* individuals either die before breeding or are completely infertile. Then the only source of gametes having the *a* gene is the heterozygote, *Aa*. The gene frequency of *A* among the survivors when $s = 1$ is

$$p_1 = \frac{p_0}{1 - sq_0^2} = \frac{p_0}{1 - q_0^2}$$

Because $1 - q_0^2 = (1 - q_0)(1 + q_0)$ and $p_0 = 1 - q_0$, the new gene frequency is

$$p_1 = \frac{(1 - q_0)}{(1 - q_0)(1 + q_0)} = \frac{1}{1 - q_0}$$

and because $q_1 = 1 - p_1$,

$$\begin{aligned} q_1 &= 1 - 1/(1 + q_0) \\ &= q_0/(1 + q_0) \end{aligned}$$

In each subsequent generation the *aa* progeny die or fail to reproduce, and the $f(a)$ continues to decrease. When $s = 1$, the gene frequencies in the *t* generation can be predicted as

$$p_t = \frac{1 + (t - 1)q_0}{1 + tq_0}$$

or, more easily, as $p_t = 1 - q_t$, where

$$q_t = \frac{q_0}{1 + tq_0}$$

The last equation can be algebraically rearranged to predict the number of generations, *t*, required for $f(a)$ to decrease from q_0 to q (for $q > 0$):

$$t = 1/q - 1/q_0$$

as shown in Example 7-9.

Example 7-9

Achondroplasia is a genetic defect in cattle in which calves are born with all four limbs amputated near the elbow and hock joints. This defect is the result of a homozygous recessive condition and is lethal. Assume $f(A) = p_0 = .8$, and $f(a) = q_0 = .2$; hence

Genotype	Initial frequency	Relative fitness
AA	.64	1
Aa	.32	1
aa	.04	0

The frequency of A in the survivors when $s = 1$ is

$$p_1 = 1/(1 + q_0)$$

Hence, with $q_0 = .2$,

$$p_1 = 1/1.2 = .833$$

and

$$q_1 = 1 - .833 = .167$$

With random mating of the surviving animals, the new progeny population becomes

Genotype	Initial frequency	Relative fitness
AA	.694	1
Aa	.278	1
aa	.028	0

The process can now be repeated for a second generation:

$$p_2 - 1/(1 + q_1) = .857$$

Note

$$p_1 - p_0 = .833 - .800 = .033$$

and

$$p_2 - p_1 = .857 - .833 = .024$$

The change in gene frequency becomes smaller with each generation because, in each succeeding generation, there are fewer *aa* genotypes against which selection can act.

By the tenth generation, $t = 10$, $f(A)$ is

$$p_{10} = \frac{1 + (t-1)q_0}{1 + tq_0} = \frac{1 + (9)(.2)}{1 + (10)(.2)} = .933$$

and the number of generations (t) required for q to decrease to .05 is

$$t = 1/q - 1/q_0 = 1/.05 - 1/.2 = 15$$

The approach to determining the influence of selection on gene frequencies for the remaining types of gene action is exactly the same:

1. For the type of gene action, determine the appropriate set of relative fitness values.
2. Calculate the proportion of the population surviving to become parents as the sum of products of fitness values and genotypic frequencies.
3. Calculate the new genotypic frequencies in the surviving population. For each genotype this is its fitness multiplied by its genotypic frequency divided by the proportion of the population surviving.
4. Calculate the gene frequencies in the surviving population from the genotypic frequencies in step 3 as $f(A \text{ in survivors}) = f(AA \text{ in survivors}) + \frac{1}{2}f(Aa \text{ in survivors})$, and $f(a \text{ in survivors}) = 1 - f(A \text{ in survivors})$.

Thus, following these steps, the formula for the gene frequencies in the survivors is obtained. For the remaining examples of gene action, the appropriate fitness values are discussed and the formula for predicting new gene frequencies presented. Each formula is derived using the four steps outlined.

No dominance

When there is no dominance between alleles at a given locus, the fitness of the heterozygote is exactly intermediate to the fitnesses of the homozygotes. If the relative fitness value of *AA* is 1 and of *aa* is $1 - s$, the fitness of the heterozygote is their average, that is, $\frac{1}{2}(1 + 1 - s) = 1 - .5s$. Then, with no dominance,

Genotype	Initial frequency	Relative fitness
AA	p_0^2	1
Aa	$2p_0q_0$	$1 - .5s$
aa	q_0^2	$1 - s$

The formula for the frequency of *A* in the survivors is

$$p_1 = \frac{p_0(1 - .5sq_0)}{1 - sq_0}$$

Because $(1 - .5sq_0) > (1 - sq_0)$, then $p_1 > p_0$. As with complete dominance, the *A* allele will eventually become fixed. The genotypic frequencies of the progeny if the survivors are randomly mated are:

$$f(AA \text{ in progeny}) = p_1^2$$
$$f(Aa \text{ in progeny}) = 2p_1q_1$$
$$f(aa \text{ in progeny}) = q_1^2$$

where $q_1 = 1 - p_1$.

Example 7-10 shows the change in gene frequency with no dominance.

Example 7-10

Assume that, in mice, the three genotypes *AA*, *Aa*, and *aa* produce three different coat colors, with *AA* being the wild type color. Let $p_0 = .6$, $q_0 = .4$, and $s = .5$, so that with no dominance,

Genotype	Initial frequency	Relative fitness
AA	.36	1
Aa	.48	$(1 - .5s) = .75$
aa	.16	$(1 - s) = .50$

Then

$$p_1 = \frac{p_0(1 - .5sq_0)}{1 - sq_0} = .675$$

and

$$q_1 = 1 - p_1 = .325$$

Incomplete dominance

With incomplete dominance the heterozygote has a relative fitness value greater than the average of the homozygotes but less than the favored homozygote. If *AA* is the favored homozygote, then

Genotype	Initial frequency	Relative fitness
AA	p_0^2	1
Aa	$2p_0q_0$	$1 - s_1$
aa	q_0^2	$1 - s_2$

where s_1 takes on a value greater than zero but less than $\frac{1}{2}s_2$; for example, if $s_2 = .5$, then $0 < s_1 < .25$. If $s_1 = 0$, there is complete dominance, and $s_1 = \frac{1}{2}s_2$ represents no dominance.

With incomplete dominance, the frequency of the *A* allele in the survivors is

$$p_1 = \frac{p_0(1 - s_1q_0)}{1 - 2s_1p_0q_0 - s_2q_0^2}$$

Because $p_1 > p_0$ with incomplete dominance, the *A* allele will eventually become fixed, as with complete dominance and no dominance.

Complete dominance, selection favoring the *aa* genotype

In the previous three examples of gene action, selection favors the *A* allele which ultimately becomes fixed. Consider the case in which selection favors the *aa* genotype. In this case the $f(a)$ increases with each generation and ultimately the *a* allele becomes fixed. The change in gene frequency will be demonstrated by using complete dominance of the *A* allele to *a*. With complete dominance, the fitness of *AA* is equal to *Aa* but less than *aa*. Hence, for $0 < s < 1$,

Genotype	Initial frequency	Relative fitness
AA	p_0^2	$1 - s$
Aa	$2p_0q_0$	$1 - s$
aa	q_0^2	1

The frequency of the A allele in the survivors is

$$p_1 = \frac{p_0(1 - s)}{1 - s(1 - q_0^2)}$$

Because $[1 - s(1 - q_0^2)] > (1 - s)$, then $p_1 < p_0$. The frequency of A decreases and thus the frequency of a increases.

Overdominance

Overdominance refers to the heterozygote having greater fitness than either homozygote. For $0 < s_1 < 1$ and $0 < s_2 < 1$, overdominance may be represented as

Genotype	Initial frequency	Relative fitness
AA	p_0^2	$1 - s_1$
Aa	$2p_0q_0$	1
aa	q_0^2	$1 - s_2$

The frequency of the A allele in the survivors is

$$p_1 = \frac{p_0(1 - s_1p_0)}{1 - s_1p_0^2 - s_2q_0^2}$$

Because the heterozygote is the favored genotype, the expectation is that neither allele will become fixed. The question of whether $f(A)$ will decrease or increase is more complicated than in the previous examples. The frequency of A in the survivors, p_1, will equal the initial population frequency, p_0, when the ratio $(1 - s_1p_0)/(1 - s_1p_0^2 - s_2q_0^2)$ is equal to 1. This ratio equals 1 when $p_0 = (s_2/s_1)q_0$. When this is the case, $p_1 = p_0$, and selection would not be expected to change gene frequency. If $p_0 > (s_2/s_1)q_0$, then $f(A)$ would decrease, and, if $p_0 < (s_2/s_1)q_0$, $f(A)$ would increase.

If $p_0 = (s_2/s_1)q_0$, then $p_1 = p_0$; with no change in gene frequency, the population is in equilibrium. For populations not in equilibrium, $f(A)$ will increase or decrease to equilibrium. The frequency of $f(A)$ at equilibrium will be

$$p_E = \frac{s_2}{s_1 + s_2}$$

7-5 Gene Frequencies in Small Populations

One of the assumptions of the Hardy-Weinberg law is a large population of breeding animals. In small populations, fluctuations in gene frequency may occur from one generation to the next by chance. Unlike migration, mutation, and selection, which cause directional changes in gene frequency, fluctuations due to small population size are random in both direction and magnitude. These changes in gene frequency are referred to as *random drift* or *random genetic drift.*

The concept of sampling is important in the study of random drift. Assume an infinitely large number of genes with $f(A) = p_0$ and $f(a) = q_0$. From this pool of genes, a sample of n genes is drawn at random. The expansion $(pA + qa)^n$ gives the probabilities that given numbers of each allele are obtained in the sample of n genes (see Chapter 5). For example, if eight genes are drawn at random from a gene pool where $p_0 = \frac{3}{4}$ and $q_0 = \frac{1}{4}$, the following probabilities are obtained for the possible numbers of A alleles in the sample of eight:

Number of A genes	Probability
0	.000015
1	.000366
2	.003845
3	.023071
4	.086517
5	.207642
6	.311462
7	.266968
8	.100113

The most probable event is that six of the eight genes sampled are A alleles. The probability is .31, which means 31 percent of the time a sample would occur in which the gene frequency is the same, $\frac{3}{4}$, as in the gene pool. Conversely, 69

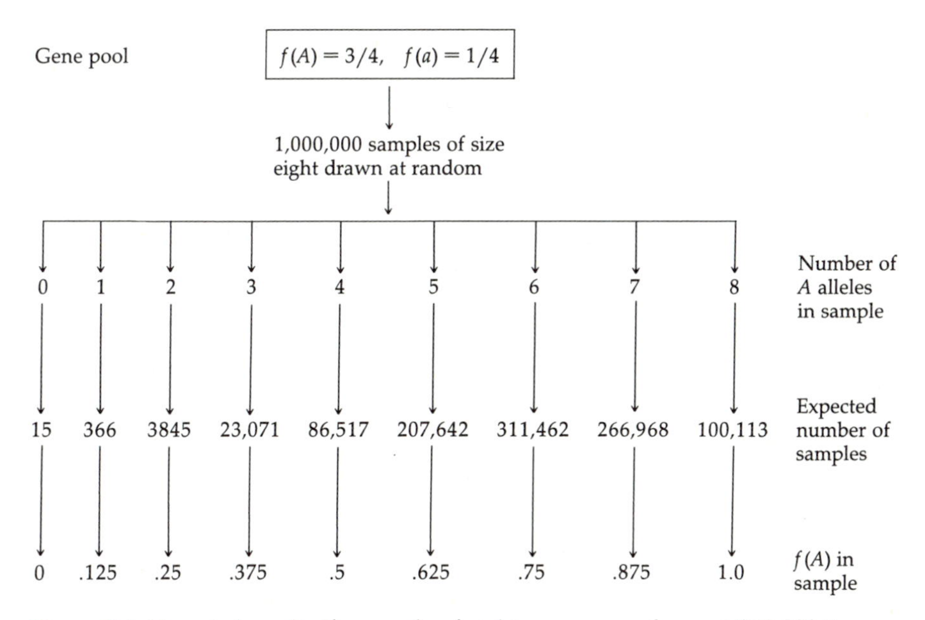

Figure 7-4 Expected results if a sample of eight genes were drawn 1,000,000 times from a gene pool where $f(A) = \frac{3}{4}$.

percent of the time the gene frequency in the sample would be different from that in the gene pool by the chance of sampling. One of every 10 samples would be expected to contain all *A* alleles, and the potential exists—although small—that all eight genes would be the *a* allele. Figure 7-4 shows the expected results if eight genes were drawn 1,000,000 times from the gene pool. The figure illustrates that sampling can cause random change in gene frequency and also may result in the loss of one of the alleles. The probability that *A* or *a* is fixed in the sample (that is, the other allele is lost) is dependent on the gene frequencies in the pool from which the sample is drawn and the number of genes sampled:

$$P(A \text{ fixed}) = P(\text{all } A \text{ alleles in a sample of } n \text{ alleles}) = p_0^n$$

and

$$P(a \text{ fixed}) = P(\text{all } a \text{ alleles in a sample of } n \text{ alleles}) = q_0^n$$

The example of sampling eight genes from a large gene pool shows how sampling can influence gene frequency. However, no consideration was given to how the gene pool was established; that is, the frequency of parent genotypes was not considered. The effect of the frequency of parent genotypes will now be examined for the following small population where, for each sex:

Genotype	Initial frequency
AA	5
Aa	2
aa	1

Assume each female and male is mated and that one progeny results from each of the eight matings. Of the 16 genes represented in the progeny, 10 must be allele *A* (one from each of the five males and five females being *AA*), and two must be allele *a* (one from each *aa* male and female). Sampling cannot influence the number of genes received by the progeny from these homozygous parents. The four genes coming from the heterozygotes, however, are subject to sampling with $P(A) = .5$ and $P(a) = .5$. The possible outcomes are that the progeny receive from none to four *A* alleles from the four heterozygous parents with the following probabilities:

Number of *A* genes from heterozygotes	Probability	Total number of *A* genes in progeny	Resulting $f(A)$ in progeny
0	.0625	$10 + 0 = 10$	.6250
1	.2500	$10 + 1 = 11$	.6875
2	.3750	$10 + 2 = 12$	.7500
3	.2500	$10 + 3 = 13$	.8125
4	.0625	$10 + 4 = 14$	.8750

The frequency of the *A* gene in the parents was .75, whereas in the progeny it can range from .625 to .875, although the expected or average frequency would be .75. Fixation cannot occur as long as there is one parent of each homozygous type. However, sampling may cause the loss of one of the homozygous types. For example, if progeny of heterozygous parents received by chance only *A* alleles from the heterozygous parents, 14 of the 16 genes in the progeny would be the *A* allele. For this case, there would be an *aa* progeny only if the *aa* × *aa* mating had occurred. If the *aa* × *aa* mating did not occur and the progeny of the heterozygous parents received all *A* alleles, then the population of progeny would be

Genotype	Number
AA	6
Aa	2
aa	0

This population would then be susceptible to the loss of the a allele in the next generation.

Consider a population of size n with the following genotypes and number of animals:

Genotype	Number
AA	n_{AA}
Aa	n_{Aa}
aa	n_{aa}

The frequency of A in the progeny could range from n_{AA}/n to $(n_{AA} + n_{Aa})/n$, with the most probable frequency being $(n_{AA} + \frac{1}{2}n_{Aa})/n$. The probability of a change in the frequency of A of a certain magnitude can be calculated from the binomial expansion as shown in Chapter 5.

7-6 Summary

When the three assumptions of the Hardy-Weinberg law—random mating, a large population, and the absence of forces acting to change gene frequencies—are not valid, changes in gene and genotypic frequencies may occur from one generation to the next.

Nonrandom mating occurs when rules are imposed which dictate the mating strategy; for example, the inter se mating of like phenotypes, or positive assortative mating. Nonrandom mating alters genotypic frequencies from one generation to the next. Positive assortative mating does not change gene frequencies, but negative assortative mating may.

Migration refers to the movement of individuals from one breeding population (population 2) to another (population 1). The gene frequencies in population 1 after migration are

$$q_1' = q_1 + m(q_2 - q_1)$$

and

$$p_1' = 1 - q_1'$$

where m is the proportion of immigrants in population 1.

A mutation is a sudden heritable change in the genetic material. Mutation rate is the probability that a particular gene mutates and the frequency of mutation is the product of the mutation rate and the gene frequency. When mutations are in one direction, that is, from B to b but not from b to B, the B allele eventually will be lost from the population. If u is the mutation rate, then the frequency of the B allele in the nth generation is

$$p_n = p_0(1 - u)^n$$

For mutations in both directions, B to b at rate u and b to B at rate v, three possibilities exist:

1. $up > vq$, and $f(B)$ will decrease;
2. $up < vq$, and $f(B)$ will increase; or
3. $up = vq$, and $f(B)$ will be stable, or in equilibrium.

The gene frequencies at equilibrium are

$$q_E = u/(u + v)$$

and

$$p_E = v/(u + v)$$

Selection causes differential fitness among phenotypes. Table 7-1 summarizes the influence of selection on gene frequencies for several cases, including dominance with selection favoring the dominant phenotype, no dominance, incomplete dominance, overdominance, and dominance with selection favoring the recessive phenotype.

In small populations, chance changes in gene frequencies may occur both in magnitude and direction. The change is referred to as random drift and is a result of sampling. The probability that genes will be lost or fixed in small populations depends on population size, gene frequencies, and genotypic frequencies.

Table 7-1 Influence of selection on gene frequency with p_0 and q_0 representing initial gene frequencies

Gene action	Genotype	Relative fitness	New frequency of *B*	Ultimate fate of *B* allele
Dominance (favoring B_)	*BB* *Bb* *bb*	1 1 $1 - s$	$p_1 = \frac{p_0}{1 - sq_0^2}$	Fixed
Dominance (favoring *bb*)	*BB* *Bb* *bb*	$1 - s$ $1 - s$ 1	$p_1 = \frac{p_0(1 - s)}{1 - sp_0(1 + q_0)}$	Lost
No dominance	*BB* *Bb* *bb*	1 $1 - .5s$ $1 - s$	$p_1 = \frac{p_0(1 - .5sq_0)}{1 - sq_0}$	Fixed
Incomplete dominance	*BB* *Bb* *bb*	1 $1 - s_1$ $1 - s_2$	$p_1 = \frac{p_0(1 - s_1q_0)}{1 - 2s_1p_0q_0 - s_2q_0^2}$	Fixed
Overdominance	*BB* *Bb* *bb*	$1 - s_1$ 1 $1 - s_2$	$p_1 = \frac{p_0(1 - s_1p_0)}{1 - s_1p_0^2 - s_2q_0^2}$	Reaches equilibrium at $p_E = \frac{s_2}{s_1 + s_2}$

CHAPTER 8

Probability of Detecting Carriers of Recessive Genes

It is often important to know whether a particular animal showing a dominant phenotype is homozygous for the dominant allele or heterozygous, especially if the recessive homozygote has an undesirable characteristic. The recessive genotype may be simply a coat color which does not meet the standard for a breed or, more seriously, it may be a lethal or semilethal characteristic.

In domestic species, males have the potential to produce many more progeny in their reproductive lifetimes than females. An extreme example is the use of artificial insemination in cattle, which allows a male to have thousands of progeny. It is important that such a male not carry a recessive allele for a characteristic such as mulefoot. Breeders need a system of screening genotypes to prevent carriers of an undesirable allele from having many progeny. Two potential sources of information exist which may determine the probability of heterozygosity—the animal's pedigree and its progeny. As this chapter will show, there is never complete assurance that a particular animal is homozygous for a dominant allele, that is, not a carrier of the recessive allele.

8-1 Pedigree Information

Pedigree information may help determine whether an animal is a known or likely carrier of a recessive allele. An animal showing the phenotype of the dominant allele (the dominant phenotype) is known to be a carrier if either parent had a homozygous recessive genotype.

Progeny of a known carrier are likely to be carriers. For example, assume that a breeder of horned Hereford cattle (genotype *pp*) begins using polled bulls which in test matings have not been found to transmit the horned gene. The bulls are assumed to have the genotype *PP*. All of the F_1 polled progeny are carriers because they had dams with the *pp* genotype. If the F_1 progeny are inter se mated, the expected phenotypic distribution in the F_2 is 3 : 1 polled to horned. The expected genotypic ratio is 1 *PP* : 2 *Pp* : 1 *pp*. All polled F_2 progeny must be considered to be possible carriers of the *p* gene even though one of three is *PP*.

8-2 Progeny Testing

A *progeny test* is the mating of an animal to obtain progeny for observation. The breeder uses the information obtained from the progeny to help determine the animal's genotype. Although both males and females may be progeny tested, the examples in this chapter are for progeny testing males. For several of the tests, large numbers of progeny are required so that testing is effective only for males.

Assume a male is suspected of carrying a recessive allele. His genotype will be designated as $A\_$. There are several alternatives for progeny testing this male. Each test depends on the type of females available for mating. In all cases the following procedure is used:

1. Assume the suspected animal is in fact heterozygous.
2. Calculate the probability that the dominant phenotype will be observed in a single progeny.
3. Calculate the probability that if *n* progeny are observed, all will show the dominant phenotype. This is calculated as the probability of observing the dominant phenotype in a single progeny, raised to the *n*th power.
4. Calculate the probability that at least one recessive phenotype will be observed. The sum of the probabilities of all events must be 1. Therefore, the probability of at least one recessive phenotype is 1 minus the probability of all *n* progeny showing the dominant phenotype. The probability that at least one progeny will show the recessive genotype is called the *probability of detection* and refers to detecting the recessive gene, if present, in the suspected animal.

Homozygous recessive females

The probability of a progeny showing the dominant phenotype from the mating of *Aa* with *aa* is $\frac{1}{2}$. The probability that all *n* progeny will show the dominant phenotype—that is, none show the recessive *aa* phenotype—from the mating of *Aa* to *aa* females is $\frac{1}{2}^n$. The probabilities that all progeny will show the phenotype associated with the dominant allele are, for example, the following:

Number of progeny	P(all progeny being *Aa* from mating $Aa \times aa$)
1	1/2
2	1/4
3	1/8
4	1/16
5	1/32
6	1/64

If even one progeny has the recessive phenotype, the genotype of the male must be *Aa*; hence, the probability of detection is

$$\text{P(detection)} = 1 - \text{P(all progeny are } Aa \text{ from mating } Aa \times aa)$$

For the test matings with one to six progeny, the probabilities of detection are

Number of progeny	P(detection)
1	$1-(1/2)^1 = 1/2 \quad = .500$
2	$1-(1/2)^2 = 3/4 \quad = .750$
3	$1-(1/2)^3 = 7/8 \quad = .875$
4	$1-(1/2)^4 = 15/16 = .938$
5	$1-(1/2)^5 = 31/32 = .969$
6	$1-(1/2)^6 = 63/64 = .984$

Assume the test matings produce six progeny. The probability of detection of $\frac{63}{64}$ indicates that 63 of 64 heterozygotes tested are expected to have at least one progeny with the recessive phenotype. Conversely, one of 64 heterozygotes tested is expected to "pass" the test and remain undetected.

If all progeny show the phenotype for the dominant allele, the tentative conclusion is that the suspect animal is homozygous for that allele. The conclusion obviously is correct when the suspect is homozygous but incorrect when the animal is heterozygous and not detected in the test matings. The probability of detection can be interpreted as the confidence level associated with the conclusion of homozygosity. If only a single progeny is born and shows the phenotype for the dominant allele, there is little confidence in the conclusion that the suspect is homozygous. If the suspect is, in fact, heterozygous, then the conclusion of homozygosity will be wrong 50 percent of the time when based on one progeny.

The usual procedure in setting up a test program for a suspected carrier is to determine the desired level of confidence in the conclusion of homozygosity then calculate the number of progeny required to obtain that confidence level. For example, if a 95 percent or greater confidence level is desired, the equation to be solved for the number of progeny required is

$$.95 = 1 - (1/2)^n$$

or

$$.05 = (1/2)^n$$

This equation can be solved by taking the log of both sides:

$$\log .05 = n \log(1/2)$$

so that

$$n = 4.32$$

To obtain a 95 percent or greater level of confidence requires five or more progeny.

Known carrier females

For some characteristics (for example, lethals such as dwarfism in cattle), homozygous females are not available. However, a group of known carriers, that is, females that have had progeny showing the recessive phenotype, may be available. The male to be tested is assumed to be heterozygous, and the probability of one progeny being $A\_$ from an Aa by Aa mating is $\frac{3}{4}$. The probability that all n progeny show the dominant phenotype given the mating of heterozygotes is $(\frac{3}{4})^n$. The probabilities of detection for different numbers of progeny are

Number of progeny	P(all progeny are $A\_$ from $Aa \times Aa$ matings)	P(detecting the recessive gene in suspect)
1	.75	.25
2	.56	.44
5	.24	.76
10	.06	.94
20	.003	.997

Because the probability of a dominant phenotype from an *Aa* by *Aa* mating is greater than the probability of a dominant phenotype from an *Aa* by *aa* mating, more progeny are needed to achieve the same confidence level. If a 95 percent level for probability of detection is desired, then

$$\text{P(detection)} = 1 - \text{P(one progeny is } A\_ \text{ from an } Aa \times Aa \text{ mating)}^n$$

Thus,

$$.95 = 1 - (3/4)^n$$

and

$$.05 = (3/4)^n$$

Again, taking the log of both sides,

$$\log .05 = n \log .75$$

so that

$$n = \log .05/\log .75$$

and

$$n = 10.4 \text{ progeny}$$

Therefore, 11 progeny would be needed to reach the .95 level of confidence.

Daughters of known carriers

Assume a bull is known to be a carrier of a recessive gene, for example, the allele for mulefoot. Due to the low frequency of the gene in the population, all his mates are assumed to be *MM*, homozygous for the normal allele; therefore, his daughters are expected to be $\frac{1}{2}MM$ and $\frac{1}{2}Mm$. The expected $f(m)$ in his daughters is .25, which is higher than the frequency in the population. This group of females can be used to test suspected carriers of the mulefoot gene. The expected genotypic frequencies of progeny from mating a suspected carrier to daughters of a known carrier can be obtained as the probabilities of union between gametes. If the male is a carrier, the probability of his progeny receiving an *M* gene from him is $\frac{1}{2}$. If the probability the progeny receive an *m* gene from the females is the expected $f(m) = \frac{1}{4}$, then the probabilities of union are

Gametes from		
Male	**Females**	**P(union)**
M	M	3/8
M	m	1/8
m	M	3/8
m	m	1/8

The probability of a normal progeny is $\frac{7}{8}$, and the probability of all n progeny being normal is $(\frac{7}{8})^n$. The probability of detection is

$$\text{P(detection)} = 1 - (7/8)^n$$

A 95 percent level of detection requires 23 progeny.

In the first two cases, using homozygous recessive females or females known to be heterozygous, the probabilities of detection were precise. In this case, however, the probability of detection assumes $f(MM) = .5$ and $f(Mm) = .5$ in the daughters of the known carrier such that $f(m) = .25$. The genotypic frequencies of the daughters can be different from .5, due to sampling. Then the expected probability of detection will not represent the true probability of detection for that particular sample of daughters.

Daughters of the suspect

A male's own daughters may be used as a test population. Sire to daughter matings represent intensive inbreeding as will be discussed in Chapter 15. Assume all initial mates of the sire are AA for a particular locus; then, for a male heterozygous at that locus, his daughters' genotypic frequencies are expected to be $f(AA) = .5$ and $f(Aa) = .5$. For a particular locus, the probability of detection is the same for mating the male to his daughters as for mating him to daughters of known carriers.

There are, however, two important differences. First, a longer period of time is required to test a male with matings to his daughters because the daughters must reach breeding age. Second, using daughters of a known carrier is a test for only one specific characteristic. There are, however, potentially many undesirable recessive alleles that may be present in an animal's genotype. Mating a sire to his daughters is a test for recessive alleles at all loci, not just one locus, with each having the same expected probability of detection. This form of inbreeding is an effective method for screening an individual's genotype.

Mating a suspect randomly to the female population

Mating a male to females randomly chosen from a population is a method of detecting recessive genes. Assume in the female population the following genotypic frequencies:

Genotype	Frequency
AA	p^2
Aa	$2pq$
aa	q^2

The probability of drawing an *a* gene at random from the female gene pool is q. The expected genotypic frequencies of progeny that result from mating a suspected carrier to this population of females are the probabilities of the union of gametes carrying the *A* and *a* genes, as shown:

Gametes from		
Male	**Females**	**P(union)**
A	*A*	$(1/2)p$
A	*a*	$(1/2)q$
a	*A*	$(1/2)p$
a	*a*	$(1/2)q$

The probability of an $A\_$ genotype in the progeny is

$$1 - \mathrm{P}(aa \text{ progeny}) = 1 - .5q$$

The probability that all n progeny will be $A\_$ is $(1 - .5q)^n$. The probability of detection is then calculated as

$$\mathrm{P(detection)} = 1 - (1 - .5q)^n$$

The probability of detection depends on $f(a)$ in the female population and on the number of progeny. The number of progeny required to reach a 95 percent level of detection is

$$n = \log .05/\log(1 - .5q)$$

and, for varying $f(a)$,

$f(a)$	**n required for P(detection) = .95**
.8	6
.5	11
.3	18
.1	58
.01	598

For a high $f(a)$ few progeny are needed, but for a low $f(a)$ many are needed to obtain a probability of detection of .95.

The assumption to this point has been that the homozygous recessive genotypes survive and are capable of reproducing. For lethals or recessive characteristics which result in culling, the homozygous recessive is not available for mating. Assume the following frequencies of genotypes in the female population:

Genotype	**Frequency**	**Fitness values**
AA	p^2	1
Aa	$2pq$	1
aa	q^2	0

The fitness values reflect the fact that *aa* genotypes do not enter the breeding population or are not successful if they do. As described in Chapter 7, $f(A)$ in the breeding female population is calculated as

$$f(A) = 1/(1 + q)$$

Therefore, $f(a)$ in the breeding females is

$$f(a) = q/(1 + q)$$

Again, the following table uses the probabilities of union of gametes to obtain the expected genotypic frequencies in the progeny.

Gametes from		
Male	Females	P(union)
A	A	$.5/(1+q)$
A	a	$.5q/(1+q)$
a	A	$.5/(1+q)$
a	a	$.5q/(1+q)$

The probability of an A genotype in the progeny is $(1+.5q)/(1+q)$, and the probability of detection with n progeny is

$$\text{P(detection)} = 1 - [(1+.5q)/(1+q)]^n$$

For a 95 percent detection level, the numbers of progeny required for different genotype frequencies are:

$f(a)$	n required
.8	8
.5	17
.3	24
.1	64
.01	604

8-3 Multiple Birth Species

If k is the number of progeny per mating and m is the number of matings, then $km = n$, the total number of progeny. Again, the assumption is made that the suspect individual is Aa. If homozygous recessive females are available for test matings, the probability that all progeny show the phenotype of the dominant allele is

$$\begin{aligned}\text{P(all progeny are } Aa \text{ from } Aa \times aa \text{ matings)} &= [(1/2)^k]^m \\ &= (1/2)^{km} = (1/2)^n\end{aligned}$$

as was observed with single-bearing species. If the genotypes of all tester females are the same, there is no difference between single and multiple birth species in calculating the probability of detection. If there is more than one genotype in the tester female population, the number of progeny born per mating must be considered because the probability of an $A\_$ progeny is different for each genotype of female.

Daughters of a known carrier

The expected genotypic frequencies in daughters of a known carrier are 50 percent *AA* and 50 percent *Aa*. If the suspect to be tested is heterozygous, the probability of all progeny showing the phenotype associated with the dominant allele may be partitioned as follows into probabilities for each type of mating:

$$\begin{aligned} &\text{P}[n \text{ progeny are } A\_ \text{ from } Aa \times (.5AA + .5Aa)] \\ &\quad = [.5\text{P}(1A\_ \text{ progeny from an } Aa \times AA \text{ mating})^k \\ &\qquad + .5\text{P}(1A\_ \text{ progeny from an } Aa \times Aa \text{ mating})^k]^m \end{aligned}$$

The first probability is 1 because all progeny receive an *A* gene from the females, and the second probability is .75; therefore,

$$\text{P}[n \text{ progeny are } A\_ \text{ from } Aa \times (.5AA + .5Aa)] = [.5 + .5(.75)^k]^m$$

The following are probabilities of detection for varying litter sizes:

Litter size	P[n progeny are $A\_$ from $Aa \times (.5AA + .5Aa)$]
1	$(.875)^m$
2	$(.781)^m$
3	$(.711)^m$
8	$(.550)^n$

The probability of detection is calculated as with single-bearing species. For example, with eight progeny per litter,

$$\text{P(detection)} = 1 - (.55)^m$$

If a certain level of detection is desired for a particular litter size, k, the number of matings required is calculated, not the number of progeny.

8-4 Summary

To prevent carriers of undesirable genes from having many progeny requires a system of genetic screening. Pedigree information may aid in identifying carriers and likely carriers. Progeny testing may help determine an animal's geno-

Table 8-1 Summary of progeny tests for different tester populations of mates

Tester population	P(detection)	Progeny required for P(detection) of .95	.99
aa	$1-(1/2)^n$	5	7
Aa	$1-(3/4)^n$	11	16
Daughters of known carriers ($.5AA + .5Aa$)	$1-(7/8)^n$	23	35
Daughters of suspect	$1-(7/8)^n$	23	35
Population at random (aa viable), $q = .1$	$1-(1-.5q)^n$	58	90
Population at random (aa not viable), $q = .1$	$1-[(1+.5q)/(1+q)]^n$	64	99

type. Table 8-1 summarizes probabilities of detection from progeny testing with different genotypes of mates, under the assumption that the tested animal is in fact a heterozygote.

For multiple birth species, if k is the number of progeny per litter and m is the number of matings, then $km = n$ is the total number of progeny. If the tester population includes just one genotype, aa or Aa, the probability of detection is the same as for single-bearing species:

$$\mathrm{P}(\text{detection from } aa \text{ mates}) = 1 - (1/2)^n$$

and

$$\mathrm{P}(\text{detection from } Aa \text{ mates}) = 1 - (.75)^n$$

If more than one genotype is present in the tester population, the number born per litter must be considered. As an example, if the tester population consists of daughters of known carriers,

$$\begin{aligned}&\mathrm{P}\{\text{all progeny are } A\_ \text{ from } Aa \times [(1/2)AA + (1/2)Aa]\}\\&\quad = [(1/2)\mathrm{P}(\text{one progeny is } A\_ \text{ from an } Aa \times AA \text{ mating})^k\\&\qquad + (1/2)\mathrm{P}(\text{one progeny is } A\_ \text{ from an } Aa \times Aa \text{ mating})^k]^m\\&\quad = [.5 + .5(.75)^k]^m\end{aligned}$$

CHAPTER 9

Genetic Value and Artificial Selection

Artificial selection is selection based on a set of rules dictating which individuals will become parents. Selection, either artificial or natural, changes gene frequencies. The prediction of change in gene frequencies is identical for both artificial and natural selection. For example, the allele for black coat color in Angus and Holstein cattle is dominant to the allele for red coats. If the rule for selection is to keep only black animals, this selection is synonymous with having fitness values of 1 for the homozygous and heterozygous genotypes for black, and zero for the recessive genotype for red. The gene frequency of the dominant allele for black after one generation of selection will be $1/(1 + q_0)$ as with natural selection. Artificial selection differs from natural selection in that animals removed from the population in artificial selection may be capable of surviving and reproducing; they are not, however, allowed the opportunity.

Artificial selection is the tool used by humans to dramatically alter the appearance and productivity of domesticated animals. Much of the selection practiced by livestock breeders is for performance traits, that is, traits for which the phenotype is a measurement. Breeders practice selection to increase the pounds of milk produced by dairy cattle, to increase the weight of lambs, calves, and young pigs at weaning, and also to improve reproductive traits like litter size. The genotype for these traits influences the magnitude of the measurement for the phenotype. Breeders are interested in the result of artificial selection on the average performance of the progeny for the particular trait.

This chapter will discuss genotypic values, population means and variances, and the impact of artificial selection on future progeny performance for traits that are measured. The single locus model will be used as an introduction to the study of quantitative characteristics which are influenced by genes at many loci.

9-1 Genotypic Values and Population Means

Genotypic values and population means will be defined using the following example. Assume the phenotype of interest is the amount of hormone in a milliliter of serum, and that the amount is influenced by the genotype at only one locus. The population is assumed to be in equilibrium with the following gene frequencies and phenotypic measurements:

Genotype	Frequency	Phenotype
B_1B_1	p^2	$P_{11} = 12$ units
B_1B_2	$2pq$	$P_{12} = 10$ units
B_2B_2	q^2	$P_{22} = 6$ units

The alleles are distinguished by subscripts, with B_1 increasing the amount of hormone. The subscripts for the phenotype (P) indicate which alleles comprise the genotype represented by P. The genotype of the animal influences the measurement obtained for the phenotype. Thus, the genotype has a certain value measured in units of the trait. The *genotypic value* is defined as the deviation of the phenotype from the average of the two homozygous phenotypes, P_{11} and P_{22}. The average of the two homozygotes is denoted by m, that is, $m = \frac{1}{2}(P_{11} + P_{22})$, and the genotypic value is denoted by V. For this example, $m = 9$ units and the genotypic values for each of three genotypes are

Genotype	Genotypic value (V)
B_1B_1	$V_{11} = P_{11} - m = a = 3$ units
B_1B_2	$V_{12} = P_{12} - m = d = 1$ unit
B_2B_2	$V_{22} = P_{22} - m = -a = -3$ units

Because m is defined as the average of phenotypes for the two homozygous genotypes, the genotypic value of B_2B_2 (that is, $-a$) is the negative of that for B_1B_1 (that is, a). The advantage of assigning genotypic values of a, d, and $-a$ is to obtain general formulas for means and variances of the population.

The relationship of d to a defines the kind of dominance such that:

Relationship of d to a	Relationship of phenotypes	Gene action
$d > a$	$P_{12} > P_{11}$	Overdominance
$d = a$	$P_{12} = P_{11}$	Complete dominance of B_1
$0 < d < a$	$m < P_{12} < P_{11}$	Incomplete dominance of B_1
$d = 0$	$P_{12} = m$	No dominance
$0 > d > -a$	$m > P_{12} > P_{22}$	Incomplete dominance of B_2
$d = -a$	$P_{12} = P_{22}$	Complete dominance of B_2
$d < -a$	$P_{12} < P_{22}$	Overdominance

The relationship of genotypic values to the kinds of dominance is also shown in Figure 9-1.

Each phenotype can be expressed by the following equation (or model):

$$P_{ij} = m + V_{ij}$$

This model assumes no environmental influences on the phenotype (discussed in Chapter 11) and that the measurement of P is made without error.

Population mean

A population parameter of interest for traits which are measured is the phenotypic mean (or average), which is denoted by the symbol μ (Greek mu). The

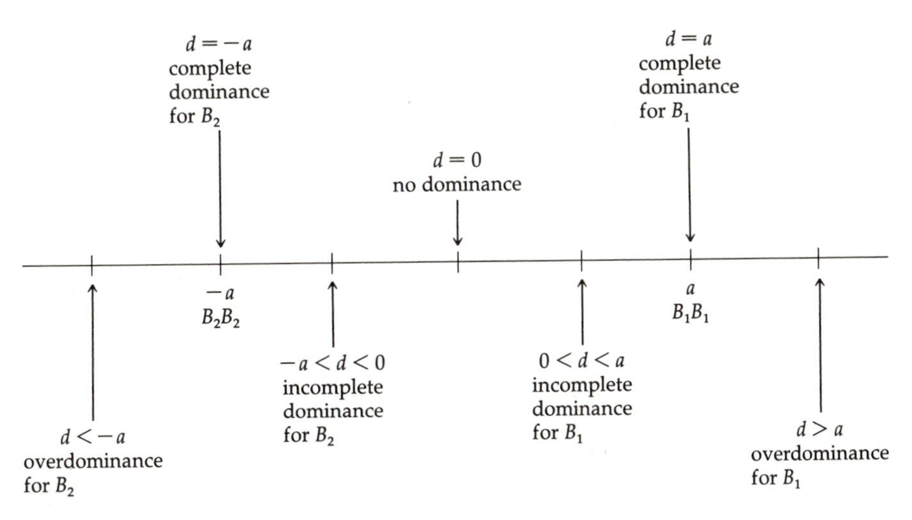

Figure 9-1 Illustration of kinds of dominance.

mean is calculated as the sum of all phenotypic measurements divided by the number of phenotypes observed. For the single locus model, an equivalent calculation is

$$\mu = f(B_1B_1)P_{11} + f(B_1B_2)P_{12} + f(B_2B_2)P_{22}$$

For example, μ for a population in Hardy-Weinberg equilibrium is

$$\mu = p^2P_{11} + 2pqP_{12} + q^2P_{22}$$

An alternative computing formula for obtaining the mean is based on substituting $m + V_{ij}$ for each phenotype, for example, $m + a$ for P_{11}. For a population in equilibrium the mean is

$$\begin{aligned}\mu &= p^2(m + a) + 2pq(m + d) + q^2(m - a)\\ &= m(p^2 + 2pq + q^2) + a(p^2 - q^2) + 2pqd\end{aligned}$$

Because $p^2 + 2pq + q^2 = 1$, and $(p^2 - q^2) = (p + q)(p - q) = p - q$, the mean is

$$\mu = m + [a(p - q) + 2pqd]$$

The population mean, written in this way, is a fixed portion, m, plus the average of genotypic values, $[a(p - q) + 2pqd]$. The latter part may be changed with selection, which changes gene frequencies. In the hormone example, increasing $f(B_1)$ would increase the mean hormone level in the population. For example, when $m = 9$, $a = 3$, and $d = 1$,

If p	Then μ
.2	6.92
.5	9.50
.8	11.12

9-2 Selection and Selection Response

The first step in developing a selection program is to define the goal to be achieved by selection. If, in the hormone example, the goal is to raise the population mean, one strategy would be to cull all homozygous B_2B_2 animals from the breeding population. If initially $p = .5$, then

Genotype	Initial frequency	Genotypic value
B_1B_1	.25	$a = 3$
B_1B_2	.50	$d = 1$
B_2B_2	.25	$-a = -3$

The mean for this generation, μ, is 9.5. If all B_2B_2 animals are culled, the gene frequency in the surviving animals is $p_1 = 1/(1 + q) = \frac{2}{3}$. Random mating of the surviving animals would produce progeny with the following genotypic frequencies:

Genotype	Frequency
B_1B_1	4/9
B_1B_2	4/9
B_2B_2	1/9

The mean of the progeny, μ_1, is

$$\mu_1 = m + a(p_1 - q_1) + 2p_1q_1d$$
$$\mu_1 = 9 + 3(2/3 - 1/3) + 2(2/3)(1/3)(1)$$
$$= 10.44$$

Response to selection, denoted as $\Delta\mu$, is the change in the population mean from the parent to the progeny generation:

$$\Delta\mu = \mu_1 - \mu = .94$$

Because m is a constant, the change in the mean was a result of increasing the average genotypic value by increasing the frequency of B_1.

In this example, if the selection rule had been to keep only B_1B_1 animals, the progeny mean would have been 12.0. The B_1 gene would become fixed in the progeny so that no further response to selection could occur. The trait would be at its genetic limit.

9-3 Breeding Value

The difference in the phenotypes of animals in the single locus example is a function of the genotypic values. For example,

$$P_{11} - P_{22} = (m + a) - (m - a) = 2a$$

With complete dominance, $P_{11} = P_{12}$; hence, phenotypically the individuals are equal. However, parents do not pass their genotype on to their progeny but rather pass on only a random sample of one gene to each locus of the progeny. The critical question is which parental genotype will produce progeny with the highest average? The answer to this question will define an animal's breeding value.

The term *breeding value* is self-descriptive, referring to the value of an animal in a breeding program. Breeding value is a measure of the animal's expected progeny performance relative to the population mean. For the single locus example, the breeding value for each genotype is calculated as twice the difference of the expected progeny mean from the population mean. The reason for doubling the progeny deviation is that the progeny contain only a sample one-half of the parent's genes. The progeny deviation itself represents the *transmitting ability* of the parent, which is one-half the breeding value.

Assume a male with genotype B_1B_1 is mated to the following female population:

Genotype	Frequency	Genotypic value
B_1B_1	p^2	a
B_1B_2	$2pq$	d
B_2B_2	q^2	$-a$

The expected genotypic frequencies of the progeny can be determined from the probabilities of union between alleles from the male and the female gene pool. With a B_1B_1 male, the probability of a B_1 gene is 1 and of a B_2 gene is 0. In the female gene pool, the probability of drawing a B_1 gene is p and a B_2 gene is q. Hence, the frequency of B_1B_1 in the progeny is expected to be P(B_1 from males)P(B_1 from females) $= p$, and $f(B_1B_2)$ is expected to be P(B_1 from males)P(B_2 from females) $= q$. The progeny genotypic frequencies and genotypic values are as follows:

Genotype	Frequency	Genotypic value
B_1B_1	p	a
B_1B_2	q	d
B_2B_2	0	$-a$

The expected mean of the progeny of the B_1B_1 male, denoted μ_{11}, is the sum of the products of genotypic frequencies and corresponding phenotypic values:

$$\begin{aligned}\mu_{11} &= f(B_1B_1)P_{11} + f(B_1B_2)P_{12} + f(B_2B_2)P_{22}\\ &= p(m + a) + q(m + d) + 0(m - a)\\ &= m + pa + qd\end{aligned}$$

The breeding value of the B_1B_1 male, denoted BV_{11}, is twice the deviation of his progeny mean from the population mean, that is,

$$\begin{aligned}BV_{11} &= 2(\mu_{11} - \mu)\\ &= 2[m + pa + qd - m - a(p - q) - 2pqd]\\ &= 2(qd + qa - 2pqd)\\ &= 2[qa + qd(1 - 2p)]\\ &= 2q[a + d(q - p)]\end{aligned}$$

Likewise, the breeding values of B_1B_2 and B_2B_2 males are

$$\begin{aligned}BV_{12} &= 2(\mu_{12} - \mu)\\ &= (q - p)[a + d(q - p)]\end{aligned}$$

and

$$\begin{aligned}BV_{22} &= 2(\mu_{22} - \mu)\\ &= -2p[a + d(q - p)]\end{aligned}$$

The term, $[a + d(q - p)]$, appearing in the breeding value of each genotype is usually denoted by α (Greek alpha). Hence, breeding values are

Genotype	Breeding value
B_1B_1	$2q\alpha$
B_1B_2	$(q - p)\alpha$
B_2B_2	$-2p\alpha$

The term α is often referred to as the *average effect of a gene substitution.*

The breeding value is a function of gene frequencies and genotypic values. Gene frequencies are likely to vary from population to population, as are the breeding values. Two cases using the hormone example will be examined: first, $f(B_1) = .4$ in Example 9-1, and second, $f(B_1) = .8$ in Example 9.2.

Example 9-1

Assume $p = .4$, the frequency of B_1, and that the population is

Genotype	Frequency	Genotypic value
B_1B_1	.16	$a = 3$
B_1B_2	.48	$d = 1$
B_2B_2	.36	$-a = -3$

The mean of this population when $m = 9$ is

$$\begin{aligned}\mu &= m + a(p - q) + 2pqd \\ &= 8.88\end{aligned}$$

For this population, the average effect of a gene substitution is

$$\begin{aligned}\alpha &= a + d(q - p) \\ &= 3.2\end{aligned}$$

The breeding values are

Genotype	Breeding value
B_1B_1	$2q\alpha = 3.84$
B_1B_2	$(q - p)\alpha = .64$
B_2B_2	$-2p\alpha = -2.56$

Both B_1B_1 and B_1B_2 parents will increase the population mean since both breeding values are positive.

Example 9-2

Assume now that $p = .8$ and the population is

Genotype	Frequency	Genotypic value
B_1B_1	.64	3
B_1B_2	.32	1
B_2B_2	.04	-3

For this population,

$$\begin{aligned}\mu &= m + a(p - q) + 2pqd \\ &= 11.12\end{aligned}$$

Also

$$\begin{aligned}\alpha &= a + d(q - p) \\ &= 2.4\end{aligned}$$

Therefore, the breeding values are

Genotype	Breeding value
B_1B_1	$2q\alpha = .96$
B_1B_2	$(q - p)\alpha = -1.44$
B_2B_2	$-2p\alpha = -3.84$

All breeding values are smaller than those obtained in Example 9-1 in which $p = .4$. Only B_1B_1 parents are expected to increase the population mean.

Comparing the breeding values obtained in these two examples demonstrates that the breeding values are specific to the population in question. That is, breeding values depend on genotypic frequencies.

The difference between breeding values is additive; that is, the difference between BV_{11} and BV_{12} is the same as the difference between BV_{12} and BV_{22} and is equal to α.

$$\begin{aligned}BV_{11} - BV_{12} &= 2q\alpha - (q - p)\alpha \\ &= \alpha\end{aligned}$$

and

$$\begin{aligned}BV_{12} - BV_{22} &= (q - p)\alpha - (-2p\alpha) \\ &= \alpha\end{aligned}$$

The breeding value of an individual is then referred to as its *additive genetic merit.* Replacing (substituting) a B_1 gene in a B_2B_2 genotype gives a B_1B_2 genotype and an increase in breeding value of α (the average effect of the gene substitution). Likewise, replacing the B_2 gene in the heterozygote with a B_1 gene increases its breeding value by α.

9-4 Dominance Deviation

In Example 9-1 genotypic values and breeding values were

Genotype	Genotypic value	Breeding value
B_1B_1	3	3.84
B_1B_2	1	.64
B_2B_2	−3	−2.56

The breeding value is the value of each genotype as a parent. Because a parent passes on, to each progeny, one or the other of its genes, the breeding value represents the sum of the value of each allele in the genotype. The question arises as to why the genotypic values and the breeding value differ. The difference between the genotypic value, V_{ij}, and the breeding value, BV_{ij}, for B_1B_1 can be represented as

$$\begin{aligned} V_{11} - BV_{11} &= a - 2q\alpha \\ &= a - 2q[a + d(q - p)] \\ &= [a(p - q) + 2pqd] - 2q^2d \end{aligned}$$

and for B_1B_2 as

$$\begin{aligned} V_{12} - BV_{12} &= d - (q - p)\alpha \\ &= [a(p - q) + 2pqd] + 2pqd \end{aligned}$$

and for B_2B_2 as

$$\begin{aligned} V_{22} - BV_{22} &= -a - (-2p\alpha) \\ &= [a(p - q) + 2pqd] - 2p^2d \end{aligned}$$

The term $[a(p - q) + 2pqd]$ is the mean genotypic value of the population and appears in the difference $V_{ij} - BV_{ij}$ for each genotype. The remaining term in each difference is referred to as the dominance deviation, denoted as D_{ij}. That is, $D_{11} = -2q^2d$, $D_{12} = 2pqd$, and $D_{22} = -2p^2d$. The dominance deviation is defined as the value of the gene combination in the genotype. Hence, the genotypic value, V_{ij}, can be represented as the following sum:

$$V_{ij} = \text{mean genotypic value} + BV_{ij} + D_{ij}$$

A summary table of the different values for each genotype is

Genotype	Genotypic value	Breeding value	Dominance deviation
B_1B_1	a	$2q\alpha$	$-2q^2d$
B_1B_2	d	$(q-p)\alpha$	$2pqd$
B_2B_2	$-a$	$-2p\alpha$	$-2p^2d$

Earlier the phenotype of an animal was represented as

$$P_{ij} = m + V_{ij}$$

This model may now be written as

$$P_{ij} = m + [a(p-q) + 2pqd] + BV_{ij} + D_{ij}$$

Because $m + [a(p-q) + 2pqd]$ is μ, the phenotype is represented as

$$P_{ij} = \mu + BV_{ij} + D_{ij}$$

Example 9-3 shows phenotypes partitioned numerically into their component parts.

Example 9-3

For the hormone example assume $p = .6$, the frequency of B_1, so that

$$\mu = m + a(p-q) + 2pqd$$
$$= 10.08$$

and

$$\alpha = a + d(q-p)$$
$$= 2.8$$

Hence,

Genotype	BV_{ij}	D_{ij}
B_1B_1	2.24	$-.32$
B_1B_2	$-.56$	.48
B_2B_2	-3.36	$-.72$

Then,

$$\begin{aligned} P_{11} &= \mu + BV_{11} + D_{11} \\ &= 10.08 + 2.24 - .32 \\ &= 12 \\ P_{12} &= \mu + BV_{12} + D_{12} \\ &= 10.08 - .56 + .48 \\ &= 10 \\ P_{22} &= \mu + BV_{22} + D_{22} \\ &= 10.08 - 3.36 - .72 \\ &= 6 \end{aligned}$$

For the hormone example these are the phenotypes for each genotype.

Figure 9-2 is a plot of the phenotypes and $\mu + BV$ for each genotype for the hormone example with $p = .6$. The genotypes are ordered by the number of B_1 genes in the genotype, zero, 1, or 2. The slope of the line is the average effect of a gene substitution, $\alpha = 2.8$, for this example. The difference between the phenotype and $\mu + BV$ is the dominance deviation. If d were zero, then all dominance deviations would be zero and the trait would be completely additive, $P_{ij} = \mu + BV_{ij}$.

9-5 Variances

The mean of a population is the average of the phenotypes. The actual observations vary about the mean, as in Example 9-1 in which the mean was 8.88 while the actual observations were either 12, 10, or 6. The variation of the observations about the mean can be quantified and is termed the variance. The variance is usually denoted by σ^2 (Greek sigma squared).

The phenotypic variance, denoted σ_P^2, is calculated for the single locus model as

$$\sigma_P^2 = f(B_1B_1)(P_{11} - \mu)^2 + f(B_1B_2)(P_{12} - \mu)^2 + f(B_2B_2)(P_{22} - \mu)^2$$

which is the weighted average of the deviations squared from the mean, $(P_{ij} - \mu)^2$. Because $P_{ij} - \mu = BV_{ij} + D_{ij}$ (the sum of the breeding value and the dominance deviation), the phenotypic variance can be written as

$$\sigma_P^2 = p^2(BV_{11} + D_{11})^2 + 2pq(BV_{12} + D_{12})^2 + q^2(BV_{22} + D_{22})^2$$

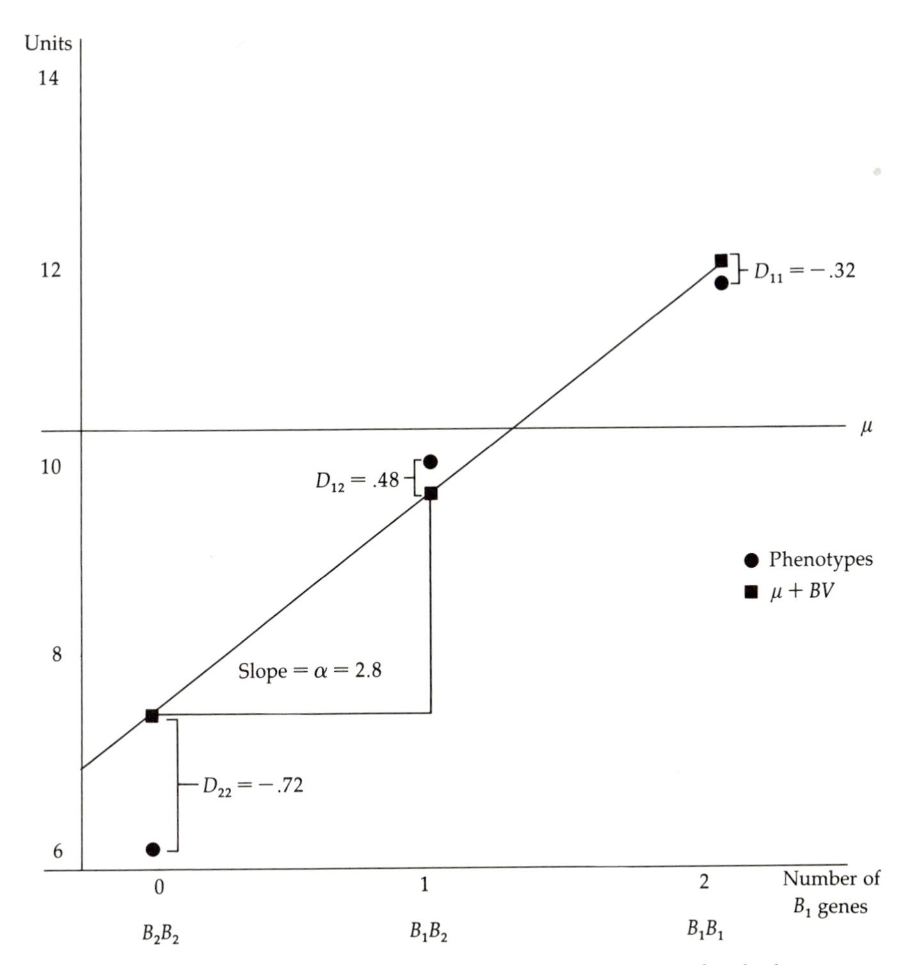

Figure 9-2 Plot of the phenotypes and $\mu + BV$ for each genotype for the hormone example with $f(B_1) = .6$.

Section 9-6 contains the algebraic reduction of the equation for σ_P^2. The exercise in algebra is tedious, but shows, in summary, that the phenotypic variance is

$$\sigma_P^2 = 2pq\alpha^2 + (2pqd)^2$$

The term $2pq\alpha^2$ is the weighted sum of the squared breeding values and represents that portion of σ_P^2 that is attributed to variation in breeding values among animals in the population. The variation among breeding values is called the *additive genetic variance* and is denoted as either σ_{BV}^2 or σ_A^2. The term

$(2pqd)^2$ is the weighted sum of squared dominance deviations among animals in the population. The variation among dominance deviations is called the dominance variance and is denoted as σ_D^2. Thus,

$$\sigma_P^2 = \sigma_A^2 + \sigma_D^2$$

Together, $\sigma_A^2 + \sigma_D^2$ is the genetic variance, denoted σ_G^2, for a trait controlled by genes at a single locus. In this case $\sigma_P^2 = \sigma_G^2$ because no nongenetic causes of variation have been considered.

9-6 Phenotypic Variance for a Single Locus Model Assuming Equilibrium

$$\begin{aligned}\sigma_P^2 &= p^2(BV_{11} + D_{11})^2 + 2pq(BV_{12} + D_{12})^2 + q^2(BV_{22} + D_{22})^2 \\ &= p^2(BV_{11}^2 + D_{11}^2 + 2BV_{11}D_{11}) + 2pq(BV_{12}^2 + D_{12}^2 + 2BV_{12}D_{12}) \\ &\quad + q^2(BV_{22}^2 + D_{22}^2 + 2BV_{22}D_{22}) \\ &= (p^2BV_{11}^2 + 2pqBV_{12}^2 + q^2BV_{22}^2) + (p^2D_{11}^2 + 2pqD_{12}^2 + q^2D_{22}^2) \\ &\quad + 2(p^2BV_{11}D_{11} + 2pqBV_{12}D_{12}^2 + q^2BV_{22}D_{22})\end{aligned}$$

The first term in parentheses represents the weighted sum of squared breeding values; the second term, the weighted sum of squared dominance deviations; and the final term, the weighted sum of cross products. The weighted sum of squared breeding values reduces to $2pq\alpha^2$ as follows:

$$\begin{aligned}p^2BV_{11}^2 + 2pqBV_{12}^2 + q^2BV_{22}^2 &= p^2(2q\alpha)^2 + 2pq[(q-p)\alpha]^2 + q^2(-2p\alpha)^2 \\ &= 4p^2q^2\alpha^2 + 2pq\alpha^2(q-p)^2 + 4q^2p^2\alpha^2 \\ &= 8p^2q^2\alpha^2 + 2pq\alpha^2(q^2 + p^2 - 2pq) \\ &= 8p^2q^2\alpha^2 + 2pq^3\alpha^2 + 2p^3q\alpha^2 - 4p^2q^2\alpha^2 \\ &= 4p^2q^2\alpha^2 + 2pq^3\alpha^2 + 2p^3q\alpha^2 \\ &= 2pq\alpha^2(p^2 + 2pq + q^2) \\ &= 2pq\alpha^2\end{aligned}$$

The weighted sum of squared dominance deviations reduces to $(2pqd)^2$ as follows:

$$\begin{aligned}p^2D_{11}^2 + 2pqD_{12}^2 + q^2D_{22}^2 &= p^2(-2q^2d)^2 + 2pq(2pqd)^2 + q^2(-2p^2d)^2 \\ &= 4p^2q^4d^2 + 8p^3q^3d^2 + 4p^4q^2d^2 \\ &= 4p^2q^2d^2(q^2 + 2pq + p^2) \\ &= 4p^2q^2d^2 \\ &= (2pqd)^2\end{aligned}$$

Finally, the weighted sum of cross products of breeding values and dominance deviations equals zero:

$$
\begin{aligned}
2p^2BV_{11}D_{11} + 4pqBV_{12}D_{12} + 2q^2BV_{22}D_{22} &= 2p^2(2q)(-2q^2d) + 4pq[(q-p)\alpha](2pqd) \\
&\quad + 2q^2(-2p\alpha)(-2p^2d) \\
&= -8p^2q^3(\alpha d) + 8p^2q^2 \times (q-p)(\alpha d) + 8p^3q^2(\alpha d) \\
&= 8p^2q^2(-q\alpha d + q\alpha d - p\alpha d + p\alpha d) \\
&= 0
\end{aligned}
$$

Therefore,

$$\sigma_P^2 = 2pq\alpha^2 + (2pqd)^2$$

which is the weighted sum of squared breeding values and squared dominance deviations.

9-7 Summary

Phenotypes for many traits can be measured quantitatively. The average of the phenotypes of the two homozygotes is defined as a constant, m, associated with all phenotypes: $m = (P_{11} + P_{22})/2$. The deviations of the phenotypes from m are the genotype values, V, for genotypes at a single locus, so that

$$P_{ij} = m + V_{ij}$$

where $V_{11} = a$, $V_{12} = d$, and $V_{22} = -a$.

For a single locus, the mean of a population, μ, is the sum of products of frequencies and phenotypes:

$$\mu = f(B_1B_1)P_{11} + f(B_1B_2)P_{12} + f(B_2B_2)P_{22}$$

For populations at equilibrium the mean is

$$\mu = m + [a(p-q) + 2pqd]$$

The term in brackets is the mean genotypic value of the population which can be changed by selection.

The breeding value (BV) of an animal refers to its value in a breeding program and is defined for a single locus as twice the progeny deviation from the population mean. One-half the breeding value of an animal is defined as the transmitting ability of the animal. The breeding values of the genotypes are

Genotype	Breeding value (BV_{ij})
B_1B_1	$2q\alpha$
B_1B_2	$(q-p)\alpha$
B_2B_2	$-2p\alpha$

where α is the average effect of a gene substitution and is

$$\alpha = [a + d(q-p)]$$

The breeding values are dependent on gene frequencies and thus may vary from population to population. The difference between the breeding values of B_1B_1 and B_1B_2 is α, as is the difference between the breeding values of B_1B_2 and B_2B_2. The breeding value is therefore referred to as the additive genetic value of the genotype.

The dominance deviations of the genotypes for a single locus are

Genotype	Dominance deviation (D_{ij})
B_1B_1	$D_{11} = -2q^2d$
B_1B_2	$D_{12} = 2pqd$
B_2B_2	$D_{22} = -2p^2d$

The model for a phenotype is the sum of the common mean, μ, the breeding value, and the dominance deviation, as follows:

$$P_{ij} = \mu + BV_{ij} + D_{ij}$$

If d (the genotypic value of the heterozygote) is zero, the trait is completely additive.

The amount of variation in phenotypes about the population mean is termed the variance. For a single locus the phenotypic variance is the weighted average of squared deviations from the mean:

$$\sigma_P^2 = f(B_1B_1)(P_{11}-\mu)^2 + f(B_1B_2)(P_{12}-\mu)^2 + f(B_2B_2)(P_{22}-\mu)^2$$

At equilibrium,

$$\sigma_P^2 = 2pq\alpha^2 + (2pqd)^2$$

where $2pq\alpha^2$ is the variance among breeding values, which is denoted σ_A^2 and called the additive genetic variance, and $(2pqd)^2$ is dominance variation, denoted σ_D^2. The sum $\sigma_A^2 + \sigma_D^2$, for a single locus, is the total genetic variance, which is denoted σ_G^2.

CHAPTER 10

Relationships and Inbreeding

The *additive relationship* is the most commonly used measure of relationship. It is a measure of the fraction of like genes shared by two animals and thus is an indication of how reliable one of the relative's records will be in predicting the genetic value of the other animal. The *inbreeding coefficient* of an animal is calculated as one-half the additive relationship between the parents. *Line-breeding* is a system of mating which maintains close relationships to a particular animal. Thus, knowledge of relationships can be helpful in selecting animals on the basis of relatives' records, in arranging matings to avoid high levels of inbreeding, and in establishing lines tracing to desirable animals.

This chapter first introduces the common relationships and discusses how these relationships can be computed easily and quickly by counting the generations linking the related pair of animals. When inbreeding of an ancestor is involved, however, more precise methods of calculating relationships are needed. Two methods, the tabular method, which simultaneously accounts for relationships among all animals and which is computationally easy, and the path coefficient method, which is intuitively more understandable, will be described.

10-1 Common Relationships

An animal is often said to have so much of the blood of one animal and so much of the blood of another. Of course such statements are not literally true but may actually apply to the fraction of genes that came from each ancestor.

An animal is related to itself by 100 percent. The relationship between a parent and its offspring is said to be 50 percent. This halving of relationships and genes in common occurs with each generation. The relationship to each parent is 50 percent, to each grandparent 25 percent, to each great-grandparent 12.5 percent, and so on as illustrated in Figure 10-1. Thus, after only a few generations, any ancestor is likely to be the source of only a small fraction of the genes of its descendants. The rule of halving holds only when all the ancestors are unrelated to one another. When some are related, the problem of determining relationships is more complicated and will be discussed later in the sections on the tabular method and path coefficient method.

Relationships between individuals not in direct line of descent but with a common ancestor can also be determined essentially by a halving for each intervening animal in the pedigree, as will be discussed more completely later

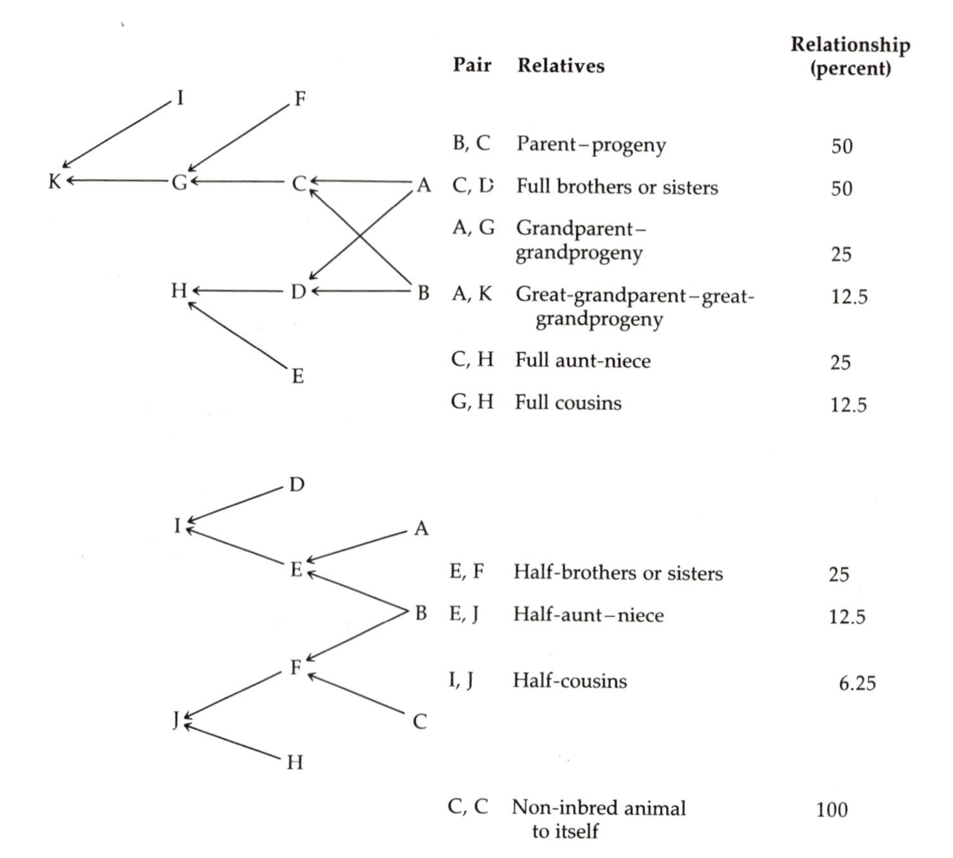

Pair	Relatives	Relationship (percent)
B, C	Parent–progeny	50
C, D	Full brothers or sisters	50
A, G	Grandparent–grandprogeny	25
A, K	Great-grandparent–great-grandprogeny	12.5
C, H	Full aunt-niece	25
G, H	Full cousins	12.5
E, F	Half-brothers or sisters	25
E, J	Half-aunt–niece	12.5
I, J	Half-cousins	6.25
C, C	Non-inbred animal to itself	100

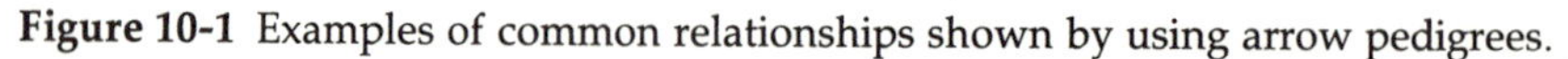
Figure 10-1 Examples of common relationships shown by using arrow pedigrees.

in this section. For example, calves by the same bull but with different dams are called *paternal half-sibs* and have an average relationship of 25 percent. *Maternal half-sibs* (same mother, different father) are also 25 percent related to each other. Full brothers and sisters, called *full sibs,* have the same sire and dam, as do fraternal twins. *Fraternal twins* are related by 50 percent to each other—25 percent through the sire and 25 percent through the dam—as are full sibs born at different times. These and some other less close relationships are also shown in Figure 10-1. *Identical twins* are genetically alike; thus they are 100 percent related to each other, just as an animal with itself.

The halving of the common hereditary material with each intervening animal in the pedigree is the basis of the method of tracing paths to calculate relationships. This method, which is commonly taught, is intuitively obvious and computationally easy when the animals are not inbred and when relationships among only a few animals are needed. Section 10-5 on the path coefficient method describes the necessary modification to account for inbreeding. The tabular method discussed in the next section is much more efficient for complicated pedigrees and when relationships among many animals must be computed. Another advantage of the tabular method is that, as new animals are born, the relationships to the older animals and among the new animals can be easily added to the table.

All methods assume that the relationship between any random pair of animals at some specified time in the past is zero. The animals of this base period are the *base population.* Thus, calculated relationships will not be precise but will be the best approximation that is possible relative to the base period.

The symbol, a_{XY}, will be used to denote the additive relationship between relatives X and Y. Animal breeders use the additive relationship to estimate breeding value; it is numerically the same as what is usually thought to be the relationship between a pair of animals. For example, if X is the parent of Y, then $a_{XY} = .50$.

Sewall Wright (1921) is responsible for the idea of tracing paths to establish the relationships among animals. Malécot (1948), however, is given credit for the definition of relationships based on probabilities of individual genes at a locus being identical by descent. The phrase, genes are *identical by descent,* means that the genes are exact replicates from a common ancestor, as contrasted with being physically the same allele or *identical in state.* Genes identical by descent are identical in state, but genes identical in state are not necessarily identical by descent.

The method of tracing paths can be used for both collateral relatives and direct relatives. *Direct relatives* are those in direct line of descent, such as parent and progeny or grandparent and grandprogeny. For example, if C, G, and K are direct descendants of A, the arrow pedigree is

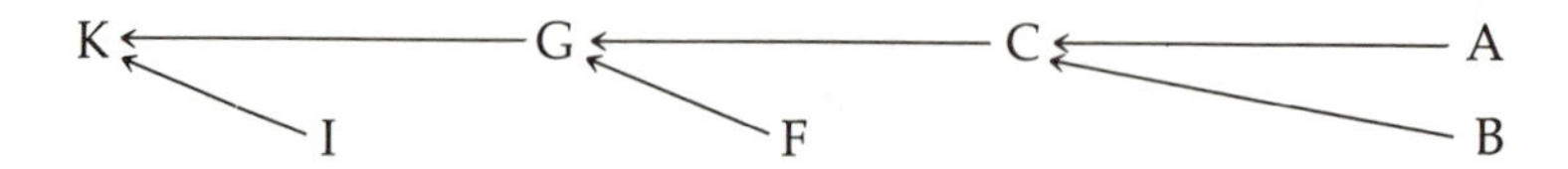

Individual A transmits a sample half of its genes to C; thus, $a_{AC} = (\frac{1}{2})^1 = \frac{1}{2}$. Equivalently, the probability that a particular gene of A at each locus will be transmitted to C is one-half. Animal C transmits a sample half of its genes to G; thus, $a_{AG} = (\frac{1}{2})(\frac{1}{2}) = (\frac{1}{2})^2 = \frac{1}{4}$, that is, the probability that a particular gene of A will be transmitted to G is one-fourth. Similarly, $a_{AK} = (\frac{1}{2})(\frac{1}{2})(\frac{1}{2}) = (\frac{1}{2})^3 = \frac{1}{8}$.

More than one path can exist between direct relatives, as illustrated below.

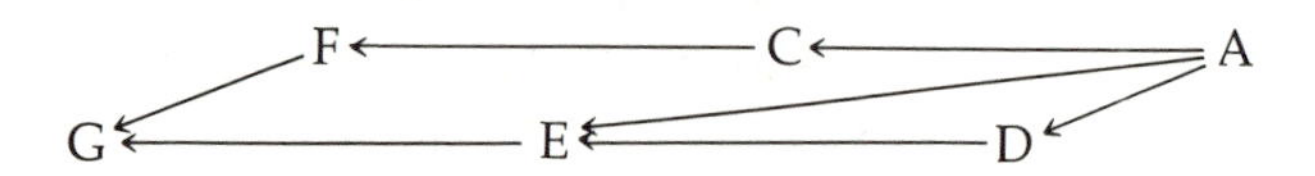

The rules for calculating the relationship between direct relatives can be illustrated with this example:

1. Count for each path the number of intervening generations from the animal to its direct ancestor. For the example, the number of generations for the three paths between G and A are: $n_{GEA} = 2$, $n_{GEDA} = 3$, and $n_{GFCA} = 3$.
2. Compute

$$a_{AG} = (1/2)^{n_{GEA}} + (1/2)^{n_{GEDA}} + (1/2)^{n_{GFCA}}$$

Thus, for this example,

$$a_{AG} = (1/2)^2 + (1/2)^3 + (1/2)^3 = 1/2$$

This simple procedure will not work to calculate a_{EG} because the parents of E are related: A is the parent of D, and, therefore E, is inbred. A more complicated rule is needed when an ancestor is inbred, which will be introduced later in this chapter. The relationship, a_{EG}, will be $\frac{23}{32}$, not $\frac{1}{2}$.

Two animals, with one or more common ancestors but which are not direct descendants are *collateral relatives*. The method of tracing paths can be illustrated for a pair of full sibs (C and D) having unrelated parents (A and B).

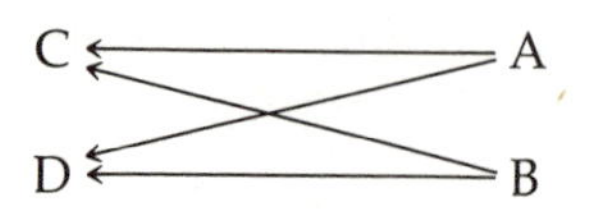

In this case, there are two common ancestors, A and B. The probability that C will receive a particular gene from A is $\frac{1}{2}$. The probability that D will receive the same gene from A is also $\frac{1}{2}$. Thus, C and D are related through A by $\frac{1}{2} \times \frac{1}{2} = (\frac{1}{2})^2 = \frac{1}{4}$. Similarly, the relationship of C and D through B also is $\frac{1}{2} \times \frac{1}{2} = (\frac{1}{2})^2 = \frac{1}{4}$. These two paths are independent so that the total relationship of C and D through both common ancestors is $\frac{1}{2} \times \frac{1}{2} + \frac{1}{2} \times \frac{1}{2} = \frac{1}{4} + \frac{1}{4}$; that is, $a_{CD} = \frac{1}{2}$.

The rule for calculating the relationship between collateral relatives is to sum the relationships due to each independent path. The relationships for the two paths in the example of full sibs are $(\frac{1}{2})^{n_{CA}+n_{AD}}$ and $(\frac{1}{2})^{n_{CB}+n_{BD}}$, where n_{CA} is the number of steps from C to common ancestor A, and n_{AD} is the number of steps from A to D. Similarly, n_{CB} and n_{BD} are the number of steps from C to B and from B to D. Each pair of n's reflects the total number of steps from C through the common ancestor to its relative D. Thus, $a_{CD} = (\frac{1}{2})^{n_{CA}+n_{AD}} + (\frac{1}{2})^{n_{CB}+n_{BD}} = (\frac{1}{2})^{1+1} + (\frac{1}{2})^{1+1} = \frac{1}{2}$.

The rules for direct relatives and collateral relatives are similar in that in both cases the number of steps from one relative to another is counted. However, with direct relatives the count is *to* the ancestor and with collateral relatives the count is *through* the common ancestor.

Relatives can be both direct and collateral simultaneously. For example, in the illustrated pedigree,

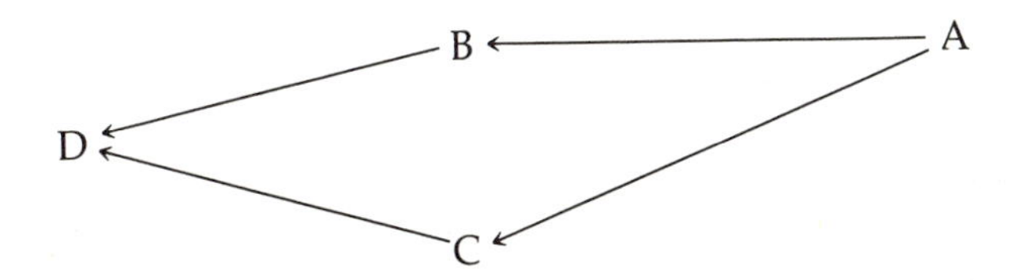

the direct relationship between B and D is $\frac{1}{2}$, and the collateral path is $B \rightarrow A \rightarrow C \rightarrow D$ with collateral relationship $(\frac{1}{2})^3 = \frac{1}{8}$ for a total relationship, $a_{BD} = \frac{1}{2} + \frac{1}{8} = \frac{5}{8}$.

These rules can be applied to the common kinds of relatives shown in Figure 10-1 to obtain the relationships shown in the figure. The pedigrees are shown by arrow diagrams which are easier to follow than are pedigrees which list two parents, four grandparents, and eight great-grandparents.

The *inbreeding coefficient* of an animal is defined as the probability that both genes at a locus are identical by descent. The only way this can happen is if the parents are related, that is, the parents have a gene in common by descent. The symbol for inbreeding of animal X is F_X. The inbreeding coefficient is

$$F_X = (1/2)a_{S_X D_X}$$

where $a_{S_XD_X}$ is the additive relationship between the sire (S_X) and dam (D_X) of X. The additive relationship of X to itself is

$$a_{XX} = 1 + F_X$$

10-2 Computing Additive Relationships with the Tabular Method

Relationships among, and inbreeding coefficients of, all animals in a herd or flock can be calculated with the tabular method by following a few simple rules. The tabular method at first may appear quite complicated, but two or three examples will show that many relationships can be calculated simply and quickly. Later in the chapter the more traditional but more difficult path coefficient method will be described.

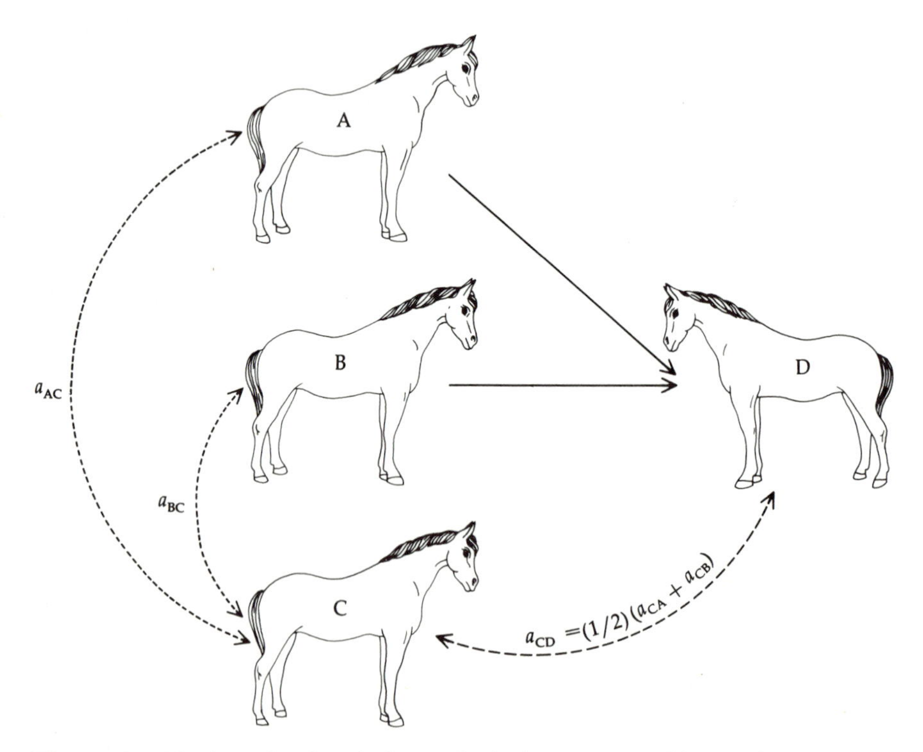

Figure 10-2 The basis for the tabular method of computing additive relationships is that the relationship between C and D is one-half the relationship between C and the sire of D plus one-half the relationship between C and the dam of D.

The procedure for calculating a table of additive relationships among all animals in a herd is based on the fact that if two animals are related, then one or both of the parents of one must also be related to the other animal of the pair. In fact, as illustrated in Figure 10-2, if C and D are the two animals and A and B are the parents of D, the additive relationship between animals C and D is one-half the relationship between C and A plus one-half the relationship between C and B; that is, $a_{CD} = \frac{1}{2}a_{CA} + \frac{1}{2}a_{CB}$. Thus, relationships determined sequentially from oldest to youngest animals can be computed by halving previously calculated relationships. The other basis of the method is that the coefficient of inbreeding of an animal is one-half the additive relationship between its parents, as diagrammed in Figure 10-3. The relationship of the parents will have been computed earlier and will already be in the table.

The tabular method—as well as the path tracing method described in Section 10-5—require the assumption that the relationships among animals in the base population are zero if unknown. The relationships that are known can be included in the table.

The obvious outcome of the tabular method is the construction of a table that, when finished, gives the additive relationship of any animal to any other in the table. An example is Table 10-1 which was constructed for animals related as shown in Figure 10-4. Animals A and B are base animals. One parent of C is not known; no entry is made for unknown parents. Unknown parents are ignored. The table gives the additive relationships among all pairs of the

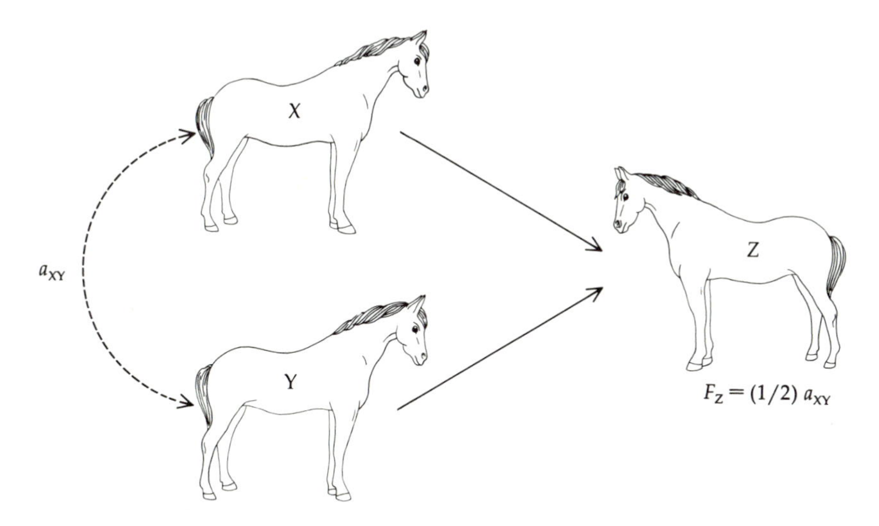

Figure 10-3 The inbreeding coefficient for an animal (F_Z for animal Z) is one-half the relationship between the parents (a_{XY} for X and Y).

Table 10-1 Relationship table for animals in Figure 10-4

	Animals			
	—	—	A–?	A–B
Animals	A	B	C	D
A	1	0	1/2	1/2
B	0	1	0	1/2
C	1/2	0	1	1/4
D	1/2	1/2	1/4	1

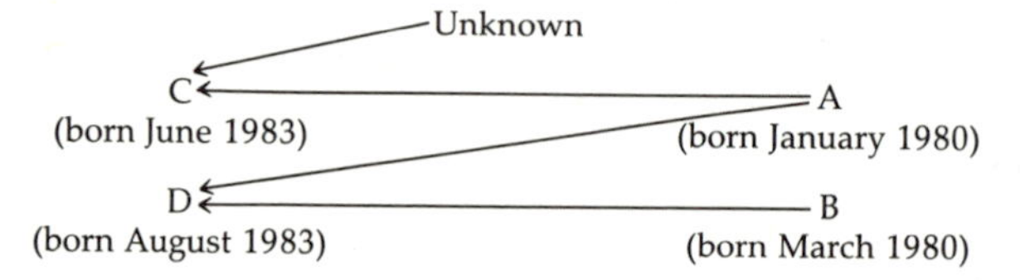

Figure 10-4 Arrow chart of pedigree to illustrate the rules for calculating relationships shown in Table 10-1.

four animals. For example, to find the additive relationship between C and D, go to row C and then across to column D. The additive relationship is $\frac{1}{4}$.

The rules for computing additive relationships and inbreeding coefficients

The following rules for computing the relationship table are illustrated for the pedigree shown in Figure 10-4.

Rule 1 Determine which animals are to be included. Put them in order by date of birth, oldest first; that is, A, B, C, and D. This step is very important so that a progeny does not appear before its parents in the table.

Rule 2 Write the names or numbers of the animals in order of birth across the top of the table to identify the columns and down the left side of the table to identify the rows as shown.

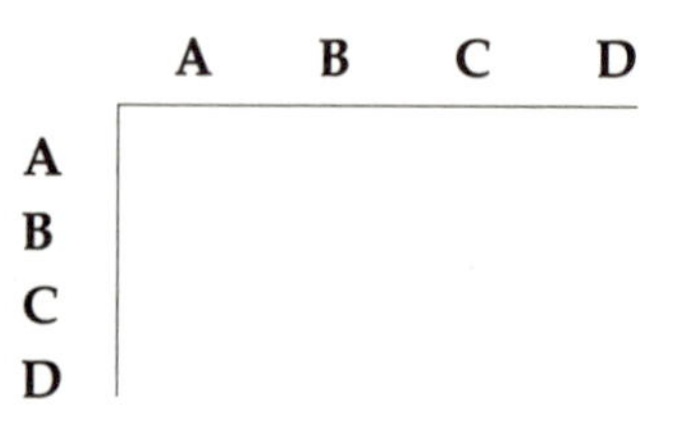

Rule 3 Write above the numbers of the animals the names or numbers of their parents, if known. If unknown, put a dash or a question mark. No entries are necessary or should be made in the table for unknown parents.

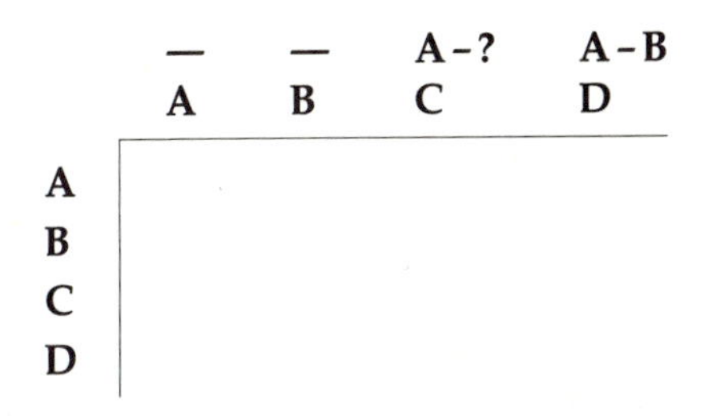

	—	—	A – ?	A – B
	A	B	C	D
A				
B				
C				
D				

Rule 4 Put a 1 in each of the diagonal cells of the table, such as row 1, column 1; row 2, column 2. The 1 is the animal's basic additive relationship to itself unless the animal is inbred. If the inbreeding coefficient of any of the first or base animals is known, add that to the diagonal for that animal. All other inbreeding coefficients will be added to the diagonals as they are computed according to rule 6. For the base generation animals, enter their relationships if known or, if not known, assume the relationships are zero.

	—	—	A – ?	A – B
	A	B	C	D
A	1	0		
B	0	1		
C			1	
D				1

Rules 5 and 6 will be repeated until all rows of the relationship table are completed.

Rule 5 Compute entries for each off-diagonal cell of row 1 according to the rule of one-half the entry in this row corresponding to the column of the first parent plus one-half the entry in this row corresponding to the column of the second parent.

	—	—	A – ?	A – B
	A	B	C	D
A	1	0	1/2	1/2
B	0	1		
C			1	
D				1

When the first row is finished, write the same values down the first column. For example, $a_{CA} = a_{AC} = \frac{1}{2}$, and $a_{DA} = a_{AD} = \frac{1}{2}$.

	— A	— B	A – ? C	A – B D
A	1	0	1/2	1/2
B	0	1		
C	1/2		1	
D	1/2			1

Rule 6 Go to the next row and begin at the diagonal, which now has a 1 in it. To that 1, add one-half of the relationship between the animal's parents, which can be found from an earlier entry in the table at the intersection of the row and column of the two parents. One-half the relationship between the parents is the inbreeding coefficient described in rule 4. The inbreeding coefficient often is zero.

In this example, F_B is assumed zero because the parents of B are unknown and assumed to be unrelated.

	— A	— B	A – ? C	A – B D
A	1	0	1/2	1/2
B	0	1 + 0		
C	1/2		1	
D	1/2			1

Continue across the row as before, computing the off-diagonal entries according to rule 5. Put the values for this row down the corresponding column.

	— A	— B	A – ? C	A – B D
A	1	0	1/2	1/2
B	0	1 + 0	(1/2)(0 + 0) = 0	(1/2)(0 + 1) = 1/2
C	1/2	0	1	
D	1/2	1/2		1

Rule 7 Repeat rules 5 and 6 in this manner until the table is complete, always remembering to do a row at a time and to put the same values in the corresponding column before going to the next row.

For the third row and column,

	— A	— B	A – ? C	A – B D
A	1	0	1/2	1/2
B	0	1	0	1/2
C	1/2	0	1 + 0	(1/2)(1/2 + 0) = 1/4
D	1/2	1/2	1/4	1

For the last diagonal $F_{DD} = \frac{1}{2}a_{AB} = \frac{1}{2}(0) = 0$.

	— A	— B	A – ? C	A – B D
A	1	0	1/2	1/2
B	0	1	0	1/2
C	1/2	0	1	1/4
D	1/2	1/2	1/4	1 + 0

In summary, the two basic steps are

1. the diagonal entries are 1 plus the inbreeding coefficient, which is one-half the relationship between the animal's parents, and
2. each off-diagonal relationship is one-half of the sum of two relationships that appear to the left in the same row corresponding to the columns of the parents of the animal.

Identical twins are handled differently. Only one needs to be included in the table because the relationship will be the same for any animal with either of the identical twins. Another method is to insert a duplicate of the row and column of one twin for the other, remembering that the relationship between them is the same as the relationship of either one to itself.

Example 10-1 will help clarify the procedure for a situation involving inbreeding.

Example 10-1 Computing relationships using the tabular method

Suppose that sire A is mated to his own daughter, C, as in Figure 10-5. The only individuals in an animal's pedigree that must be recorded are its parents. The animals are ordered by age as described in rule 1. The names of the known

Figure 10-5 Pedigree for a sire, A, mated to his daughter, C, illustrated in two ways.

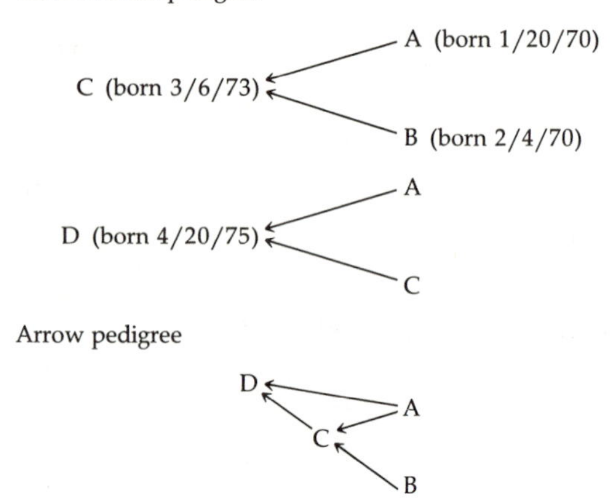

parents are written above the animals' names as shown in Table 10-2 (rules 2 and 3). The parents of A and B can be ignored unless they are known to be related to each other. A and B represent the base population for this example. Parents of C and D are written above C and D at the top of the table. Write 1's in all four diagonal cells, that is, where the (A, A), (B, B), (C, C), and (D, D) rows and columns meet (rule 4).

Begin with row 1, the row for A. B is assumed not related to A, so enter a zero under the column labeled B. If the relationship of A and B is known and is not zero, enter that instead.

The entry for row A, column C is determined according to rule 5 (one-half the entry for row A, column A plus one-half the entry for row A, column B because A and B are the parents of C). Thus, $a_{AC} = \frac{1}{2}(1) + \frac{1}{2}(0) = \frac{1}{2}$.

Table 10-2 Relationship table for the sample pedigree in Figure 10-6

	Parents			
	—	—	A–B	A–C
Animals	A	B	C	D
A	1	0	1/2	3/4
B	0	1	1/2	1/4
C	1/2	1/2	1 + (1/2)(0) = 1 + 0	3/4
D	3/4	1/4	3/4	1 + (1/2)(1/2) = 1 + 1/4

The entry for row A, column D will be $\frac{1}{2}$ the (A, A) entry plus one-half the (A, C) entry, which is $\frac{1}{2}(1) + \frac{1}{2}(\frac{1}{2}) = \frac{3}{4}$. The first row is now finished, with the additive relationships of animal A to the other three animals having been computed.

The last part of rule 5 is to write the values in row A down column A. Column A will be 1, 0, $\frac{1}{2}$, and $\frac{3}{4}$.

The second row, B, is done similarly, beginning at row B, column B. Because the parents of B are not related B is not inbred. Therefore, nothing is added to the 1 already in the (B, B) cell (rule 6).

The entry for row B, column C is $\frac{1}{2}(0) + \frac{1}{2}(1) = \frac{1}{2}$. The value for row B, column D is $\frac{1}{2}(0) + \frac{1}{2}(\frac{1}{2}) = \frac{1}{4}$. The values for row B are now written down column B: 0, 1, $\frac{1}{2}$, and $\frac{1}{4}$. The 0 and 1 were already there.

The row C, column C entry is the 1 already there plus one-half the additive relationship between the parents of C, which are A and B (rule 6). The entry for row A, column B, is zero so that $a_{AB} = 0$ and the entry for row C, column C, is $1 + \frac{1}{2}(0) = 1$.

The row C, column D entry is $\frac{1}{2}(\frac{1}{2}) + \frac{1}{2}(1) = \frac{3}{4}$. This value is also written into row D, column C, according to rule 5.

The last entry is row D, column D. One-half the additive relationship between parents A and C is added to the 1 already there: $a_{AC} = \frac{1}{2}$ as found from the row A, column C entry. The row D, column D entry is $1 + \frac{1}{2}(\frac{1}{2}) = 1 + \frac{1}{4}$. Thus, D is one-fourth or 25 percent inbred and $a_{DD} = 1 + \frac{1}{4}$. The calculations are complete as shown in Table 10-2.

With this special kind of mating, animal D is related to A, her father, by 75 percent and also by 75 percent to her mother C, but is related to her grandmother by only 25 percent. Other than for identical twins, the only way that two animals can be related by more than 50 percent is if one or both the animals is inbred. This was true of D and C and also D and A because D is inbred.

If new animals are born in the herd, their relationships to the older animals and among themselves can easily be added to the table. For example, if E is the progeny of C and D, a column and a row for E would be added to the table. Previous computations would not need to be repeated.

The mathematical definition of the additive relationship, attributed to Malécot (1948), is twice the probability of identical genes occurring in the two animals. That probability for a noninbred animal compared to itself (or an identical twin) is .5. Thus, doubling the probability gives an additive relationship of a noninbred animal with itself of 1. The probability of identical genes by descent for a completely inbred animal compared with itself is 1; therefore, the additive relationship of a completely inbred animal with itself is 2. In general, the additive relationship of an animal, say X, with itself is

$$a_{XX} = 1 + F_X$$

The maximum value of an inbreeding coefficient is 1. Therefore, the lower limit for the additive relationship of an animal with itself is 1, if noninbred, and the upper limit is 2, if the inbreeding coefficient is maximum.

Similarly, the limits for the additive relationship between two animals are zero when not related, and 2 when the parents are members of the same inbred line in which all animals are genetically identical and have an inbreeding coefficient of 1.

The additive relationship is the appropriate measure of the fraction of identical gene effects and is a relative measure of the covariance between breeding values of relatives. Thus, the additive relationship is used in devising weighting factors for records of relatives in genetic evaluation. The additive relationship seems illogical only when there is inbreeding and is the most useful measure of relationship, although the coefficient of relationship is sometimes used.

The *coefficient of relationship* is a measure of relationship that has a minimum value of zero and a maximum value of 1. The coefficient of relationship is equivalent to the correlation between the breeding values of the two animals (Section 9-3). For noninbred animals, the coefficient of relationship is the same as the additive relationship. For inbred animals, say X and Y, the coefficient of relationship is

$$\begin{aligned} R_{XY} &= a_{XY}/\sqrt{(a_{XX}\, a_{YY})} \\ &= a_{XY}/\sqrt{(1 + F_X)(1 + F_Y)} \end{aligned}$$

For example, suppose the coefficient of relationship between animals C and D in Table 10-2 is to be determined: $a_{CD} = \frac{3}{4}$, $a_{CC} = 1$, and because D is inbred, $a_{DD} = 1 + \frac{1}{4}$. Thus, $R_{CD} = \frac{3}{4}/\sqrt{1(1 + \frac{1}{4})} = .67$ rather than $a_{CD} = .75$. The coefficient of relationship between noninbred animals A and C (Table 10-2) is $R_{AC} = \frac{1}{2}/\sqrt{1(1)} = .5$, which is the same as a_{AC}.

Just as the additive relationship is a measure of the covariance between breeding values (additive genetic values), the *dominance relationship* is a measure of the covariance between dominance deviations (Section 9-4). Selection for dominance effects is generally beyond the scope of this book. Nevertheless the dominance relationship between relatives can be computed from the additive relationships among the parents of the two animals. The parents must be related if the complete genotype at a locus is identical by descent for the two animals.

Let animal X have sire X_S and dam X_D, and animal Y have sire Y_S and dam Y_D. Then the dominance relationship between X and Y can be computed as

$$d_{XY} = (a_{X_SY_S}a_{X_DY_D} + a_{X_SY_D}a_{X_DY_S})/4$$

This formula holds unless X or Y is inbred and may be a good approximation even if X or Y is inbred.

The dominance relationship for an animal with itself is always 1:

$$d_{XX} = 1$$

The dominance relationship is zero for most relatives.

The most common relatives with a nonzero dominance relationship are full sibs. As an example of the computing formula, assume the pedigree for full sibs, C and D, as illustrated:

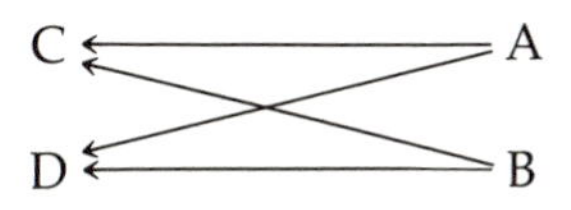

The additive relationship table contains the relationships needed to compute the dominance relationship between C and D.

	—	—	A – B	A – B
	A	**B**	**C**	**D**
A	1	0	1/2	1/2
B	0	1	1/2	1/2
C	1/2	1/2	1	1/2
D	1/2	1/2	1/2	1

Note that A is the sire of both C and D, and B is the dam of both C and D. That is, the dominance relationship between full sibs C and D is

$$\begin{aligned} d_{CD} &= (a_{AA}a_{BB} + a_{AB}a_{BA})/4 \\ &= [(1)(1) + (0)(0)]/4 \\ &= 1/4 \end{aligned}$$

10-3 Inbreeding

Inbreeding occurs in the progeny of related parents. The preceding section described an easy method of calculating the most commonly used measure of inbreeding, *Wright's coefficient of inbreeding,* named after its originator Sewall Wright.

The inbreeding coefficient is defined as the probability that two genes at a locus are identical by descent (Malécot, 1948). The inbreeding coefficient can be interpreted in either of two ways: (1) as the expected fraction of an animal's loci which have the pair of genes identical by descent, or, equivalently, (2) as the fraction of animals with the same pattern of related ancestors having genes identical by descent at a particular locus.

If identity of genes could be known precisely, as in the following hypothetical example, the inbreeding coefficient could be determined exactly. Suppose an animal has five loci with alleles tagged by identity by descent from common ancestors, rather than physical type, as follows:

$$\text{Genotype by ancestry } (a_1a_8\ b_6b_6\ c_7c_8\ d_4d_4\ e_1e_6)$$

For example, a_1 comes from ancestor 1, and c_8 from ancestor 8. The fraction of loci with genes identical by descent is two out of five and $F = .4$. For a real animal, the genes cannot be identified by their ancestry. Thus, the probabilities of identity by descent must be calculated in some way as, for example, by the tabular method described in the previous section. Another approach to defining inbreeding using gene frequencies rather than probabilities of identity by descent is described in Section 10-4.

10-4 Inbreeding and Changes in Genotypic Frequencies

A single locus with two alleles is sufficient to show an alternative definition of inbreeding. Assume for alleles A and a with frequencies p and $q, q = 1 - p$, that the population is initially in Hardy-Weinberg equilibrium, as shown in the second column of Table 10-3. The inbreeding coefficient, F, for a population can be defined as the fractional decrease in the frequency of heterozygotes which, in turn, results in an increase in frequencies of the homozygotes. Thus, the genes of a fraction F of the $2pq$ heterozygotes will have been transferred to homozygotes. The frequency of Aa becomes

$$f(Aa) = 2pq - 2pqF = 2pq(1 - F)$$

Table 10-3 Summary of frequencies of heterozygotes and homozygotes with inbreeding

Genotype	Initial frequency with random mating	Frequency if inbreeding coefficient is F
AA	p^2	$p^2 + pqF = p^2(1 - F) + pF$
Aa	$2pq$	$2pq - 2pqF = 2pq(1 - F)$
aa	q^2	$q^2 + pqF = q^2(1 - F) + qF$
Frequency of A	p	p
Frequency of a	q	q

Of the $2 \times 2pqF$ genes moved from the heterozygous subpopulation to the homozygous subpopulations, one-half must be A alleles and one-half must be a alleles; that is, $2pqF$ are A alleles and $2pqF$ are a alleles. The $2pqF$ A alleles exist only in pairs in the homozygous state. Thus, because of the pairing, the number of new AA homozygotes formed is pqF. The $2pqF$ a alleles also form pqF new homozygotes of type aa. These homozygotes are in addition to the original p^2 AA and the q^2 aa homozygotes which do not result from inbreeding. Thus with inbreeding the frequencies of AA and aa become

$$f(AA) = p^2 + pqF$$

and because $q = 1 - p$,

$$f(AA) = p^2(1 - F) + pF$$

and

$$f(aa) = q^2 + pqF$$

and similarly,

$$f(aa) = q^2(1 - F) + qF$$

as shown in Table 10-3.

For example, assume initial frequencies of $p = .6$ and $q = .4$, and that positive assortative mating has resulted in an inbreeding coefficient of $F = .7$. Then the frequencies initially and after inbreeding are

Genotype	Initial frequency	Frequency after inbreeding
AA	.36	$.36 + .24(.7) = .528$
Aa	.48	$.48(1 - .7) = .144$
aa	.16	$.16 + .24(.7) = .328$
	1.00	1.00

Examination of Table 10-3 reveals that inbreeding

1. decreases the frequency of heterozygotes by $2pqF$;
2. increases the frequency of each homozygote by the same frequency pqF, although the proportional increase depends on p^2 and q^2;
3. does not change gene frequency; and also that
4. when inbreeding is maximum ($F = 1$, so that $1 - F = 0$) the population is completely homozygous, with a fraction p being AA, and a fraction q being aa; and
5. when inbreeding is minimum ($F = 0$) and the Hardy-Weinberg conditions exist, the frequencies are the usual Hardy-Weinberg frequencies.

Inbreeding depression

Table 10-3 also illustrates a likely reason for the phenomenon of inbreeding depression. If the effect of a homozygote, say *aa*, is deleterious to the animal, then with inbreeding the frequency of such animals increases by pqF for that locus. The frequency of deleterious homozygotes at other loci would also increase at a rate dependent on the allelic frequencies at those loci. Such effects would be averaged over many loci so that even if only some loci have deleterious genes, inbreeding depression would still occur. If the gene effects are additive for all loci then there would be no inbreeding depression because the average additive value depends only on the effects of the *A* or *a* alleles (see Chapter 9) which do not change in frequency with inbreeding. In the case of additive effects, any detrimental effects of increasing the frequency of *aa* homozygotes is exactly offset by the beneficial effects of increasing the frequency of *AA* homozygotes.

The increased chance of obtaining a homozygote, which could be a deleterious recessive, can be calculated if the frequency of the gene is known. For example, assume the frequency of the recessive gene for mulefoot in cattle is $q = \frac{1}{100}$. If Hardy-Weinberg conditions exist, then $q^2 = \frac{1}{10000}$ is the risk of a mulefoot calf. If full brothers and sisters (additive relationship of $\frac{1}{2}$) are mated, their progeny would have inbreeding coefficients of $F = \frac{1}{4}$. With inbreeding the proportion of homozygous recessives becomes $q^2(1 - F) + qF$, as shown in Table 10-3. Because, in this example, $q = \frac{1}{100}$ and $F = \frac{1}{4}$, then the expected frequency of homozygous recessives would be $\frac{103}{40000}$, or approximately $\frac{26}{10000}$. Thus, the chance is 26 times greater that a mulefoot calf would be born from matings of brothers and sisters than from matings of unrelated animals.

The inbreeding coefficient is a measure of homozygosity. Increased or decreased homozygosity can result in various phenotypic effects. The consequences of inbreeding as a mating system are discussed in Chapter 15, but three applications and another property of inbreeding are summarized below.

Applications of inbreeding

1. Inbreeding increases the chance of expression of deleterious recessive genes, which would allow culling of affected and carrier animals and thereby reduce the frequency of the detrimental genes. However, the cost must be balanced against the potential gain.
2. Line crosses resulting from matings between inbred lines would have mostly heterozygous loci and thus might be superior to noninbred animals if there is some form of dominant gene action (see Chapter 15).
3. Inbreeding can be used to fix a desirable type (if the reproductive rate is sufficient to allow selection to eliminate the undesirable genes) and to achieve greater uniformity. For example, progeny will be more like an inbred parent than like a noninbred parent. The relationship of a noninbred parent and its progeny is $\frac{1}{2}$, whereas the relationship of an inbred parent and its progeny is $(1 + F)/2$. Similarly, progeny of an inbred parent will be more alike than progeny of a noninbred parent: paternal or maternal half-sibs, $(1 + F)/4$ versus $\frac{1}{4}$; full sibs, one parent inbred, $(\frac{1}{2} + F/4)$ versus $\frac{1}{2}$; and full sibs with both parents inbred (F_S and F_D) but unrelated, $[\frac{1}{2} + (F_S + F_D)/4]$ versus $\frac{1}{2}$.
4. Mating of unrelated animals always results in noninbred progeny. If the parents are inbred but are unrelated, then their progeny are not inbred.

10-5 Computing Relationships and Inbreeding Using Path Coefficients

The method of tracing paths is often taught for computing relationships and inbreeding coefficients. If many relationships and inbreeding coefficients are needed, the tabular method is easier and much quicker. However, if, as described early in this chapter, only a simple pedigree is examined, the tracing of paths is often easier. The method is based on the same principle of halving in each generation; that is, there is a one-half chance that a particular gene at a particular locus of the parent will be transmitted to its progeny. If a common ancestor is inbred, the procedure must be modified to account for the increased chance of both genes at a locus of the ancestor being identical.

The modification requires the inclusion of the term $1 + F_Z$, where Z is the direct ancestor if the relationship is between direct relatives, or the common ancestor in the path between the relatives. Examples similar to those of Section 10-1 will be used to illustrate the modification.

Direct relatives

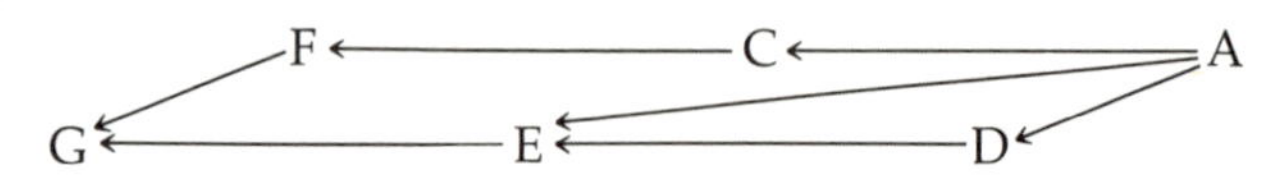

As before, to calculate the relationship between A and G, a_{AG}, count the number of generations between G and A for each separate path:

$$n_{GFCA} = 3$$
$$n_{GEA} = 2$$
$$n_{GEDA} = 3$$

Then

$$a_{AG} = (1/2)^{n_{GFCA}}(1 + F_A) + (1/2)^{n_{GEA}}(1 + F_A) + (1/2)^{n_{GEDA}}(1 + F_A)$$

If F_A is 0, then the calculation is the same as earlier:

$$a_{AG} = (1/2)^3 + (1/2)^2 + (1/2)^3 = 1/2$$

If F_A is .50, then

$$a_{AG} = (1/2)^3\,(1 + .5) + (1/2)^2\,(1 + .5) + (1/2)^3\,(1 + .5) = 3/4$$

Collateral relatives

As before, to calculate the relationship between F and E, a_{FE}, count the number of generations from F to E through the common ancestor A for the two different paths:

$$n_{FCA} + n_{AE} = 3$$
$$n_{FCA} + n_{ADE} = 4$$

so that

$$a_{FE} = (1/2)^3\,(1 + F_A) + (1/2)^4\,(1 + F_A)$$

If F_A is 0,

$$a_{FE} = (1/2)^3 + (1/2)^4 = 3/16$$

If F_A is .50,

$$a_{FE} = (1/2)^3\,(1 + .5) + (1/2)^4\,(1 + .5) = 9/32$$

If two or more common ancestors are included, the procedure is to include all the different paths through all common ancestors.

To ensure that the paths are independent, these rules must be followed:

1. an animal can occur only once in a single path;
2. a path follows direct descendants only, that is, it does not zig-zag but must go directly back to the common ancestor, from offspring to parents and so on; and
3. inbreeding of intermediate animals is ignored, with only the inbreeding coefficient of the common ancestor considered in each path.

As an example, consider the following diagrammed full sib mating system:

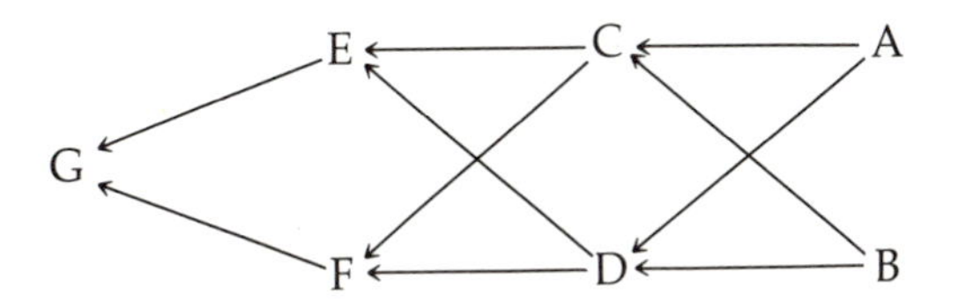

To calculate a_{GE} there is one direct path and six collateral paths through common ancestors C, D, A, and B. For the direct path,

$$n_{GE} = 1$$

For the indirect paths,

Path	n
E → (C) ← F ← G	$n_1 = 1 + 2 = 3$
E → (D) ← F ← G	$n_2 = 1 + 2 = 3$
E → C → (A) ← D ← F ← G	$n_3 = 2 + 3 = 5$
E → C → (B) ← D ← F ← G	$n_4 = 2 + 3 = 5$
E → D → (A) ← C ← F ← G	$n_5 = 2 + 3 = 5$
E → D → (B) ← C ← F ← G	$n_6 = 2 + 3 = 5$

Thus,

$$a_{GE} = (1/2)^1(1 + F_E) + (1/2)^3(1 + F_C) + (1/2)^3(1 + F_D) + (1/2)^5(1 + F_A) + (1/2)^5(1 + F_B) + (1/2)^5(1 + F_A) + (1/2)^5(1 + F_B)$$

If the common ancestors are not inbred, and if $F_E = \frac{1}{4}$ as the diagram indicates, then,

$$a_{GE} = (1/2)(1 + 1/4) + 1/8 + 1/8 + 1/32 + 1/32 + 1/32 + 1/32 = 1$$

If A and B are inbred a_{GE} will depend on F_A and F_B. For the calculation of a_{GA}, the different paths are:

$$n_{GECA} = 3$$
$$n_{GEDA} = 3$$
$$n_{GFCA} = 3$$
$$n_{GFDA} = 3$$

so that

$$a_{GA} = (1/2)^3(1 + F_A) + (1/2)^3(1 + F_A) + (1/2)^3(1 + F_A) + (1/2)^3(1 + F_A) = 4(1/2)^3(1 + F_A)$$

Inbreeding coefficients

The inbreeding coefficient for an animal is one-half the additive relationship between its parents. If X_S and X_D are parents of X, then $F_X = \frac{1}{2}a_{X_S X_D}$. Therefore, to calculate the inbreeding coefficient by the path coefficient method,

1. compute the additive relationship between the parents by tracing the paths, and
2. multiply by $\frac{1}{2}$.

As an example, consider the previous example of full sib matings and calculate the inbreeding coefficient for E, F_E.

1. Compute a_{CD}:

$$a_{CD} = (1/2)^{n_{CA} + n_{AD}}(1 + F_A) + (1/2)^{n_{CB} + n_{BD}}(1 + F_B)$$

and if $F_A = 0$ and $F_B = 0$, then,

$$a_{CD} = (1/2)^2 + (1/2)^2 = 1/2$$

2. Compute F_E:

$$F_E = (1/2)a_{CD}$$
$$= (1/2)(1/2) = 1/4$$

Similarly, to compute F_G:

1. Compute a_{EF}:

$$a_{EF} = (1/2)^2(1 + F_C) + (1/2)^2(1 + F_D) + (1/2)^4(1 + F_A) + (1/2)^4(1 + F_B) + (1/2)^4(1 + F_A) + (1/2)^4(1 + F_B)$$

2. Compute F_G:

$$F_G = 1/2a_{EF}$$

Example 10-2 illustrates the method of tracing paths to calculate additive relationships and inbreeding coefficients when one of the base animals is inbred.

Example 10-2 Computing relationships and inbreeding by tracing paths

The arrow diagram represents three generations of brother by sister matings. Assume that $F_A = 1/4$.

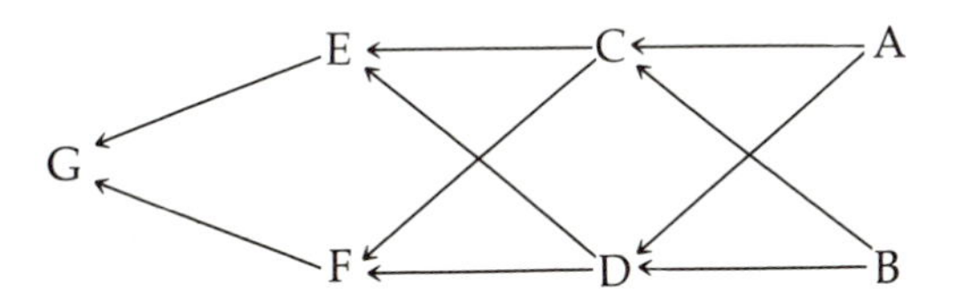

Compute the inbreeding coefficient of animal G, F_G.

1. Compute a_{EF}, the relationship between the parents of G. There are four common ancestors of E and F, that is, A and B, as well as C and D. Between E and F there are two different paths through A, two through B, one through C and one through D:

$$F \rightarrow D \rightarrow \textcircled{A} \leftarrow C \leftarrow E \qquad n_1 = 2 + 2$$
$$F \rightarrow C \rightarrow \textcircled{A} \leftarrow D \leftarrow E \qquad n_2 = 2 + 2$$
$$F \rightarrow D \rightarrow \textcircled{B} \leftarrow C \leftarrow E \qquad n_3 = 2 + 2$$
$$F \rightarrow C \rightarrow \textcircled{B} \leftarrow D \leftarrow E \qquad n_4 = 2 + 2$$
$$F \rightarrow \textcircled{C} \leftarrow E \qquad n_5 = 1 + 1$$
$$F \rightarrow \textcircled{D} \leftarrow E \qquad n_6 = 1 + 1$$

Thus,

$$a_{EF} = (1/2)^{2+2}(1 + 1/4) + (1/2)^{2+2}(1 + 1/4) + (1/2)^{2+2} + (1/2)^{2+2} + (1/2)^{1+1} + (1/2)^{1+1} = 25/32.$$

2. The inbreeding coefficient of G is

$$F_G = (1/2)a_{EF} = 25/64$$

10-6 Practice Problem in Computing Relationships and Inbreeding

The following is an example of systematic inbreeding whereby sire A is mated back to his daughters in each succeeding generation. The arrow charts (Figure 10-6) are drawn in two ways: (*A*) on the left corresponding to usual pedigrees, and (*B*) on the right in a form necessary for tracing paths. Pedigrees should be redrawn as on the right if the method of tracing paths is used.

Compute the relationships among the five animals as well as the inbreeding coefficients of D and E. Assume that the parents of A and B are unrelated. Use both the tabular method and the path coefficient method. Time yourself to determine which is most efficient. The solutions are given in the tabular form.

Solution to practice problem

	—	—	A–B	A–C	A–D
	A	B	C	D	E
A	1	0	1/2	3/4	7/8
B	0	1	1/2	1/4	1/8
C	1/2	1/2	1	3/4	5/8
D	3/4	1/4	3/4	1 + 1/4	1
E	7/8	1/8	5/8	1	1 + 3/8

Computations [relationships are given (row, column)]:

(A, C) = (1/2)(1) + (1/2)(0) = 1/2   (C, A) = 1/2
(A, D) = (1/2)(1) + (1/2)(1/2) = 3/4   (D, A) = 3/4
(A, E) = (1/2)(1) + (1/2)(3/4) = 7/8   (E, A) = 7/8

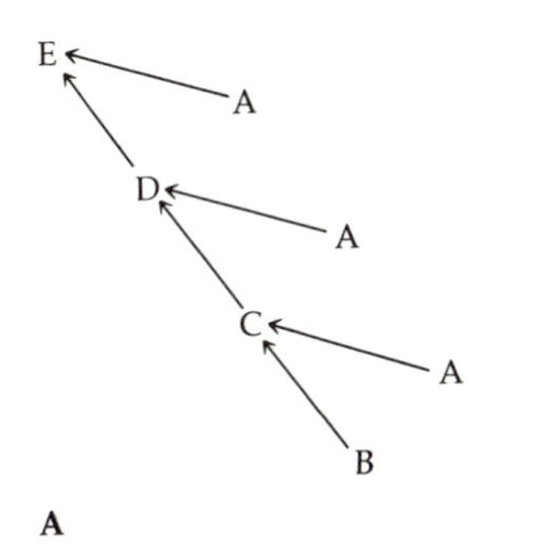

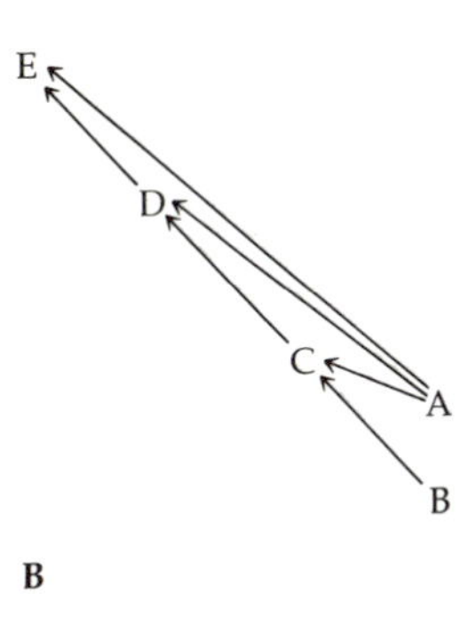

Figure 10-6 Pedigree for practice problem.

$$
\begin{aligned}
(B, C) &= (1/2)(0) + (1/2)(1) = 1/2 && (C, B) = 1/2\\
(B, D) &= (1/2)(0) + (1/2)(1/2) = 1/4 && (D, B) = 1/4\\
(B, E) &= (1/2)(0) + (1/2)(1/4) = 1/8 && (E, B) = 1/8\\
(C, C) &= 1 + (1/2)(A, B) = 1 + (1/2)(0) = 1;\ F_C = 0 &&\\
(C, D) &= (1/2)(1/2) + (1/2)(1) = 3/4 && (D, C) = 3/4\\
(C, E) &= (1/2)(1/2) + (1/2)(3/4) = 5/8 && (E, C) = 5/8\\
(D, D) &= 1 + (1/2)(A, C) = 1 + (1/2)(1/2) &&\\
&= 1 + 1/4;\ F_D = 1/4 &&\\
(D, E) &= (1/2)(3/4) + (1/2)(1 + 1/4) = 1 && (E, D) = 1\\
(E, E) &= 1 + (1/2)(A, D) = 1 + (1/2)(3/4) &&\\
&= 1 + 3/8;\ F_E = 3/8 &&
\end{aligned}
$$

10-7 Summary

Additive relationships among animals are needed for weighting records of relatives for genetic evaluation and for computing inbreeding coefficients.

If ancestors are not inbred, the additive relationship between relatives X and Y can be computed as follows:

1. determine the number of different paths from X to Y through each common ancestor;
2. count the number of intervening generations for each path, n_1, n_2, . . . ; and
3. compute

$$a_{XY} = (1/2)^{n_1} + (1/2)^{n_2} + \cdots$$

The inbreeding coefficient for an animal is one-half the additive relationship between its parents; if X is the progeny of X_S and X_D,

$$F_X = (1/2)a_{X_S X_D}$$

If ancestors are inbred, the method of counting generations is modified to include the inbreeding coefficient of the common ancestor (which may be the same for more than one path). If $F_1, F_2, \ldots$ are the inbreeding coefficients of the common ancestor for each path,

$$a_{XY} = (1/2)^{n_1}(1 + F_1) + (1/2)^{n_2}(1 + F_2) + \cdots$$

The tabular method of computing additive relationships and inbreeding coefficients is more efficient than the method of tracing paths for determining relationships among many animals and for complex pedigrees. The method is based on two principles:

1. additive relationships among younger animals can be determined from relationships among older animals (appearing earlier in the table), and
2. the inbreeding coefficient of an animal is one-half the additive relationship of its parents which, because the parents are older, appears earlier in the table.

The additive relationship of an animal X with itself is

$$a_{XX} = 1 + F_X$$

The dominance relationship, d_{XY}, is the probability of genotypes at a locus being identical by descent. If X and Y are not inbred, the dominance relationship can be calculated from the additive relationships among the parents of X and Y:

$$d_{XY} = (1/4)(a_{X_S Y_S}a_{X_D Y_D} + a_{X_S Y_D}a_{X_D Y_S})$$

Wright's coefficient of relationship between X and Y is

$$R_{XY} = a_{XY}/\sqrt{a_{XX}a_{YY}}$$

The upper and lower limits for additive relationships are zero and 2.

The limits for the inbreeding coefficient are zero and 1.

The limits for the additive relationship of an animal with itself are 1 (noninbred) and 2 (completely inbred).

The limits for dominance relationships are zero and 1.

The limits on Wright's coefficient of relationship, which is the same as the correlation between breeding values, are zero and 1.

Further Readings

Falconer, D. S. 1981. *Introduction to Quantitative Genetics,* 2d ed. Longman, Inc., New York.

Li, C. C. 1955. *Population Genetics.* University of Chicago Press, Chicago.

Lush, J. L. 1945. *Animal Breeding Plans,* 3d ed. Iowa State College Press, Ames.

Malécot, G. 1948. *Les Mathématiques de l' Hérédité.* Masson and Cie, Paris.

Turner, H. N., and S. S. Y. Young. 1969. *Quantitative Genetics in Sheep Breeding.* Cornell University Press, Ithaca, N.Y.

Van Vleck, L. D. 1983. *Notes on the Theory and Application of Selection Principles for the Genetic Improvement of Animals.* Department of Animal Science, Cornell University, Ithaca, N.Y.

Wright, S. 1921. Systems of Mating. *Genetics* **6:**111–178.

CHAPTER 11

Introduction to Quantitative Traits

11-1 Quantitative Traits

In previous chapters, the traits considered were simply inherited or qualitative characteristics. There are, however, many traits of livestock for which simple models of inheritance do not explain the differences in phenotypes. These are called quantitative traits and have two basic characteristics:

1. the phenotype is influenced by genes at many loci, and
2. the phenotype is influenced in part by environmental factors (as was the case with some qualitative traits).

In many instances, the phenotype is a measurement and may be a certain value over some continuous range. Two examples are 305-day milk yield in dairy cattle and 205-day weaning weight in beef cattle. Other traits, however, may also be considered as quantitative traits although the observations, or phenotypes, fall into discrete categories: for example, survival to a given age, in which case the observation is zero if the animal died or 1 if it lived, and litter size for which fractional observations are not possible. For both these traits, the genotype comprises genes at many loci and the environment can influence the observation. Such traits are sometimes referred to as "quasi-continuous" characteristics.

Genotype

Although the genotype for a quantitative trait consists of genes at many loci, the fundamental principles of Mendelian inheritance apply to each locus, as do the various types of inter- and intralocus gene actions. However, the influence of genes at each individual locus on the phenotype is usually quite small relative to the influence of the total genotype; hence, segregation at these loci cannot be observed nor can a unique contribution be assigned to each gene. The study of quantitative traits requires an understanding of the concept of variation and knowledge of parameters and statistical techniques.

Chapter 9 examined the concept of numerical values, or measurements, for phenotype (P). The measurements were represented with the model

$$P = \mu + A + D$$

where μ represents the mean, A repesents the breeding value (additive genetic merit), and D the dominance deviation for a single locus. Together, A and D represent the total genotypic contribution (G) to the phenotype; hence

$$P = \mu + G$$

The variance among phenotypes is

$$\sigma_P^2 = \sigma_G^2$$

which may also be partitioned into its component parts:

$$\sigma_P^2 = \sigma_A^2 + \sigma_D^2$$

For a single locus with arbitrary genotypic values and known gene frequencies, breeding values, dominance deviations, and their respective variances could be calculated. For quantitative traits, where each locus contributes a relatively small amount to the total phenotype and gene frequencies are not known, these calculations at each locus are not possible. The breeding values—the important component of the total genotypic value relative to selection—must be estimated in total. *Breeding value,* which is the additive genetic value, represents the sum of additive genetic contributions from each locus. Similarly, *dominance genetic value* represents the sum of all dominance deviations even though a particular contribution cannot be assigned to each locus.

When more than one locus is involved in the expression of a phenotype, there may be effects due to joint occurrences of alleles at more than one locus. These interloci gene interactions are called epistasis, as described in Chapter 2. If I represents the sum of all epistatic effects, then

$$G = A + D + I$$

The genetic variance also has an additional term for epistatic effects:

$$\sigma_G^2 = \sigma_A^2 + \sigma_D^2 + \sigma_I^2$$

Environmental influences

The second characteristic of quantitative traits is the possible influence of many relatively small and random environmental effects on the expression of the phenotype. These factors do not depend on the genotype of the animals and are an additional source of variation. Also, any measurement, such as size or weight, is subject to measurement error which is extraneous to the individual's genotype. Including these factors in the model for the phenotype gives

$$P = \mu + G + E$$

The environmental (and measurement error) contribution, E, represents a random deviation, either plus or minus, about the fixed portion, μ (see Chapter 9), and is the sum of all random environmental factors and errors of measurement which are not genetic.

The equation

$$P = \mu + G + E$$

shows the problem confronting a breeder wishing to select genetically superior animals to be parents. The phenotype is the only information available on which to make selection decisions. The phenotypic model shows that an above average phenotype may occur because an individual is genetically superior or because an individual receives, by chance, a better than average environmental contribution. Any information which increases the accuracy of identifying superior animals from phenotypic observations (for example, progeny testing) becomes extremely useful in selection programs.

The variance of phenotypes in a population can be expressed as

$$\sigma_P^2 = \sigma_G^2 + \sigma_E^2 + 2\,\mathrm{Cov}(G, E)$$

The environmental variance, σ_E^2, is included in the equation. The additional term, $\mathrm{Cov}(G, E)$, is the covariance between an animal's genotype and the effects of its environment (see Section 11-2). The covariance indicates how two variables vary together and may be negative, zero, or positive. If, for example, better than average genotypes in the population receive better than average treatment (such as better feed and better health care), the covariance between G and E would be positive. However, the assumption made throughout the remainder of the discussion of quantitative traits is that the genotype of an animal and its environment are independent so that $\mathrm{Cov}(G, E) = 0$. Thus, the phenotypic variance is

$$\sigma_P^2 = \sigma_G^2 + \sigma_E^2$$

11-2 Genetic Parameters

Heritability

Heritability is an extremely important population parameter that is used both for the estimation of breeding values for quantitative characteristics and for predicting the response expected from various selection schemes. Heritability can be defined in the broad sense or in the narrow sense. *Heritability in the broad sense,* denoted h_B^2, is defined as the ratio of the genetic variance to the phenotypic variance:

$$h_B^2 = \sigma_G^2/\sigma_P^2$$

Heritability in the broad sense describes what proportion of the total variation is due to differences among genotypes of individuals in the population. Since $\sigma_P^2 \geqq \sigma_G^2 \geqq 0$, then $0 \leqq h_B^2 \leqq 1$. *Heritability in the narrow sense,* denoted as h^2, is defined as the ratio of the additive genetic variance to the phenotypic variance:

$$h^2 = \sigma_A^2/\sigma_P^2$$

Thus, h^2 is the proportion of the total variance that is due to differences among the breeding values of individuals in the population. Since $\sigma_G^2 \geqq \sigma_A^2$, then $0 \leqq h^2 \leqq h_B^2 \leqq 1$.

Heritability is specific to the population and trait under consideration. If either the genetic or environmental variance for the same trait in two populations differs, h^2 will also be different.

Repeatability

A second parameter of interest, *repeatability,* represents the degree of association between measurements on the same animal for traits which are measured more than once. Examples of traits which can be measured more than once are litter size, lactation milk yield, and fleece weight. An assumption made in the definition of repeatability is that the genotype for the trait is the same each time the trait is measured. For example, the effects of genes which influence production of a dairy heifer in the first lactation are assumed to be the same genetic effects which influence production in subsequent lactations.

As with heritability, repeatability is defined as a ratio of variance components. The ratio, however, depends on the variances of two types of environmental influences: permanent environmental (PE) and temporary environmental (TE). *Permanent environmental effects* are those which influence all observations made on the individual. For example, the feeding regime used to raise young dairy heifers, if extreme (poor feeding or excessive feeding) can influence mammary development, hence becoming a permanent effect influencing all lactations. A *temporary environmental effect* is one which influences only a single observation on the individual; whether or not an individual receives a particularly favorable or unfavorable influence is simply by chance for each observation. Other nonrandom environmental effects may influence records, for example, herd feeding programs. Records may be "adjusted" for these identifiable effects and will not be considered here. The E term is partitioned into

$$E = PE + TE$$

The environmental variance now also has two parts,

$$\sigma_E^2 = \sigma_{PE}^2 + \sigma_{TE}^2$$

Thus,

$$\sigma_P^2 = \sigma_G^2 + \sigma_{PE}^2 + \sigma_{TE}^2$$

Repeatability (r) is the ratio

$$r = \frac{\sigma_G^2 + \sigma_{PE}^2}{\sigma_G^2 + \sigma_{PE}^2 + \sigma_{TE}^2}$$

The numerator contains the variance of genotypic and permanent environmental effects which are constant from one record to the next. The denominator is again the total phenotypic variance, σ_P^2. Also, $r \geqq h_B^2$ because of the additional variance component, σ_{PE}^2, in the numerator. Repeatability is used in predicting future records for an animal that has one or more previous records.

11-3 Statistics

Quantitative genetic principles rely heavily on the concepts of variation of and covariation among variables. This section addresses the estimation of phenotypic variances and covariances, their use in estimating simple regression and correlation coefficients, and the relationship of these statistics to heritability and repeatability.

Variance

In Chapter 9, the variance of a trait was defined as

$$\sigma_P^2 = \frac{\sum_i (P_i - \mu)^2}{N}$$

where N is the number of observations. The variance is the average of squared deviations from the mean, μ, and although the deviations may be negative, their square is positive. Hence, any variance is greater than, or equal to, zero.

In the equation defining σ_P^2, μ was assumed known. However, in real situations when only a sample of possible observations is obtained, the mean is not known and must be estimated from the data. The estimate of μ is denoted as $\hat{\mu}$ and is obtained as

$$\hat{\mu} = \frac{\left(\sum_i P_i\right)}{N}$$

If $\hat{\mu}$ is used to estimate the variance among observations, the equation becomes

$$\hat{\sigma}_P^2 = \frac{\sum_i (P_i - \hat{\mu})^2}{N - 1}$$

The denominator is reduced by 1 since $\hat{\mu}$ used in the equation is dependent on the same sample of records used to estimate the variance. The notation $\hat{\sigma}_P^2$ denotes an estimate of the population variance, σ_P^2. An alternative and algebraically equivalent equation is

$$\hat{\sigma}_P^2 = \frac{\sum_i P_i^2 - \frac{\left(\sum_i P_i\right)^2}{N}}{N-1}$$

This formula uses the sum of squares and the sum of the phenotypes, which are usually much easier to calculate than the sum of squared deviations from $\hat{\mu}$.

The square root of the variance is called the *standard deviation.* The standard deviation is used along with the mean to describe the normal distribution of observations, as will be discussed in a later section.

Covariance

The covariance defines how two random variables vary together. For two variables, X and Y, the covariance is denoted as σ_{XY} and is estimated as

$$\hat{\sigma}_{XY} = \frac{\sum_i (X_i - \hat{\mu}_X)(Y_i - \hat{\mu}_Y)}{N-1}$$

X and Y may be two traits measured on the same animal or two components influencing the same trait, as seen previously when considering the covariance between G and E for a particular set of phenotypes. The equation shows (1) that if above average observations of X occur in conjunction with above average observations of Y, then σ_{XY} is positive, (2) that if above average values of X are associated with below average values of Y (or vice versa), then σ_{XY} is negative, and (3) that if the sum of products of deviations is zero, then $\sigma_{XY} = 0$. If two variables are independent, their covariance is zero.

As with the variance, a comparable formula exists which does not require deviating observations from their means:

$$\hat{\sigma}_{XY} = \frac{\sum_i X_i Y_i - \frac{\left(\sum_i X_i\right)\left(\sum_i Y_i\right)}{N}}{N-1}$$

The calculations of variances and covariances are shown in Example 11-1.

Example 11-1

Assume the following observations have been obtained for birth weight (B) and weaning weight (W) for a group of beef calves:

Animal	Birth weight (pounds)	Weaning weight (pounds)
1	60	430
2	70	500
3	68	489
4	62	430
5	65	440
6	75	525
7	60	460
8	65	400
9	62	425
10	63	410

The mean of each trait, the variance of each trait, and the covariance between B and W are to be estimated. The following must be calculated:

the sums:

$$\sum_{i=1}^{10} B_i = 650$$

$$\sum_{i=1}^{10} W_i = 4{,}500$$

the sums of squares:

$$\sum_{i=1}^{10} B_i^2 = 42{,}456$$

$$\sum_{i=1}^{10} W_i^2 = 2{,}039{,}750$$

and the sum of products:

$$\sum_{i=1}^{10} B_i W_i = 293{,}855$$

For B:

$$\hat{\mu}_B = \frac{\sum_{i=1}^{10} B_i}{N} = \frac{650}{10} = 65$$

$$\hat{\sigma}_B^2 = \frac{\sum_{i=1}^{10} B_i^2 - \frac{\left(\sum_{i=1}^{10} B_i\right)^2}{N}}{N-1}$$

$$= \frac{42{,}456 - \frac{(650)^2}{10}}{9}$$

$$= 22.9$$

For W:

$$\hat{\mu}_W = \frac{\sum_{i=1}^{10} W_i}{\mathrm{N}} = \frac{4{,}500}{10} = 450$$

$$\hat{\sigma}_W^2 = \frac{\sum_{i=1}^{10} W_i^2 - \frac{\left(\sum_{i=1}^{10} W_i\right)^2}{N}}{N-1}$$

$$= \frac{2{,}039{,}750 - \frac{(4500)^2}{10}}{9}$$

$$= 1638.9$$

For $\mathrm{Cov}(B, W)$:

$$\mathrm{Cov}(B, W) = \frac{\sum_{i=1}^{10} B_i W_i - \frac{\sum_{i=1}^{10} B_i \sum_{i=1}^{10} W_i}{N}}{N-1}$$

$$= \frac{293{,}855 - \frac{(650)(4500)}{10}}{9}$$

$$= 150.6$$

Rules for determining variances and covariances of sums of variables

Earlier in this chapter the phenotypic variance for the model

$$P = \mu + G + E$$

was defined as

$$\sigma_P^2 = \sigma_G^2 + \sigma_E^2 + 2\,\mathrm{Cov}(G, E)$$

Certain rules can be invoked for finding the variance of the sum $\mu + G + E$. These rules and other related rules are as follows:

1. The variance of a constant is zero; that is, a constant, such as μ, does not vary.
2. The variance of a random variable, say X, is σ_X^2.
3. The variance of a constant times a random variable, say c times X, is $c^2\sigma_X^2$, the constant squared times the variance of the random variable.
4. The covariance between a constant and a random variable is zero; that is, Cov (c, X) = 0.
5. The covariance between two random variables, say X and Y, is σ_{XY}. Also, the covariance of a random variable with itself is its variance; that is, $\mathrm{Cov}(X, X) = \sigma_X^2$.
6. If two random variables are independent, then their covariance is zero.
7. The covariance between two random variables each multiplied by a constant, say c_1X and c_2Y, is $c_1c_2\sigma_{XY}$, the product of the constants times the covariance.
8. The variance of a sum is the sum of the variances of each variable plus two times the sum of all possible covariances; for example, the variance of $X + Y$ is $\sigma_X^2 + \sigma_Y^2 + 2\sigma_{XY}$.
9. The covariance of a random variable with a sum of random variables is the sum of covariances. For example, $\mathrm{Cov}(X, Y_1 + Y_2 + Y_3) = \mathrm{Cov}(X, Y_1) + \mathrm{Cov}(X, Y_2) + \mathrm{Cov}(X, Y_3)$.

If V denotes a variance, using the example $P = \mu + G + E$, then

$$V(P) = V(\mu + G + E)$$

which by rule 8 is

$$V(P) = V(\mu) + V(G) + V(E) + 2[\mathrm{Cov}(\mu, G) + \mathrm{Cov}(\mu, E) + \mathrm{Cov}(G, E)]$$

By rules 1 and 4, $V(\mu)$, Cov(μ, G), and Cov(μ, E) are zero; hence,

$$V(P) = V(G) + V(E) + 2\text{Cov}(G, E)$$

which by rules 2 and 5 can be rewritten as

$$V(P) = \sigma_G^2 + \sigma_E^2 + 2\sigma_{G,E}^2$$

If G and E are independent, $\sigma_{G,E} = 0$ (rule 6). Then $V(P) = \sigma_G^2 + \sigma_E^2$.

Example 11-2 uses the rules above to obtain the variance of the estimated mean, $\hat{\mu}$.

Example 11-2

What is the variance of $\hat{\mu}$ estimated from a set of independent observations? First, if $V(\mu) = 0$, why does a variance exist for $\hat{\mu}$? μ is a population parameter and thus is a constant for that population. However, $\hat{\mu}$ is an estimate of μ from a sample of records, and since the samples may vary, $\hat{\mu}$ may also vary even though the true but unknown μ is constant.

Since

$$\hat{\mu} = (1/N) \sum_i P_i$$

then

$$V(\hat{\mu}) = V\left[(1/N) \sum_i P_i\right]$$

For a sample, $1/N$ is a constant; therefore, by rule 3,

$$V(\hat{\mu}) = (1/N^2)\, V\left(\sum_i P_i\right)$$

By rule 8,

$$V(\hat{\mu}) = (1/N^2)\,[V(P_1) + V(P_2) + \cdots + V(P_N) + 2(\text{all covariances})]$$

Since all observations are stated to be independent, all covariances are zero according to rule 6, and

$$V(\hat{\mu}) = (1/N^2)[V(P_1) + V(P_2) + \cdots + V(P_N)]$$

The variance of any P_i is σ_P^2, rule 2. Thus,

$$V(\hat{\mu}) = (1/N^2)\,(N\sigma_P^2)$$

so that

$$V(\hat{\mu}) = \sigma_P^2/N$$

Hence, the variance of $\hat{\mu}$ is equal to the phenotypic variance divided by the sample size when the observations are independent and $V(\hat{\mu})$ approaches zero as $N \rightarrow \infty$.

Correlation

The covariance indicates how two random variables vary together and can be negative, zero, or positive. However, the covariance is in units of the traits measured; for example, the covariance between height and weight will be in terms of inches and pounds while the covariance between milk yield and fat percent will be in terms of pounds and percent. Comparing covariances among various traits with different units of measurement becomes difficult and many times is meaningless. The relative degree of association between various traits can be compared using a standardized statistic, the correlation. The correlation between two random variables, say X and Y, is defined as

$$r_{XY} = \sigma_{XY}/\sigma_X\sigma_Y$$

Since variances are in terms of units squared, standard deviations are in the units of measurement; hence, in the ratio, the units in the covariance are cancelled out by the units in the standard deviations. A correlation must be between -1 and 1. The closer the correlation is to either -1 or 1, the closer the association between traits. Two traits with a negative correlation of $-.7$ are more closely associated than two traits with a positive correlation of .5 since $-.7$ is closer to -1 than .5 is to 1. In the first case, as one trait increases, the second tends to decrease. In the second case, the traits tend to increase or decrease together. If two traits are independent, their covariance is zero; hence, r_{XY} is also zero.

The definition of a correlation is general enough to consider the correlation between an individual's breeding value and its phenotype. The covariance between the breeding value, A, and phenotype, P, is

$$\text{Cov}(A, P) = \text{Cov}(A, \mu + A + D + I + E)$$

From rule 9,

$$\text{Cov}(A, P) = \text{Cov}(A, \mu) + \text{Cov}(A, A) + \text{Cov}(A, D) + \text{Cov}(A, I) + \text{Cov}(A, E)$$

Under the assumption of independence (rule 6),

$$\text{Cov}(A, P) = \text{Cov}(A, A)$$

Using rule 5,

$$\text{Cov}(A, P) = \sigma_A^2$$

Hence, the correlation between breeding value and phenotype is

$$r_{A,P} = \frac{\sigma_A^2}{\sigma_P \sigma_A} = \frac{\sigma_A}{\sigma_P}$$

This correlation is the square root of heritability, since $h^2 = \sigma_A^2/\sigma_P^2$.

Regression

In many cases breeders must predict the value of one variable given the value of another. Two examples are: predicting second lactation performance from the first lactation record and predicting yearling weight from weaning weight. The statistic used to make such prediction is the *regression coefficient.* If random variable Y is to be predicted from variable X, the regression coefficient of Y on X, denoted $b_{Y \cdot X}$ (called "b-Y-dot-X") is

$$b_{Y \cdot X} = \text{Cov}(X, Y)/\sigma_X^2$$

First, what are the units associated with $b_{Y \cdot X}$? If X is measured in inches and Y in pounds, the covariance is inches by pounds and the variance is inches squared. The regression coefficient, therefore, is in terms of pounds per inch. The regression coefficient gives the expected change of trait Y in pounds for a one-inch change in trait X. $b_{Y \cdot X}$ can be negative, zero, or positive depending on $\text{Cov}(X, Y)$. Figure 11-1 summarizes the units associated with means, variances, standard deviations, covariances, regression, and correlation.

The regression equation

$$\hat{Y}_i = \hat{\mu}_Y + b_{Y \cdot X}(X_i - \hat{\mu}_X)$$

Units for:			
	Means	μ_H: inches	μ_W: pounds
	Variances	σ_H^2: inches^2	σ_W^2: pounds^2
	Standard deviations	σ_H: inches	σ_W: pounds
	Covariance	Cov(H, W): inches by pounds	
	Correlation	$r = \frac{\mathrm{Cov}(H, W)}{\sigma_H \sigma_W} : \frac{\text{inches by pounds}}{\text{inches} \cdot \text{pounds}} = \text{none}$	
	Regression of weight on height	$b_{W \cdot H} = \frac{\mathrm{Cov}(H, W)}{\sigma_H^2} : \frac{\text{inches by pounds}}{\text{inches}^2} = \frac{\text{pounds}}{\text{inches}}$	

Figure 11-1 Summary of units associated with parameters (H = height in inches; W = weight in pounds).

can be used to predict Y when X is known. $X_i - \hat{\mu}_X$ represents for a particular animal the deviation of the variable X_i from the estimated mean of X. Since the $b_{Y \cdot X}$ is the expected change in variable Y per unit change in variable X, $b_{Y \cdot X}(X_i - \hat{\mu}_X)$ represents the difference in Y expected for that particular X, as compared to the Y expected for an average X. If $X_i = \hat{\mu}_X$ (that is, $X_i - \hat{\mu}_X = 0$), the predicted value of Y is $\hat{\mu}_Y$.

The equation to predict Y can be rewritten as

$$\begin{aligned} \hat{Y}_i &= (\hat{\mu}_Y - b_{Y \cdot X} \hat{\mu}_X) + b_{Y \cdot X} X_i \\ &= \hat{\alpha} + b_{Y \cdot X} X_i \end{aligned}$$

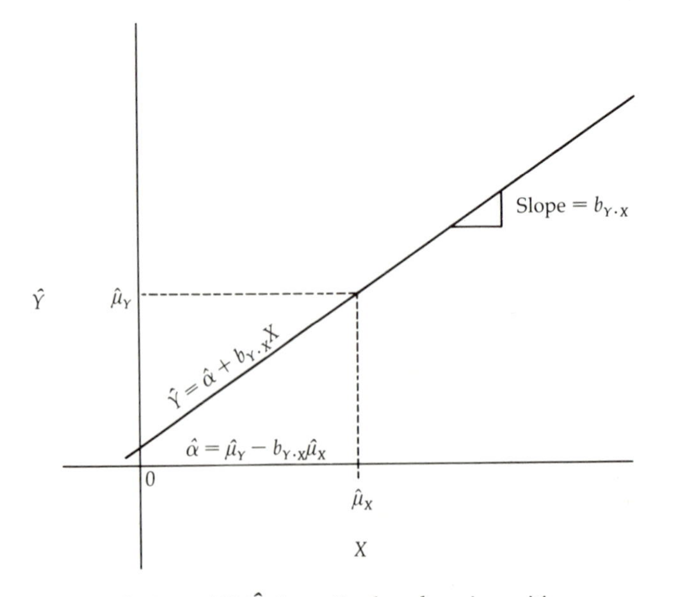

Figure 11-2 Prediction of Y, $\hat{Y}$, from X when $b_{Y \cdot X}$ is positive.

Here $\hat{\alpha}$ represents the estimated intercept of Y (the value of Y when X_i is zero). Figure 11-2 represents the equation to predict Y from X. The slope of the line is $b_{Y \cdot X}$, that is, the change in Y for change of one unit of X.

Genetic parameters can be defined as regression coefficients. For example, the regression of an animal's breeding value on its phenotype is

$$b_{A \cdot P} = \text{Cov}(A, P)/\sigma_P^2 = \sigma_A^2/\sigma_P^2$$

This regression coefficient is equal to h^2. The equation to predict the breeding value from the animal's phenotype is

$$\hat{A}_i = \hat{\mu}_A + b_{A \cdot P}(P_i - \hat{\mu}_P)$$

If $\hat{\mu}_A = 0$, as defined in Chapter 9, and since $b_{A \cdot P} = h^2$,

$$\hat{A}_i = h^2(P_i - \hat{\mu}_P)$$

Example 11-3

In Example 11-1 the variances of birth weight and weaning weight as well as their covariances were calculated from observations on 10 animals. Using these estimated parameters, the correlation between these traits and the regression of weaning weight on birth weight can be obtained.

Correlation:

$$\begin{aligned} \hat{r}_{B,W} &= \frac{\widehat{\text{Cov}}(B, W)}{\hat{\sigma}_B \hat{\sigma}_W} \\ &= \frac{150.6}{\sqrt{22.9}\,\sqrt{1638.9}} \\ &= .78 \end{aligned}$$

Regression of weaning weight on birth weight:

$$\begin{aligned} \hat{b}_{W \cdot B} &= \frac{\widehat{\text{Cov}}(B, W)}{\hat{\sigma}_B^2} \\ &= \frac{150.6}{22.9} \\ &= 6.6 \end{aligned}$$

If an animal's birth weight were 72 pounds, its predicted weaning weight would be

$$\begin{aligned}\hat{W} &= \hat{\mu}_W + \hat{b}_{W \cdot B}(BW - \hat{\mu}_B) \\ &= 450 + 6.6\,(72 - 65) \\ &= 496.2 \text{ pounds}\end{aligned}$$

11-4 Estimating Genetic Variances

To estimate heritability (or to predict response to selection), an estimate of the additive genetic variance, σ_A^2, is needed. Methods of estimating σ_A^2 involve the covariance between records of relatives. The use of records on a parent and its offspring and the use of sib records will be discussed.

From N pairs of parent and offspring records, the covariance between the parent phenotype, P_i, and the offspring phenotype, O_i, can be calculated as

$$\widehat{\text{Cov}}(P, O) = \frac{\sum_i (P_i - \hat{\mu}_P)(O_i - \hat{\mu}_O)}{N - 1}$$

To estimate σ_A^2 the fraction of σ_A^2 contained in $\widehat{\text{Cov}}(P, O)$ must be known, that is, the expectation of Cov(P, O) is needed. This expectation is equal to $\frac{1}{2}\sigma_A^2$. In general, the covariance between relatives' records is $a_{ij}\sigma_A^2$ if nonadditive genetic variation is ignored, where a_{ij} is the additive relationship between relatives i and j (see Chapter 10). A parent and its offspring are related by $\frac{1}{2}$; hence, the calculated covariance can be equated to its expectation:

$$\widehat{\text{Cov}}(P, O) = (1/2)\sigma_A^2$$

and

$$\hat{\sigma}_A^2 = 2\widehat{\text{Cov}}(P, O)$$

where $\hat{\sigma}_A^2$ denotes the estimate of σ_A^2.

A covariance can be calculated from records on N pairs of half-sibs and equated to its expectation. The covariance between half-sibs is $\frac{1}{4}\sigma_A^2$ since a_{ij} is $\frac{1}{4}$; hence, an estimate of the additive genetic variance is

$$\hat{\sigma}_A^2 = 4\widehat{\text{Cov}}(\text{half-sibs})$$

Certain types of records may not be as useful in estimating σ_A^2. For example, the covariance between records on N pairs of full sibs contains a fraction of the dominance variance as well as a fraction of the additive genetic variance. The covariance between full sib records equated to its expectation would be

$$\widehat{\text{Cov}}(\text{full sibs}) = (1/2)\sigma_A^2 + (1/4)\sigma_D^2$$

and could be used to estimate σ_A^2 only if $\sigma_D^2 = 0$. If this were the case, then

$$\hat{\sigma}_A^2 = 2\widehat{\text{Cov}}(\text{full sibs})$$

Often sib data consists of N groups of two or more sibs rather than N pairs of sibs. For such data sets, a different statistical technique, the analysis of variance, is used to partition the phenotypic variance. This method will not be covered but it is described in most statistical texts.

11-5 The Normal Distribution

If a large number of observations for most quantitative traits were plotted, the frequencies of observations would show the characteristic bell-shaped curve of the normal distribution, as shown in Example 11-4. Hence, it is important to recognize the features of the normal distribution.

The sum of all frequencies must equal 1; therefore, the area represented under the curve equals 1. The normal distribution is symmetric about μ, and the frequency of observations between μ and some point $\mu + t$ is the same as the frequency of observations between μ and $\mu - t$. Table 11-1 shows the

Table 11-1 The frequency of observations between μ and $\mu + t$ (or between μ and $\mu - t$)

t	**Frequency of observations between μ and $\mu + t$ (or between μ and $\mu - t$)**
$.5\sigma$	.192
σ	.341
1.5σ	.433
2σ	.477
2.5σ	.494
3σ	.499

frequency of observations between μ and t, where t is increased by .5 standard deviations.

From Table 11-1, the frequency of observations between $\mu - \sigma$ and $\mu + \sigma$ is calculated as .682, between $\mu - 2\sigma$ and $\mu + 2\sigma$ as .954, and between $\mu - 3\sigma$ and $\mu + 3\sigma$ as .998. These frequencies are twice the values in Table 11-1 because of the symmetry of the normal distribution.

Other frequencies can be calculated as shown in Examples 11-4 and 11-5.

Example 11-4

What is the frequency of observations greater than $\mu + \sigma$, $f(X > \mu + \sigma)$, represented by the shaded area in the following curve:

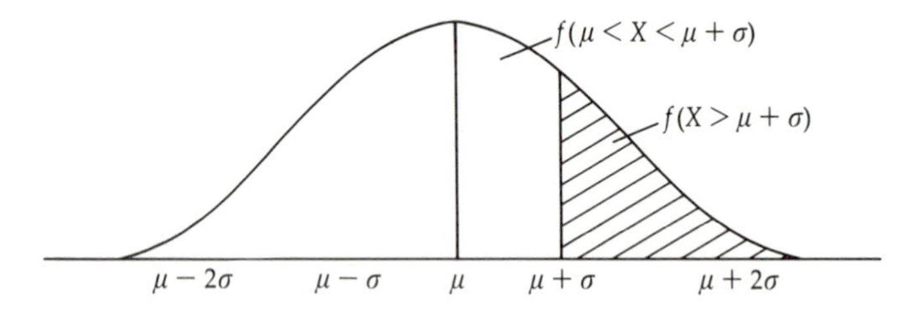

The frequency of observations greater than μ, $f(X > \mu)$, is .5. From Table 11-1, the frequency of observations between μ and $\mu + \sigma$, $f(\mu < X < \mu + \sigma)$ is .341. Examining the curve shows that

$$
\begin{aligned}
f(X > \mu + \sigma) &= f(X > \mu) - f(\mu < X < \mu + \sigma) \\
&= .5 - .341 = .159
\end{aligned}
$$

The frequency of observations below $\mu + \sigma$ is

$$
\begin{aligned}
f(X < \mu + \sigma) &= f(X < \mu) + f(\mu < X < \mu + \sigma) \\
&= .5 + .341 = .841
\end{aligned}
$$

Example 11-5

What is $f(X > \mu - 2\sigma)$?

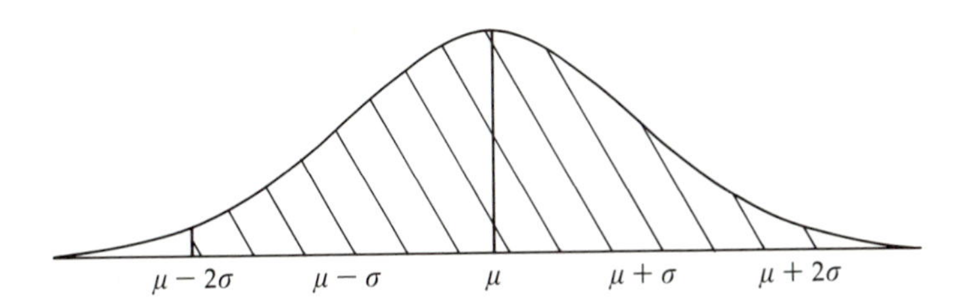

From the diagram it can be seen that

$$\begin{aligned} f(X > \mu - 2\sigma) &= f(X > \mu) + f(\mu - 2\sigma < X < \mu) \\ &= .5 + .433 \\ &= .933 \end{aligned}$$

11-6 Summary

Quantitative traits are

1. influenced by alleles at many loci, each having a relatively small contribution to the genetic merit of the individual, and
2. influenced in part by environmental factors.

The genetic merit of an individual may be partitioned into its breeding value (A), dominance genetic value (D), and epistatic genetic value (I). Thus,

$$G = A + D + I$$

The model for a phenotype is

$$P = \mu + G + E$$

where E represents the environmental factors and any measurement error. The variance among phenotypes, σ_P^2, can be partitioned into the genetic variance and environmental variance:

$$\sigma_P^2 = \sigma_G^2 + \sigma_E^2$$

Heritability in the broad sense, h_B^2, is the ratio of genetic variance to the phenotypic variance:

$$h_B^2 = \sigma_G^2 / \sigma_P^2$$

Heritability in the narrow sense, h^2, is the ratio of the variance among breeding values, σ_A^2, to the phenotypic variance:

$$h^2 = \sigma_A^2/\sigma_P^2$$

Permanent environmental effects, *PE*, are those common to all records on an individual, while temporary environmental effects, *TE*, are those specifically influencing each record. Repeatability is the degree of association between two records for the same trait measured on the same animal, and is defined as

$$r = (\sigma_G^2 + \sigma_{PE}^2)/\sigma_P^2$$

The variance of a trait measures the amount of variation of observations about the mean and is defined as

$$\sigma_P^2 = \frac{\sum_i (P_i - \mu)^2}{N}$$

If μ is not known and is estimated from a sample of observations, the estimate of the variance is

$$\hat{\sigma}_P^2 = \frac{\sum_i (P_i - \hat{\mu})^2}{N-1}$$

or

$$\hat{\sigma}_P^2 = \frac{\sum_i P_i^2 - \frac{\left(\sum_i P_i\right)^2}{N}}{N-1}$$

The covariance defines how two random variables vary together and may be estimated as

$$\hat{\sigma}_{XY} = \frac{\sum_i (X_i - \hat{\mu}_X)(Y_i - \hat{\mu}_Y)}{N-1}$$

or

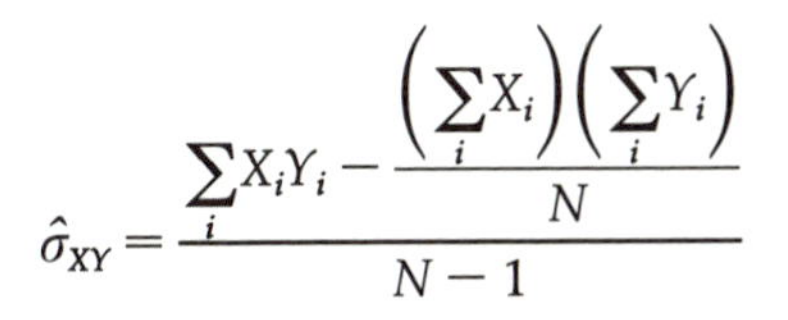

$$\hat{\sigma}_{XY} = \frac{\sum_i X_i Y_i - \frac{\left(\sum_i X_i\right)\left(\sum_i Y_i\right)}{N}}{N-1}$$

The correlation between two random variables is defined as

$$r_{XY} = \sigma_{XY}/\sigma_X \sigma_Y$$

and must be in the range $-1 \leqq r_{XY} \leqq 1$.

The regression coefficient for the regression of a random variable Y on X is defined as

$$b_{Y \cdot X} = \sigma_{XY}/\sigma_X^2$$

and is used to predict the variable Y when X is known. The regression equation for predicting Y from X is

$$\hat{Y}_i = \hat{\mu}_Y + b_{Y \cdot X}(X_i - \hat{\mu}_X)$$

Estimates of variance among breeding values, σ_A^2, may be obtained from the covariance between records on relatives. For parent-offspring records,

$$\hat{\sigma}_A^2 = 2\widehat{\text{Cov}}(\text{parent, offspring})$$

For half-sibs,

$$\hat{\sigma}_A^2 = 4\widehat{\text{Cov}}(\text{half-sib records})$$

The expectation of the covariance between full sibs is

$$\text{Cov(full sibs)} = (1/2)\sigma_A^2 + (1/4)\sigma_D^2$$

and therefore may not be used to estimate σ_A^2 unless σ_D^2 is zero.

The normal distribution is symmetric about its mean, μ. The frequency of observations expected between two points can be obtained by the use of Table 11-1.

CHAPTER 12

Prediction of Breeding Values

The goal of animal breeders is rapid genetic improvement, for which accurate prediction of breeding value is the most crucial factor. The breeder can rank the animals and cull those with the poorest evaluations while selecting those with the best evaluations as replacements. Accurate evaluations require proper application of heritability, repeatability, and relationships to weight records of the animal and its relatives. The selection index procedure described in this chapter maximizes the accuracy of predicting breeding value, thus maximizing genetic progress from selection.

12-1 The Model

Most economically important traits of animals are influenced by genes at many loci, each with a relatively small effect as described in the previous chapter. The phenotypic expression of such economic traits usually is measured quantitatively, for example, milk production per lactation, growth rate of beef steers per day, litter size in pigs, reproductive interval in sheep, and racing speed in horses. These traits are quantitative in two ways—in terms of measurement and in number of genes involved. The sum of all the effects of all of an animal's genes can be thought of as the genetic value of the animal. Chapter 11 presented the genetic value as comprising three parts:

1. the sum of effects of substitution values of individual genes, or the *additive genetic value;*

2. the sum of dominance effects of the paired genes at each locus, or the *dominance genetic value*; and
3. the sum of effects due to combinations of genes at different loci, or the *epistatic genetic value.*

Selection is usually for parents that will produce superior progeny. Therefore this chapter will describe the prediction of additive genetic value, that is, prediction of breeding value. Additive value is the only part of total genetic value that can be effectively selected for because

1. a parent cannot contribute dominance effects to its progeny inasmuch as dominance effects depend on a particular pair of genes at a locus, and each parent can contribute only one gene at each locus; and
2. epistatic effects depend on combinations of genes at different loci, which, by independent segregation, do not usually stay together from one generation to the next.

Sometimes the breeder is more concerned with predicting the next record of an animal rather than its breeding value. Therefore, prediction of producing ability will also be described.

The symbolic model

As shown in the previous chapter, the model for the phenotypic expression of a quantitative trait is

$$P_{ij} = \mu + G_i + PE_i + TE_{ij}$$

where P_{ij} is the jth record of the ith animal;

μ is a constant level of performance for all animals which can be thought of as the average value of genes that all animals in the population have in common plus the average level of management;

G_i is the sum of the genetic values ($G_i = A_i + D_i + I_i$, where A_i, D_i, and I_i are the sums of all the additive, dominance, and epistatic effects associated with the complete genotype of animal i);

PE_i is the sum of effects of *environmental* factors which *permanently* influence the performance of animal i (for example, the fetal environment or feed availability early in life); and

TE_{ij} is the sum of random *environmental* effects which affect only the jth record of animal i and are thus *temporary.*

For traits which can be measured only once, PE_i and TE_{ij} can be considered as E_i—the sum of all environmental effects. Thus,

$$P_i = \mu + G_i + E_i$$

is the model for traits which are measured only once. This model is sufficient for most purposes although there are some exceptions which will be discussed when appropriate.

The problem of selection

The model illustrates the problem of genetic selection. Only the phenotype, P_i, can be observed, yet animals are to be selected that have preferred genetic values, G_i. The environmental effects, E_i, however, prevent direct measurement of G_i. The relative sizes of G_i and E_i, neither of which can be measured, determine how accurately P_i estimates G_i. The key factors that predict how successful selection is likely to be in improving the average genetic value will be discussed in Chapter 13. The use of records of the animal and its relatives to predict breeding value will now be described.

Genetic evaluation for a single trait logically involves records of the animal and its relatives, all properly weighted. Weighting is similar to averaging. Records of animals more closely related to the animal being evaluated receive more weight than records of distantly related animals because closely related animals have more genes and gene effects that are the same. The weights will be given for some simple cases. The general procedure for determining the weights is given in Section 12-11.

Adjustments for identified nonrandom environmental effects

First, and very importantly, before the records are weighted by the amount of genetic information each contributes, the records must be adjusted for all management or other identifiable nongenetic effects that influence the phenotype. In the model,

$$P_i = \mu + G_i + E_i$$

the μ term can be thought of as representing major and identifiable nonrandom environmental effects such as herd management level, the age of the animal when the record was made, the year, and the season of the year when the record was made. In that sense, μ might be different for each animal and each record of each animal. For example, a more complicated model might be

$$P_i = \text{management effect} + \text{sex of animal effect} + \text{age of mother effect} + \text{year effect} + \text{season effect} + G_i + E_i + \mu^*$$

where μ^* now represents the average genetic level of animals plus the average of unidentified environmental effects in the population. Perfect adjustments for such factors by deviating the record from identifiable nongenetic effects will be assumed in this text. The model for an adjusted record is

$$X_i = P_i - \mu = G_i + E_i$$

In terms of the more complicated model the adjusted record is

$$\begin{aligned} X_i &= P_i - \text{management effect} - \text{sex effect} - \text{age of mother effect} \\ &\quad - \text{year effect} - \text{season effect} - \mu^* \\ &= G_i + E_i \end{aligned}$$

12-2 The Selection Index Procedure

The procedure used to find the appropriate weights for records of the animal and its relatives is called the *selection index procedure.* The word "index" refers to the numerical value obtained for each animal which is used to rank (or index) the animal relative to others. Section 12-11 describes the equations used to find the best weights for each of the records. The index, denoted as $\hat{A}_i$ (an estimate of A_i), for the prediction of the additive genetic value for animal i, is

$$\hat{A}_i = b_1X_1 + b_2X_2 + \cdots + b_NX_N$$

where b_1 is the weight for X_1, the record on relative 1; b_2 is the weight for X_2, the record on relative 2; and b_N is the weight for X_N, the record on the Nth relative. Section 12-11 generalizes the procedure so that the X's can be averages of records of any relatives or averages of records of the same kind of relative (for example, half-sib progeny).

What properties should the index that predicts additive genetic value have? Certainly the index should be as similar as possible to the underlying true additive genetic value. Animals also should be ranked as accurately as possible for their unknown breeding values. The average true additive genetic value of animals selected as having the largest estimated breeding values should be as large as possible. These properties of the selection index will now be defined more precisely.

Properties of selection index

1. The selection index maximizes $r_{A,\hat{A}}$, which is the correlation between true A_i and $\hat{A}_i$, the prediction of A_i. Thus, no other evaluation procedure can have a higher correlation with true breeding value. This correlation is often called the *accuracy of evaluation.*
2. The selection index minimizes the average squared prediction error, that is, minimizes the average of all $(A_i - \hat{A}_i)^2$. The difference, $A_i - \hat{A}_i$, is *prediction error.*
3. Because the superiority of the selected group is maximized the genetic gain is faster with this procedure than with any other. The index for all animals must have the same $r_{A,\hat{A}}$ for this property to hold true when all animals have the same kinds of records included in the prediction of breeding value.
4. The probability of correctly ranking pairs of animals for their true breeding values is maximized.
5. Another property, although less important, is that the procedure is unbiased; that is, the average of $A_i - \hat{A}_i$ for all animals is zero. In other words, the average of additive genetic values for all possible animals with the same predicted additive genetic value is that predicted value.

Table 12-1 illustrates property 5 for two samples of simulated records. The simulation illustrates that:

1. Prediction is "on the average." Animals with the same phenotype have different genetic values. With large samples, the average additive genetic value is $h^2(P - \mu)$ and the average environmental effect is $(1 - h^2)(P - \mu)$.
2. Genetic and environmental proportions of phenotypic difference from μ are not h^2 and $1 - h^2$ except on the average; that is, usually $A_i \neq h^2(P_i - \mu)$, and $E_i \neq (1 - h^2)(P_i - \mu)$.
3. There is a range of additive genetic values for animals with the same predicted additive genetic value.

Properties 1 and 2 of the selection index are equivalent and do not require the records to have any special distribution. The third and fourth properties require a joint normal distribution between A_i and $\hat{A}_i$ which is usually close to being true.

Obviously, A_i is never known. However, statistical and quantitative genetic theory, which will not be described here, allows maximizing $r_{A,\hat{A}}$ and minimizing the average of squared prediction errors, as well as calculating the theoretical $r_{A,\hat{A}}$. If it were possible to know A_i so that $r_{A,\hat{A}}$ and the average of

Table 12-1 Illustration of how true additive genetic values vary for animals with the same predicted additive genetic value

Animal*	Simulated record $\mu + (A) + (E) = P$	Prediction error $\hat{A} - (A)$
1	10000 + (925) + (75) = 11000	250 − (925) = −675
2	10000 + (298) + (702) = 11000	250 − (298) = −48
3	10000 + (−343) + (1343) = 11000	250 − (−343) = 593
4	10000 + (536) + (464) = 11000	250 − (536) = −286
5	10000 + (787) + (213) = 11000	250 − (787) = −537
6	10000 + (−270) + (1270) = 11000	250 − (−270) = 520
7	10000 + (−48) + (1048) = 11000	250 − (−48) = 298
8	10000 + (267) + (733) = 11000	250 − (267) = −17
9	10000 + (−192) + (1192) = 11000	250 − (−192) = 442
10	10000 + (675) + (325) = 11000	250 − (675) = −425
	Averages 264 736	(−14)
Animal†		
11	10000 + (192) + (−1192) = 9000	−250 − (192) = −442
12	10000 + (−843) + (−157) = 9000	−250 − (−843) = 593
13	10000 + (−311) + (−689) = 9000	−250 − (−311) = 61
14	10000 + (−389) + (−611) = 9000	−250 − (−389) = 139
15	10000 + (226) + (−1226) = 9000	−250 − (226) = −476
16	10000 + (−298) + (−702) = 9000	−250 − (−298) = 48
17	10000 + (−605) + (−395) = 9000	−250 − (−605) = 355
18	10000 + (10) + (−1010) = 9000	−250 − (10) = −260
19	10000 + (−1097) + (97) = 9000	−250 − (−1097) = 847
20	10000 + (597) + (−1597) = 9000	−250 − (597) = −847
	Averages −252 −748	2

* The records of the first sample of 10 animals are all 11,000. Representative A and E terms were determined by simulation with $\mu = 10{,}000$. With heritability of .25, all have predicted additive genetic value of 250; $\hat{A} = .25(11{,}000 - 10{,}000) = 250$.

† The records of the second sample of 10 animals (animals 11–20) are all 9,000. Representative A and E terms were determined by simulation with $\mu = 10{,}000$. With heritability of .25, all have predicted additive genetic value of −250; $\hat{A} = .25(9{,}000 - 10{,}000) = -250$.

$(A_i - \hat{A})^2$ could be calculated, then the calculated $r_{A,\hat{A}}$ and average $(A_i - \hat{A}_i)^2$ would be the same as the theoretical values if the number of animals were large enough that the sample of A_i's were representative of the whole population.

The accuracy of the evaluation will be indicated for the simpler selection index predictions. When only one weight is needed for the index, $\hat{A}_i = b_1X_1$, and $r_{A,\hat{A}}$ is a simple function of b_1, as will be indicated for the simpler indexes. Table 12-2 gives accuracy values for some more complicated indexes. The calculation of accuracy in general is shown in Section 12-11. Accuracy of evaluation is a component of genetic improvement and will be discussed in detail in Chapter 13. Accuracy of evaluation is also a determinant of the range in which the true breeding value is expected to be found (Section 12-8).

Perfect adjustments for identified nongenetic and management effects are not known in the real world. A *best linear unbiased prediction* (BLUP) procedure can be used to adjust for those effects and to predict additive genetic values simultaneously. In symbolic terms, the practical result of the best linear unbiased prediction procedure is that $\hat{\mu}$, the "best" estimate of the identified nongenetic effects represented by μ, is used rather than μ in adjusting the phenotypic record. For example, if there is only one record on an animal, then $\hat{A}_i = h^2(P_i - \hat{\mu})$ rather than $\hat{A}_i = h^2(P_i - \mu)$, as shown in the next section. This distinction is rather subtle and a complete description of BLUP is beyond the scope of this book. Most of the properties of BLUP, however, are essentially the same as the properties of the simpler selection index procedure which is used in this book.

12-3 Prediction of Transmitting Ability

The next sections will be concerned primarily with prediction of additive genetic value. Many applications of genetic evaluation, however, refer to estimated transmitting ability. *Transmitting ability* is defined as one-half of additive genetic value and corresponds to the average additive effects of the sample half of the genes that are transmitted to the progeny. Transmitting ability, thus, is equivalent to the value of genes carried by an average gamete—sperm or ovum. Estimated transmitting ability (ETA) is the predicted additive genetic value divided by two:

$$\text{ETA}_i = \hat{A}_i/2$$

Accuracy of evaluation is exactly the same for ETA_i as for $\hat{A}_i$ because ranking on ETA_i or $\hat{A}_i$ is the same.

In practical situations breeders often use such terms as *predicted difference* and *sire comparison* for dairy bull evaluation and *expected progeny difference* for beef bull evaluation instead of estimated transmitting ability even though they all are estimates of one-half of additive genetic value.

The general idea of prediction of additive genetic value has now been presented. What remains to be done is to describe specifically the most usual kinds of prediction—from an animal's own records and from records of progeny.

12-4 Prediction of Additive Genetic Value and Producing Ability from Own Records

Breeding value

If only an adjusted single record, X_i, is available, where X_i is the deviation of the phenotypic record, P_i, from the fixed effects denoted by μ, then:

$$\hat{A}_i = h^2(P_i - \mu)$$
$$\hat{A}_i = h^2 X_i$$

In general the weighting factor for prediction is the same as the regression coefficient, which for this case can be written as derived in Chapter 11 as

$$\begin{aligned} b &= \mathrm{Cov}(A_i, X_i)/\mathrm{Var}(X_i) \\ &= [\mathrm{Cov}(A_i, A_i) + \mathrm{Cov}(A_i, E_i)]/\mathrm{Var}(X_i) \\ &= \sigma_A^2/\sigma_P^2 \\ &= h^2 \end{aligned}$$

Thus, when prediction of additive genetic value is from an animal's own record, the weighting factor is heritability of the trait, h^2, as illustrated in Figure 12-1.

If $\overline{X}_{i,n}$ is the average of n adjusted records on animal i (the bar over the X indicates average, the first subscript identifies the animal, and the second subscript the number of records), a similar but more complicated derivation results in the weighting factor for the average of the n records:

$$b = \frac{\mathrm{Cov}(A_i, \text{average of } n \text{ records})}{\mathrm{Var}(\text{average of } n \text{ records})}$$

After considerable algebra,

$$b = \frac{nh^2}{1 + (n-1)r}$$

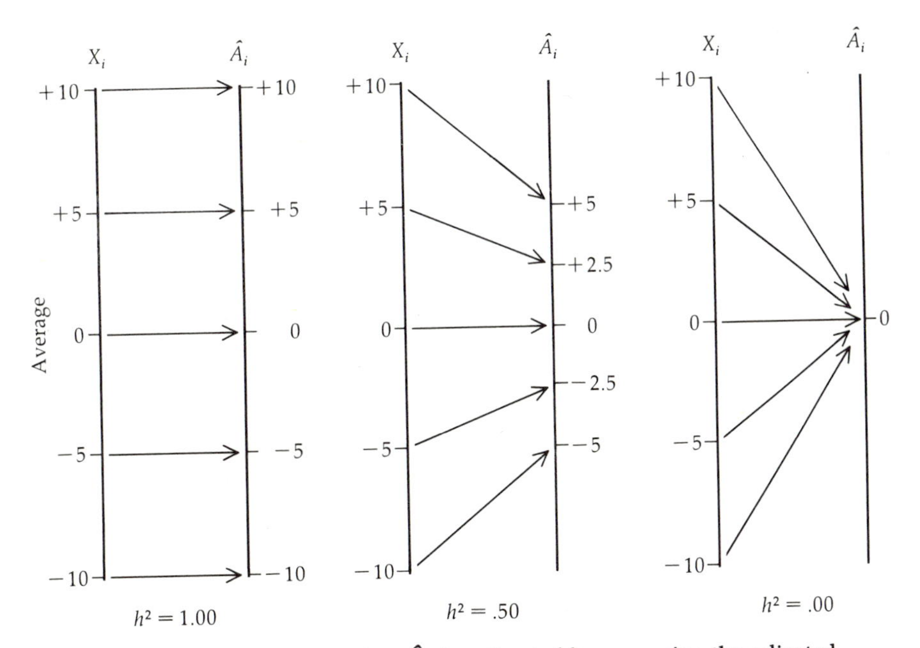

Figure 12-1 Additive genetic value, $\hat{A}_i$, is estimated by regressing the adjusted phenotypic record towards the adjusted phenotypic average of zero. The regression factor is heritability when prediction is from a single record of the animal. Note that the adjusted phenotypic record is the difference from average, $X_i = P_i - \mu$.

so that

$$\hat{A}_i = b\overline{X}_{i,n}$$

Now the weight, b, depends on heritability, repeatability (r, which was defined in Chapter 11), and number of records. Note that the symbol for repeatability, r, has no subscripts and should not be confused with the symbol for the correlation coefficient which always has two subscripts: for example, $r_{A,\hat{A}}$ is the correlation between additive genetic value and predicted additive genetic value.

The formula for the accuracy of evaluation is shown in Section 12-11 or can be derived from the definition of a correlation coefficient and application of the rules for variances and covariances of sums (Chapter 11). The derivation is somewhat complicated, but when prediction of additive genetic value is from the animal's own records, the surprising result is that accuracy is the square root of the weighting factor; that is,

$$r_{A,\hat{A}} = \sqrt{b} = \sqrt{nh^2/[1 + (n-1)r]}$$

Producing ability

Selection for breeding value is for animals which will transmit the greatest genetic superiority to their progeny. Sometimes, for traits such as milk yield, the breeder may want to select animals which will have the largest future records rather than the best progeny. Estimating a future record is the same as estimating breeding value *except* when the animal has previous records. Prediction of a future record as a difference from management level actually is for $G_i + PE_i$, the effects in the model which are repeated in every record of an animal, rather than for A_i alone. Because the permanent environmental effect, PE_i, has been expressed in any previous records of the animal, the previous records aid in predicting $G_i + PE_i$. The combined term, $G_i + PE_i$, has many names, including *producing ability, real producing ability, most probable producing ability,* and the *animal effect* (the part of the model which is expressed in every record of the animal). This text will use the term producing ability with the symbol, PA_i.

The prediction of producing ability is similar to that for A_i from $\overline{X}_{i,n}$, the average of n records on animal i, except that repeatability replaces heritability in the numerator of the weight which reflects the covariance between the previous records and producing ability:

$$b = \frac{nr}{1 + (n-1)r}$$

and

$$\widehat{PA}_i = b\overline{X}_{i,n}$$

When $n = 1$ (one previous record), then the weight reduces to r and

$$\widehat{PA}_i = r\overline{X}_{i,1}$$

Because PA_i predicts the difference from the management level in which the future record will be made, the future record can be predicted as

$$\hat{R}_i = M + \widehat{PA}_i$$

where M is the expected management level.

The accuracy of predicting PA_i is

$$\sqrt{b} = \sqrt{\frac{nr}{1 + (n-1)r}}$$

The numerical computations for estimating additive genetic value, producing ability and a future record are illustrated in Example 12-1.

Example 12-1

Following is a prediction of the additive genetic value, producing ability, and a future record from a record or the average of records on the animal to be evaluated. Units of measurement of the trait will not be repeated in the examples. If it makes the example easier, consider the trait as measured in pounds. Variances will be pounds squared.

Assume:

heritability, $h^2 = .25$
repeatability, $r = .50$
phenotypic variance, $\sigma_P^2 = 4{,}000{,}000$, so that
additive genetic variance, $\sigma_A^2 = h^2\sigma_P^2 = 1{,}000{,}000$
management levels
1. $M = 14{,}000$ where records made
2. $M = 16{,}000$ where future records will be made

The following two animals are to be evaluated:

Animal 1 has one record, $P_{1,1} = 15{,}000$.
Animal 2 has three records averaging, $\overline{P}_{2,3} = 14{,}800$.

Breeding value, A, producing ability, PA, and a future record, R, are to be predicted for each.

Evaluation of animal 1, with own record, $P_{1,1} = 15{,}000$:

$$\text{Adjusted record, } X_{1,1} = 15{,}000 - 14{,}000 = 1{,}000$$
$$\hat{A}_1 = h^2 X_{1,1} = .25(1000) = 250$$
$$PA_1 = \{nr/[1 + (n-1)r]\}\,(X_{1,1}) = .50(1000) = 500$$
$$\hat{R}_1 = M + PA_1 = 16{,}000 + 500 = 16{,}500$$

Evaluation of animal 2, with average of 3 records, $\overline{P}_{2,3} = 14{,}800$:

$$\text{Adjusted average, } \overline{X}_{2,3} = 14{,}800 - 14{,}000 = 800$$
$$\hat{A}_2 = \{nh^2/[1 + (n-1)r]\}\,(\overline{X}_{2,3}) = .375(800) = 300$$
$$PA_2 = \{nr/[1 + (n-1)r]\}\,(\overline{X}_{2,3}) = .75(800) = 600$$
$$\hat{R}_2 = M + PA_2 = 16{,}000 + 600 = 16{,}600$$

This example shows the effect of repeatability and number of records both in the prediction of additive genetic value and in prediction of producing ability. Animal 1 with one record has a larger average record (average of one record) than animal 2 (with 3 records), but because of the influence of the number of records and repeatability on the weighting factors, the weights for animal 2 are larger and more than compensate for the slightly larger record of animal 1.

12-5 Prediction of Additive Genetic Value from Progeny Records

In many species, particularly for those traits where records can be obtained only on females, the breeder obtains a progeny test (also called a progeny proof) to predict the additive genetic value of the sire. The following formula corresponds to evaluation of a sire from progeny related only through him (paternal half-sib progeny). As before, the weight is the covariance between the sire's additive genetic value and the average of records of his progeny divided by the variance of the progeny average. The algebra is complicated but the final result is as follows. If $\overline{X}_{i,p}$ is the average of single records on each of p progeny of sire i,

$$\hat{A}_i = b\overline{X}_{i,p}$$

where

$$b = \frac{2ph^2}{4 + (p-1)h^2}$$

or, equivalently, after dividing numerator and denominator by h^2,

$$b = \frac{2p}{p + (4 - h^2)/h^2}$$

The last form of the weight is often convenient to use because the ratio $(4 - h^2)/h^2$ is a constant for a particular trait. The weighting factor depends on heritability and the number of progeny. As the number of progeny becomes large, the weight approaches 2.

Accuracy of evaluation is

$$r_{A,\hat{A}} = \sqrt{\frac{p}{p + (4 - h^2)/h^2}}$$
$$= \sqrt{b/2}$$

which becomes nearly perfect (unity) as the number of progeny becomes large. The model for the progeny average indicates why the accuracy would be expected to become nearly perfect. Each progeny record includes a sample half of the additive genetic value of its sire plus a random deviation from the management level made up of other random genetic effects and from random environmental effects. The sire's contribution to the average is common to all progeny, and the random deviations are all different and thus would be expected to average zero if enough were averaged. Therefore, the model for the progeny average is

$$\frac{A_i}{2} + \frac{(\text{sum of other effects for all records})}{p} \xrightarrow{\text{which approaches}} \frac{A_i}{2}$$

as p becomes very large.

Example 12-2 describes the numerical calculations for predicting additive genetic value, shows the correspondence between $\hat{A}_i$ and ETA_i, and also shows that the predicted record of a future progeny of the sire is $\hat{R}_{\text{future progeny}} = M + \hat{A}_{\text{sire}}/2$ when no information is available on the progeny's dam.

Example 12-2

Predict the additive genetic values and transmitting abilities of two beef sires and a record of a future progeny of each sire from the paternal half-sib progeny of each sire.

Assume: $h^2 = .25$, $\sigma_P^2 = 1600$ so that $\sigma_A^2 = 400$, and future management level, $M = 500$.

Sire 3 has 50 progeny that average 450 at weaning with average management level of 400.

Sire 4 has 5 progeny that average 660 with average management level of 600.

The weight for predicting the additive genetic value of a sire from p of his paternal half-sib progeny can be written as

$$b = \frac{2p}{p + (4 - h^2)/h^2}$$

and for $h^2 = .25$

$$b = \frac{2p}{(p+15)}$$

This formula is convenient to use when different sires may have different numbers of progeny.

Evaluation of sire 3 whose progeny average, $\overline{P}_{3,50} = 450$:

$$\text{Adjusted average, } \overline{X}_{3,50} = 450 - 400 = 50$$
$$\hat{A}_3 = [2p/(p+15)](\overline{X}_{3,50}) = (100/65)(50) = 76.9$$
$$\text{ETA}_3 = \hat{A}_3/2 = 76.9/2 = 38.5$$

Prediction of a record of any future progeny of sire 3:

$$\hat{R}_{\text{progeny of 3}} = M + \text{ETA}_3 = 500 + 38.5 = 538.5$$

where $M = 500$ is the estimate of the future management level.

Prediction of the additive genetic value of a progeny when it has no records and when its dam has no records is $\hat{A}_{\text{progeny}} = \hat{A}_{\text{sire}}/2 = \text{ETA}_{\text{sire}}$. The best estimate of a future record is obtained by adding the expected management level:

$$\hat{R}_{\text{progeny}} = M + \text{ETA}_{\text{sire}}$$

Evaluation of sire 4 whose progeny average, $\overline{P}_{4,5} = 660$:

$$\text{Adjusted average, } \overline{X}_{4,5} = 660 - 600 = 60$$
$$\hat{A}_4 = [2p/(p+15)](\overline{X}_{4,5}) = (10/20)(60) = 30.0$$
$$\text{ETA}_4 = \hat{A}_4/2 = 30.0/2 = 15.0$$
$$\hat{R}_{\text{progeny of 4}} = M + \text{ETA}_4 = 500 + 15.0 = 515.0$$

This example shows the importance of adjusting for management levels when records of progeny are made under different management conditions. The progeny of sire 3 averaged considerably less than progeny of sire 4 (450 versus 660), yet the adjusted averages were much more alike (50 versus 60). The importance of the number of progeny and heritability is illustrated by the weight for sire 3 of $\frac{100}{65} = 1.54$ as compared to the weight for sire 4 of $\frac{10}{20} = .50$. The weight of 1.54 for the adjusted average of progeny of sire 3 shows that a selection index weight can be greater than 1. Selection index weights for most other relatives are always less than 1.

12-6 Prediction of Additive Genetic Value from Dam and Sire Evaluation

Prediction of genetic value of an animal with no records is usually from the evaluations of its sire and dam. Because each parent contributes a sample half of its additive genetic value to its progeny,

$$\hat{A}_{\text{progeny}} = (\hat{A}_{\text{sire}} + \hat{A}_{\text{dam}})/2$$

The accuracy of evaluation is complicated to derive but is a simple function of the accuracies of the evaluations of the two parents. In fact, accuracy of evaluation,

$$r_{A_p,\hat{A}_p} = (1/2)\sqrt{r^2_{A_s,\hat{A}_s} + r^2_{A_d,\hat{A}_d}}$$

where $r_{A_s,\hat{A}_s}$ and $r_{A_d,\hat{A}_d}$ are the accuracies for the sire and dam predictions. If only one parent is known, then a zero is usually substituted for the evaluation and accuracy of the other.

When the dam has no evaluation, the predicted additive genetic evaluation of the progeny is the same as the ETA of its sire because a zero must be inserted for the evaluation of the dam:

$$\hat{A}_p = \hat{A}_s/2 + 0/2 = \text{ETA}_{\text{sire}}$$

12-7 Genetic Evaluation with Different Sources of Records

The selection index procedure maximizes the probability of correct ranking of pairs of animals with different kinds of evaluations. For example, an animal indexed from its sire and dam can be compared to an animal indexed from its own records or can be compared to an animal indexed from its progeny records. Table 12-2 summarizes the selection index weights and accuracies of evaluations described so far in this chapter, as well as weights and accuracies for more complex cases. Ranking is not based on accuracy of evaluation. The amount of information in the records used for predicting additive genetic value has already been incorporated into calculating the index weights. The accuracy of evaluation, however, determines the confidence range for the true additive genetic value.

Table 12-2 Weights and accuracy values for predicting additive genetic value from records of various relatives (h^2 is heritability; r is repeatability)

Records	Number of records	Selection index weights (b's)	Accuracy ($r_{A,\hat{A}}$)
Individual	(1)	h^2	$\sqrt{h^2}$
	(n)	$nh^2/[1+(n-1)r]$	$\sqrt{nh^2/[1+(n-1)r]}$
Progeny (p half-sibs)	(p)	$2ph^2/[4+(p-1)h^2]$	$\sqrt{ph^2/[4+(p-1)h^2]}$
Dam or sire or progeny	(1)	$h^2/2$	$\sqrt{h^2}/2$
	(n)	$nh^2/2[1+(n-1)r]$	$\sqrt{nh^2/[1+(n-1)r]}/2$
Sire and dam	(1)	$h^2/2$; $h^2/2$	$.71\sqrt{h^2}$
	(n)	$nh^2/2[1+(n-1)r]$; $nh^2/2[1+(n-1)r]$	$.71\sqrt{nh^2/[1+(n-1)r]}$
One grandparent		$h^2/4$	$\sqrt{h^2}/4$
Four grandparents		All $h^2/4$	$\sqrt{h^2}/2$
One great-grandparent		$h^2/8$	$\sqrt{h^2}/8$
Eight great-grandparents		All $h^2/8$	$.35\sqrt{h^2}$
Individual and one parent or progeny		$[h^2-(h^2/2)^2]/[1-(h^2/2)^2]$; $[h^2(1-h^2)/2]/[1-(h^2/2)^2]$	$\sqrt{(5h^2-2h^4)/(4-h^4)}$
Individual and both parents		$h^2(h^2-2)/(h^4-2)$; $h^2(h^2-1)/(h^4-2)$; . . .	$\sqrt{h^2(2h^2-3)/(h^4-2)}$

Individual and one grandparent or grandprogeny		$h^2(h^2-16)/(h^4-16)$; $4h^2(h^2-1)/(h^4-16)$	$\sqrt{h^2(2h^2-17)/(h^4-16)}$
Individual and four grandparents		$h^2(h^2-4)/(h^4-4)$; $h^2(h^2-1)/(h^4-4)$; . . .	$\sqrt{h^2(2h^2-5)/(h^4-4)}$
Parent and progeny		$2h^2/(4+h^2)$; $2h^2/(4+h^2)$	$\sqrt{2h^2/(4+h^2)}$
{Let $A=[1+(n-1)r]/n$, $D=[1+(p-1)h^2/4]/p$, and $C=AD-(h^4/16)$}			
Individual; Paternal half-sibs	(n) (p)	$[h^2D-(h^2/4)^2]/C$; $h^2(A-h^2)/4C$	$\sqrt{b_1+(b_2/4)}$
Individual; Its paternal half-sib progeny	(n) (p)	$[h^2D-(h^2/2)^2]/[C-(3h^4/16)]$; $h^2(A-h^2)/2[C-(3h^4/16)]$	$\sqrt{b_1+(b_2/2)}$
Dam; Paternal half-sibs	(n) (p)	$nh^2/2[1+(n-1)r]$; $ph^2/[4+(p-1)h^2]$	$\sqrt{(b_1/2)+(b_2/4)}$
Dam; Sire; Progeny	(1) (1) (1)	$[h^2-(h^4/16)]/[2-(h^4/64)]$; $[h^2-(h^4/16)]/[2-(h^4/64)]$; $[h^2-(h^4/8)]/[2-(h^4/64)]$	$\sqrt{(b_1+b_2+b_3)/2}$
Paternal half-sibs; Dam; Dam's paternal half-sibs	(m) (n) (p)	$mh^2/[4+(m-1)h^2]$; $h^2[D-(h^2/16)]/2C$; $h^2(A-h^2)/8C$	$\sqrt{(b_1/4)+(b_2/2)+(b_3/8)}$

12-8 Confidence Ranges on Prediction of Breeding Value

The selection index procedure provides the best, although not perfect, way to rank animals. Culling should be done on the basis of this ranking. Sometimes, however, the breeder may want to know how close the selection index estimate is to the true breeding value. That question cannot be answered exactly because the true breeding value is never known for any animal. However, a similar question can be answered. What is the range that is 80 percent (or some other percent) sure to include the true breeding value when the selection index prediction is $\hat{A}_i$? This range, called a *confidence range,* is usually a symmetrical range—the upper limit of the range is the same amount greater than $\hat{A}_i$ as the lower limit of the range is less than $\hat{A}_i$. The confidence range depends on

1. $\hat{A}_i$, the prediction of A_i,
2. $r_{A,\hat{A}}$, the accuracy of prediction of A_i,
3. $\sigma_A^2 = h^2\sigma_P^2$, the additive genetic variance, where h^2 is heritability, σ_A^2 is the additive genetic variance, and σ_P^2 is the phenotypic variance (recall that by definition $h^2 = \sigma_A^2/\sigma_P^2$ so that $\sigma_A^2 = h^2\sigma_P^2$), and
4. the desired confidence percent which corresponds to a t value from Table 12-3.

The symmetric confidence range on true A_i is from a lower limit of

$$\hat{A}_i - t\,\sqrt{(1 - r_{A,\hat{A}}^2)\sigma_A^2}$$

to an upper limit of

$$\hat{A}_i + t\,\sqrt{(1 - r_{A,\hat{A}}^2)\sigma_A^2}$$

Table 12-3 Values of t which correspond to different symmetrical confidence ranges

Chance of A in the range	t
50%	.67
60%	.84
68%	1.00
70%	1.04
80%	1.28
90%	1.65
99%	2.58

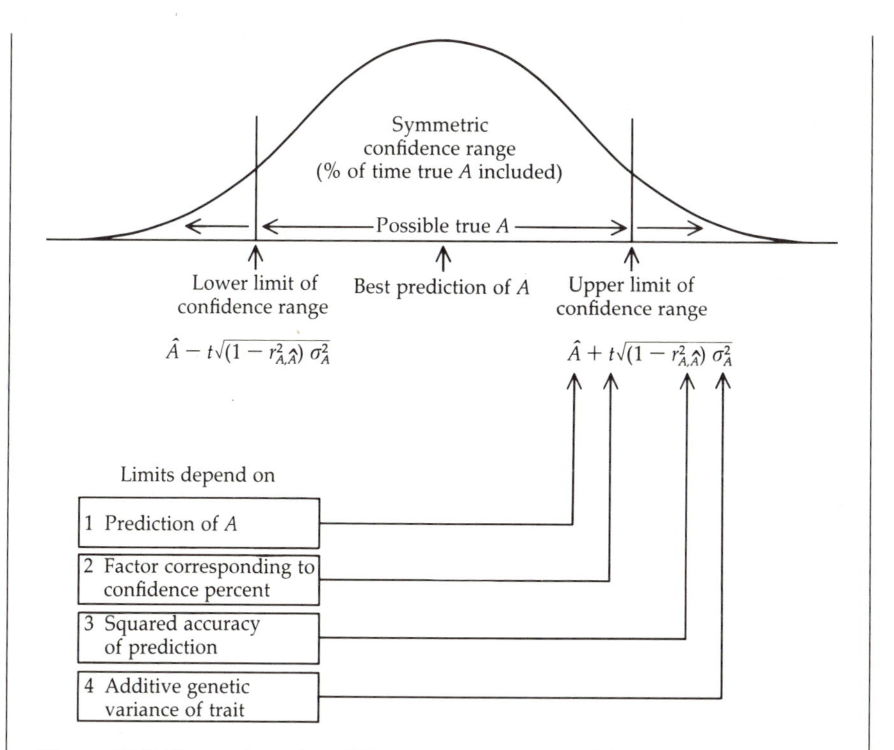

Figure 12-2 Illustration of confidence range on true additive genetic value based on prediction of additive genetic value.

If the range chosen is 80 percent, $t = 1.28$, and there is a 10 percent chance that the true A_i is greater than the upper limit and a 10 percent chance the true A_i is less than the lower limit. Figure 12-2 identifies the factors which determine the confidence range.

Confidence ranges on additive genetic value and producing ability when estimated from an animal's records

The steps in calculating the confidence range for additive genetic value when $\hat{A}_i$ is predicted from $\overline{X}_{i,n}$, the average of n records on animal i are

1. $\hat{A}_i = b\overline{X}_{i,n}$,
2. $r^2_{A,\hat{A}} = b = nh^2/[1 + (n-1)r]$,
3. σ^2_A is known or calculated from known h^2 and σ^2_P as $\sigma^2_A = h^2\sigma^2_P$,
4. look up t from Table 12-3 corresponding to the desired confidence range.

The formula for the confidence range becomes

$$\hat{A}_i + t\sqrt{(1-b)\sigma_A^2}$$

The confidence range for true producing ability is

$$\widehat{PA}_i \pm t\sqrt{(1-b)r\sigma_P^2}$$

where $b = nr/[1+(n-1)r]$ is the weight for predicting producing ability from the record or average of records of the animal, as discussed in Section 12-4. The formula is similar to the one for the confidence range for additive genetic value except that the accuracy squared, b, is for estimating producing ability, and the variance is the variance of producing ability which can be shown to be $r\sigma_P^2$.

Calculations of confidence ranges are illustrated in Example 12-3 for additive genetic value and producing ability estimated from an animal's records.

Example 12-3

What are the 80 percent confidence ranges of additive genetic value and producing ability for the animals evaluated in Example 12-1?

From Example 12-1,

$$\hat{A}_1 = .25(1000) = 250 \text{ with } b_{A,1} = .25 \text{ for } n = 1$$
$$\widehat{PA}_1 = .50(1000) = 500 \text{ with } b_{PA,1} = .50 \text{ for } n = 1$$
$$\hat{A}_2 = .375(800) = 300 \text{ with } b_{A,2} = .375 \text{ for } n = 3$$
$$\widehat{PA}_2 = .75(800) = 600 \text{ with } b_{PA,2} = .75 \text{ for } n = 3$$

In the example,

$$r_{A,\hat{A}}^2 = b_A = nh^2/[1+(n-1)r]$$
$$r_{PA,\widehat{PA}}^2 = b_{PA} = nr/[1+(n-1)r], \text{ with}$$
$$h^2 = .25$$
$$r = .50, \text{ and}$$
$$\sigma_P^2 = 4{,}000{,}000, \text{ so that } \sigma_A^2 = .25(4{,}000{,}000) = 1{,}000{,}000.$$

The 80 percent confidence ranges are:

$$\hat{A}_1 \pm t\sqrt{(1-b_{A,1})\sigma_A^2} = 250 \pm 1.28\sqrt{(1-.25)(1{,}000{,}000)}$$
$$= 250 \pm 1109$$
$$\widehat{PA}_1 \pm t\sqrt{(1-b_{PA,1})r\sigma_P^2} = 500 \pm 1.28\sqrt{(1-.5)(.5)(4{,}000{,}000)}$$
$$= 500 \pm 1280$$
$$\hat{A}_2 \pm t\sqrt{(1-b_{A,2})\sigma_A^2} = 300 \pm 1.28\sqrt{(1-.375)(1{,}000{,}000)}$$
$$= 300 \pm 1012$$
$$\widehat{PA}_2 \pm t\sqrt{(1-b_{PA,2})r\sigma_P^2} = 600 \pm 1.28\sqrt{(1-.75)(.5)(4{,}000{,}000)}$$
$$= 600 \pm 905$$

If other than 80 percent confidence ranges were desired, then the appropriate t value from Table 12-3 would replace the 1.28 used in this example.

The example shows the effect on the confidence ranges of more accurate prediction because of more records. The ranges for animal 2 are smaller than those for animal 1. For both animals the ranges are quite large—much larger than most breeders would expect.

Confidence range on additive genetic value predicted from progeny records

For evaluation of sires from the average of records of p progeny related only through the sire, the weighting factor for the average was determined as

$$b = 2p/[p + (4 - h^2)/h^2]$$

and

$$r^2_{A,\hat{A}} = b/2 = p/[p + (4 - h^2)/h^2]$$

Thus, the confidence range is $\hat{A}_i \pm t\,\sqrt{(1 - b/2)\sigma^2_A}$. Because transmitting ability is an exact one-half of additive genetic value, the estimate of transmitting ability is one-half of predicted additive genetic value. The accuracy of evaluation is the same. Therefore the confidence range for true transmitting ability is exactly one-half of the confidence range for additive genetic value. Confidence ranges based on evaluations from progeny averages are calculated in Example 12-4 for sires evaluated in Example 12-2.

Example 12-4

Calculate the 80 percent confidence ranges for additive genetic value and transmitting ability of sires 3 and 4 that were evaluated in Example 12-2.

From Example 12-2, $h^2 = .25$ and $\sigma^2_P = 1600$, so that $\sigma^2_A = .25(1600) = 400$.

For sire 3 with $p = 50$ progeny,

$$\hat{A}_3 = (100/65)(50) = 76.9 \text{ with ETA}_3 = 76.9/2 = 38.5$$

For sire 4 with $p = 10$ progeny,

$$\hat{A}_4 = (20/10)(60) = 30.0 \text{ with ETA}_4 = 15.0$$

The 80 percent confidence ranges are:

$$\hat{A}_3 \pm t\sqrt{(1 - b_3/2)\sigma_A^2} = 76.9 \pm 1.28\ \sqrt{(1 - 50/65)(400)}$$
$$= 76.9 \pm 12.3$$
$$\text{ETA}_3 \pm (1/2)t\sqrt{(1 - b_3/2)\sigma_A^2} = 38.5 \pm (.5)(1.28)\sqrt{(1 - 50/65)(400)}$$
$$= 38.5 \pm 6.1$$
$$\hat{A}_4 \pm t\sqrt{(1 - b_4/2)\sigma_A^2} = 30.0 \pm 1.28\sqrt{(1 - 5/20)(400)}$$
$$= 30.0 \pm 22.2$$
$$\text{ETA}_4 \pm (1/2)t\sqrt{(1 - b_4/2)\sigma_A^2} = 15.0 \pm (.5)(1.28)\sqrt{(1 - 5/20)(400)}$$
$$= 15.0 \pm 11.1$$

The example shows that a relatively large number of progeny results in a small confidence range. The 80 percent range with 50 progeny is ± 12.3, whereas with 10 progeny the range is nearly twice as large, ± 22.2. In fact as the number of progeny becomes large, $r^2_{A,\hat{A}}$ becomes close to 1, and the range for any confidence level shrinks to very near the prediction of additive genetic value.

Confidence range on additive genetic value for predictions based on sire's and dam's evaluations

Earlier in this chapter the accuracy of evaluating an animal with no records from the evaluation of its sire and dam was given as

$$r_{A_p,\hat{A}_p} = (1/2)\ \sqrt{r^2_{A_s,\hat{A}_s} + r^2_{A_d,\hat{A}_d}}$$

where $r_{A_s,\hat{A}_s}$ and $r_{A_d,\hat{A}_d}$ are the accuracies for the sire and dam predictions. If only one parent is known, then a zero is usually substituted for the accuracy of the unknown parent.

The confidence range is

$$\hat{A}_p \pm \text{t}\ \sqrt{(1 - r^2_{A_p,\hat{A}_p})\sigma_A^2}$$

When, for example, the dam has no evaluation, the predicted additive genetic evaluation of the progeny is the same as the ETA of its sire because a zero must be inserted for the evaluation of the dam.

$$\hat{A}_p = \hat{A}_s/2 + 0/2 = \text{ETA}_{\text{sire}}$$

The confidence range for A_{progeny} when $\hat{A}_{\text{progeny}}$ is based only on ETA_{sire}, however, is much greater than the confidence range on the ETA of the sire as can be seen by comparing the two confidence ranges after applying some simple algebra. Because

$$r^2_{A_p,\hat{A}_p} = \text{r}^2_{A_s,\hat{A}_s}/4$$

with

$$r^2_{A_s,\hat{A}_s} = p/[p + (4 - h^2)/h^2)]$$

the confidence range for the progeny is

$$\hat{A}_p \pm t\sqrt{\{1 - (1/4)p/[p + (4 - h^2)/h^2]\}\sigma_A^2}$$

whereas the confidence range for the sire's transmitting ability is

$$\text{ETA}_{\text{sire}} \pm (.5)t\sqrt{\{1 - p/[p + (4 - h^2)/h^2]\}\sigma_A^2}$$

Consequently, the confidence range may be much larger, as shown by the 80 percent range of 38.5 ± 23.0 for the additive genetic value of a progeny of sire 3 from Example 12-2 when compared to the range on the transmitting ability of sire 3 which was 38.5 ± 6.1. Even with a perfect accuracy of 1.00 for sire 3, the 80 percent confidence range of the additive genetic value of a progeny would be $38.5 \pm 1.28\sqrt{[1 - (1/4)(1)]400} = 38.5 \pm 22.2$.

12-9 Prediction of Additive Genetic Value with Inbreeding

Inbreeding may affect the prediction of genetic value in two ways. The genetic variance associated with records of inbred animals increases by $F\sigma_A^2$, which also increases the phenotypic variance by the same amount. Inbreeding may also result in a larger covariance between records of relatives. Pairs of progeny of inbred parents are more alike than pairs of progeny of noninbred parents. Furthermore, a progeny is more like an inbred parent than a progeny is like a noninbred parent. For example, if a sire is inbred, with inbreeding coefficient F_s, then the additive relationship between him and his progeny is $(1 + F_s)/2$ rather than $\frac{1}{2}$ for a noninbred sire. His progeny (assuming all have different noninbred dams) are related to each other by $(1 + F_s)/4$ rather than by $\frac{1}{4}$. Because additive relationships measure genetic effects in common, an inbred parent has more genes and gene effects in common with its progeny than does a noninbred parent.

The additional genetic effects in common should be considered in genetic evaluation. Suppose an inbred sire, i, is evaluated from the average, $\overline{X}_{i,p}$, of records of p of his progeny (all by different unrelated and noninbred dams). If F_i is his inbreeding coefficient, then the weight and index become

$$b = \left(\frac{2p}{p + \dfrac{4 - (1 + F_i)h^2}{(1 + F_i)h^2}} \right)$$

and

$$\hat{A}_i = b\overline{X}_{i,p}$$

The weight with no inbreeding ($F_i = 0$) is

$$b = \frac{2p}{p + (4 - h^2)/h^2}$$

The accuracy of evaluation is increased because the progeny have more genetic effects in common with their inbred sire than if their sire were not inbred. Thus their records are more informative. The formula for the accuracy of evaluation changes to reflect the additional information:

$$r_{A,\hat{A}} = \sqrt{\frac{p}{p + \dfrac{4 - (1 + F_i)h^2}{(1 + F_i)h^2}}}$$

rather than

$$r_{A,\hat{A}} = \sqrt{\frac{p}{p + (4 - h^2)/h^2}}$$

With more and more progeny, accuracy of evaluation becomes closer and closer to 1.00 whether or not the sire is inbred. If the sire is inbred, the accuracy is larger for the same number of progeny than if the sire were not inbred. For example, in the unlikely case the sire has $F_i = 1$, the accuracy of evaluation becomes

$$r_{A,\hat{A}} = \sqrt{\frac{p}{p + (2 - h^2)/h^2}}$$

The effect of the sire's inbreeding coefficient on accuracy of evaluation is shown in Table 12-4 for heritability of .25.

The influence of the inbreeding coefficient on predicting the additive genetic value of an animal from its own records can be illustrated easily when only one record is available. Earlier the weighting factor for such a case was defined as

$$b = \mathrm{Cov}(A_i, X_i)/\mathrm{Var}(X_i)$$

Table 12-4 Accuracy of evaluation from progeny records for inbred sires when heritability is .25

	Accuracy of evaluation when sire's inbreeding coefficient is:			
Number of progeny	.00	.25	.50	1.0
1	.25	.28	.31	.35
5	.50	.55	.58	.65
50	.88	.90	.92	.94
1000	.993	.994	.995	.997

For a noninbred animal $\text{Cov}(A_i, X_i) = \sigma_A^2$. In general, the covariance is $a_{ii}\sigma_A^2$ where a_{ii} is the additive relationship of the animal to itself. For an inbred animal, $a_{ii} = 1 + F_i$, as was developed in Chapter 10. Thus $\text{Cov}(A_i, X_i) = (1 + F_i)\,\sigma_A^2$. As stated earlier in this section, the genetic and phenotypic variances both increase with inbreeding by $F_i\sigma_A^2$. Thus, with inbreeding, $\text{Var}(X_i) = (1 + F_i)\sigma_A^2 + \sigma_E^2$, which can be rewritten as $F_i\sigma_A^2 + \sigma_P^2$. Because $\sigma_A^2 = h^2\sigma_P^2$, then

$$\begin{aligned}\text{Var}(X_i) &= F_ih^2\sigma_P^2 + \sigma_P^2 \\ &= (1 + F_ih^2)\sigma_P^2\end{aligned}$$

After performing the algebra the formula for the weighting factor becomes

$$\begin{aligned}b &= \frac{(1 + F_i)\sigma_A^2}{(1 + F_ih^2)\sigma_P^2} \\ &= \frac{(1 + F_i)h^2\sigma_P^2}{(1 + F_ih^2)\sigma_P^2} \\ &= \frac{(1 + F_i)h^2}{1 + F_ih^2}\end{aligned}$$

When $F_i = 0$,

$$b = h^2$$

as stated in an earlier section.

For evaluating an inbred animal, i, on the average, $\overline{X}_{i,n}$, of n of its records,

$$\hat{A}_i = \left(\frac{n(1 + F_i)h^2}{1 + (n - 1)r + F_ih^2}\right)\overline{X}_{i,n}$$

with

$$r_{A,\hat{A}} = \sqrt{\frac{n(1 + F_i)h^2}{1 + (n-1)r + F_i h^2}}$$

These equations are the same as for zero inbreeding except for the terms $F_i h^2$ in both the numerator and denominator. The confidence ranges either for evaluations from progeny or from an animal's own record are also modified with $(1 + F_i)\sigma_A^2$ replacing σ_A^2:

$$\hat{A}_i \pm t\sqrt{(1 - r_{A,\hat{A}}^2)(1 + F_i)\sigma_A^2}$$

12-10 Prediction of Additive Genetic Value with Environmental Correlations Between Relatives

Records of members of the same family, for example, littermates, may be more similar than can be accounted for by genes in common. A measure of this nongenetic likeness is called the *environmental correlation* and is assigned the symbol c_{FAM}^2, with the subscript identifying the kind of family relationship. The covariance due to the nongenetic likeness is defined as $c_{\mathrm{FAM}}^2\sigma_{\mathrm{P}}^2$. The square in the symbol for the environmental correlation is to emphasize that, because of how it is defined, it can never be negative. The environmental correlation reduces the weight given to records of family members and also decreases accuracy of prediction because of the confounding of genetic and environmental likeness in the records which makes the family information less informative.

Environmental correlations different from zero may exist between records of any kind of relatives. They may exist for certain kinds of relatives in some situations and not in others. For example, half-sib progeny of a sire when all are in the same herd may be treated more alike than random animals in the herd with a resulting nonzero environmental correlation. If the same half-sib progeny are spread out over many herds, with only one per herd, then an environmental correlation is unlikely. The following specific cases will assume the magnitude of the environmental correlation is the same for all possible relative pairs of the same kind.

Effect of environmental correlation on prediction from progeny records

As stated earlier, the progeny of a sire are assumed to have different mothers but to have a common environmental correlation, c^2_{PHS}, for any pair of paternal half-sibs (PHS). The average of adjusted single records of the p progeny of sire i is $\overline{X}_{i,p}$. Environmental correlations affect the variance of $\overline{X}_{i,p}$ but not the covariance between A_i and $\overline{X}_{i,p}$. The result is that the selection index weight for $\overline{X}_{i,p}$ becomes

$$b = \frac{2ph^2}{4 + (p-1)(h^2 + 4c^2_{PHS})}$$

When $c^2_{PHS} = 0$, the weight reduces to that given in Section 12-5. The effect of the positive term, $4c^2_{PHS}$, in the denominator is to make the weight smaller. As before, $r_{A,\hat{A}} = \sqrt{b/2}$, where now $r_{A,\hat{A}}$ is also smaller because of $c^2_{PHS} > 0$. Also as before, the confidence range is

$$\hat{A}_i \pm t\sqrt{(1 - r^2_{A,\hat{A}})\sigma^2_A}$$

Table 12-5 illustrates the effect of the environmental correlation on the weighting factor and on accuracy of evaluation for a situation with $h^2 = .25$ and $c^2_{PHS} = .0625$. With only a single progeny, the weights and accuracies are the same because the environmental correlation can exist only between pairs of records. What is important to notice is that the limits for both the weight and accuracy when p becomes large are greatly different with and without an environmental correlation. The accuracy of evaluation with an environmental correlation cannot reach 100 per-

Table 12-5 Illustration of the effect of an environmental correlation between paternal half-sibs on predicting additive genetic value of their sire: $h^2 = .25$ and $c^2_{PHS} = .0625$

Number of progeny	When $c^2_{PHS} = 0$		When $c^2_{PHS} = .0625$	
	b	$r_{A,\hat{A}}$	b	$r_{A,\hat{A}}$
1	.12	.25	.12	.25
3	.33	.41	.30	.39
10	.80	.62	.59	.54
50	1.54	.88	.88	.66
100	1.73	.93	.93	.69
1000	1.98	.99	.99	.70
∞	2.00	1.00	1.00	.71

cent even with an infinite number of progeny. The actual limit depends on h^2 and c^2_{PHS}.

Sib evaluation

When evaluation is based on records of an animal's littermates, the possibility of an environmental correlation must be considered. The estimated breeding value of animal i from the average of records of p full sibs, $\overline{X}_{i,p}$, is

$$\hat{A}_i = \left(\frac{ph^2}{2 + (p-1)(h^2 + 2c^2_{\text{FS}})}\right)\overline{X}_{i,p}$$

If records of p half-sibs of animal i are used for genetic evaluation, then

$$\hat{A}_i = \left(\frac{ph^2}{4 + (p-1)(h^2 + 4c^2_{\text{HS}})}\right)\overline{X}_{i,p}$$

The environmental correlations probably are not the same for paternal half-sibs as for maternal half-sibs.

Family evaluation

In some situations whole family groups might be compared with the best families selected based on their family average. An example would be that of selection for feed efficiency in poultry where the birds are fed as a group. The group is likely to be a "family" where the family members are full sibs or paternal half-sibs. If the average of single records of the p members of the family is $\overline{X}_{i,p}$, a_{FAM} is the additive relationship between family members, and c^2_{FAM} is the environmental correlation between family members, then the prediction of the average additive genetic value of the family group is

$$\hat{A}_{\text{FAM}} = \left(\frac{[1 + (p-1)a_{\text{FAM}}]h^2}{1 + (p-1)(a_{\text{FAM}}h^2 + c^2_{\text{FAM}})}\right)\overline{X}_{i,p}$$

Example 12-5 illustrates the calculations for predicting the additive genetic value of a family and for predicting the additive genetic value of an animal without a record from the average of its sibs.

Example 12-5 Prediction of additive genetic value for a family from the family average

Assume:

heritability, $h^2 = .20$
environmental correlation, $c^2_{FS} = .05$
phenotypic variance, $\sigma^2_P = 25$
genetic variance, $\sigma^2_A = 5$
management level where records made = 8

Two full sib families ($a_{FS} = \frac{1}{2}$) are to be evaluated for genetic value:

Family 1: 5 full sibs average, $\overline{P}_{1,5} = 10$; adjusted average, $\overline{X}_{1,5} = 10 - 8 = 2$.
Family 2: 3 full sibs average, $\overline{P}_{2,3} = 11$; adjusted average, $\overline{X}_{2,3} = 11 - 8 = 3$.

$$\hat{A}_{FAM} = \left(\frac{[1 + (p - 1)(a_{FS})](h^2)}{1 + (p - 1)(a_{FS}h^2 + c^2_{FS})}\right)\overline{X}_{i,p}$$

$$\hat{A}_{FAM_1} = \left(\frac{[1 + 4(1/2)](.2)}{1 + 4[(1/2)(.2) + .05]}\right)2 = .375(2) = .750$$

$$\hat{A}_{FAM_2} = \left(\frac{[1 + 2(1/2)](.2)}{1 + 2[(1/2)(.2) + .05]}\right)3 = .308(3) = .924$$

The additive genetic values of a full sib member of family 1 and a full sib member of family 2 are to be predicted. Neither has a record.

The adjusted averages are the same as in the first part of this example:

$$\overline{X}_{1,5} = 2 \text{ and } \overline{X}_{2,3} = 3$$

In general, to predict additive genetic value of another animal from its full sib average,

$$\hat{A}_i = \left(\frac{ph^2}{2 + (p - 1)(h^2 + 2c^2_{FS})}\right)\overline{X}_{i,p}$$

therefore

$$\hat{A}_1 = \left(\frac{5(.20)}{2 + 4[.20 + 2(.05)]}\right)2 = (.3125)(2) = .625$$

$$\hat{A}_2 = \left(\frac{3(.20)}{2 + 2[.20 + 2(.05)]}\right)3 = (.2308)(3) = .692$$

12-11 Selection Index with All Relatives' Records

Equations to determine weights

The general procedure for obtaining the weights for many different relatives with records in the selection index is beyond the scope of this book. A somewhat simplified version is included in this section for those who may want to develop weights for indexes which are different from those in Table 12-2.

Goal: Predict additive genetic value for animal α, that is, predict A_α.

Information available: $X_1, \ldots, X_N$, where X_i ($i = 1, \ldots, N$) is the average of n_i records on each of p_i animals all related to each other with the same additive relationship, $a_{ii'}$, and all related to α with the same additive relationship, $a_{i\alpha}$. Animals in group i with records contributing to the average, X_i, all have the same additive relationship to animals in group j with records contributing to the average, X_j, that is, a_{ij}.

Known: heritability, h^2; repeatability, r; and phenotypic variance, σ_P^2.

Index will be: $I_\alpha = \hat{A}_\alpha = b_1X_1 + b_2X_2 + \cdots + b_NX_N$.

Problem: Find the b's which are the best weights for the X's.

Solution: The selection index equations which define the b's (obtained from minimizing prediction error squared or maximizing $r_{A,\hat{A}}$) are in general:

$$\begin{array}{l}
\mathrm{Var}(X_1)b_1 + \mathrm{Cov}(X_1, X_2)b_2 + \cdots + \mathrm{Cov}(X_1, X_N)b_N = \mathrm{Cov}(X_1, A_\alpha) \\
\mathrm{Cov}(X_1, X_2)\,b_1 + \mathrm{Var}(X_2)b_2 + \cdots + \mathrm{Cov}(X_2, X_N)b_N = \mathrm{Cov}(X_2, A_\alpha) \\
\quad\vdots \qquad\qquad\qquad \vdots \qquad\qquad\qquad \vdots \qquad\qquad \vdots \\
\mathrm{Cov}(X_1, X_N)b_1 + \mathrm{Cov}(X_2, X_N)b_2 + \cdots + \mathrm{Var}(X_N)b_N = \mathrm{Cov}(X_N, A_\alpha)
\end{array}$$

The variances of the X's, or $\mathrm{Var}(X_i)$, covariances among the X's, or $\mathrm{Cov}(X_i, X_j)$, and covariances of the X's and A_α, or $\mathrm{Cov}\ (X_i, A_\alpha)$, are numerical values and thus, when the N equations with the N unknown b's are solved the selection index weights are obtained. Although these look complicated, examples with only X_1 and X_2 illustrate the procedure.

Simplified selection index equations

In general, the X's may be averages of records as well as single records, which makes the numerical determination of the variances and covariances somewhat complicated. With a few, usually noncritical, assumptions, the equations can be simplified to expressions involving h^2, r, n_i, and p_i, with a constant multiplier, σ_P^2, the phenotypic variance of the trait.

For example, the simplified selection index equations for $N = 2$ are

$$\begin{aligned} d_1\sigma_P^2 b_1 + a_{12}h^2\sigma_P^2 b_2 &= a_{1\alpha}h^2\sigma_P^2 \\ a_{12}h^2\sigma_P^2 b_1 + \quad d_2\sigma_P^2 b_2 &= a_{2\alpha}h^2\sigma_P^2 \end{aligned}$$

The diagonal coefficients are

$$d_i = \frac{\dfrac{1 + (n_i - 1)r}{n_i} + (p_i - 1)a_{ii'}h^2 + F_i h^2}{p_i}$$

The term $F_i h^2$ accounts for the added genetic variance if the animals with records are inbred where F_i is the inbreeding coefficient of each animal in group i.

When $n_i = 1$ (1 record per animal) and $F_i = 0$,

$$d_i = \frac{1 + (p_i - 1)a_{ii'}h^2}{p_i}$$

When $p_i = 1$ (1 animal in group) and $F_i = 0$,

$$d_i = \frac{1 + (n_i - 1)r}{n_i}$$

When n_i and $p_i = 1$ (1 animal in group with 1 record, that is, an animal with a single record) and $F_i = 0$, $d_i = 1$. Each diagonal coefficient, $d_i\sigma_P^2$, corresponds to Var(X_i), the variance of the average represented by X_i.

Each off-diagonal coefficient, $a_{ij}h^2\sigma_P^2$, is the covariance due to additive genetic effects in common between records of relatives i and j and does not depend on number of records or animals in either group. Each right-hand side is the additive genetic covariance between a representative record in group i and the additive genetic value of the animal being evaluated, A_α.

Because every term contains σ_P^2, σ_P^2 can be divided out of each equation and thus is not needed to find the b's.

All of the coefficients of the b's will be numerical and not symbolic because numerical values are substituted for the relationships, the p_i and n_i, and h^2 and r. The equations are then solved (possibly with a computer if N is very large) for the numerical weights (b's).

Accuracy of prediction

The accuracy of evaluation, $r_{A,\hat{A}}$, can be calculated after the b's are found as the square root of the sum of products of the b's and the corresponding right-hand sides of the selection index equations, all divided by the genetic variance, $\sigma_A^2 = h^2\sigma_P^2$:

$$r_{A,\hat{A}} = \sqrt{\frac{b_1a_{1\alpha}h^2\sigma_P^2 + b_2a_{2\alpha}h^2\sigma_P^2 + \cdots + b_Na_{N\alpha}h^2\sigma_P^2}{h^2\sigma_P^2}}$$

which simplifies to

$$r_{A,\hat{A}} = \sqrt{b_1a_{1\alpha} + b_2a_{2\alpha} + \cdots + b_Na_{N\alpha}}$$

because $h^2\sigma_P^2$ appears in all numerator terms and in the denominator. Although not obvious, the numerical limits for $r_{A,\hat{A}}$ are zero to 1. The term inside the square root operator, $r_{A,\hat{A}}^2$, cannot be negative or greater than 1.

The *final and most important step* in the selection index procedure is to apply the weights to the records to obtain the numerical genetic evaluation:

$$\hat{A}_\alpha = b_1X_1 + \cdots + b_NX_N$$

where now the b's and records are all numerical. If several animals have exactly the same kinds of records available, then the weights will be the same for the evaluation of each animal.

Assumptions made to obtain the simplified equations which are used to find the selection index weights (b's) to predict additive genetic value are:

1. selection is for additive genetic value,
2. only additive genetic effects contribute to likeness between relatives (not likely to be an important error in most cases even if not true), and
3. there are no environmental causes of likeness between relatives (for some situations, particularly with litter-bearing species, this may not be true). The equations can be modified by adding the appropriate measures of environmental likeness, c_{ij}^2 and $c_{ii'}^2$, to the $a_{ij}h^2$ and $a_{ii'}h^2$ coefficients on the left of the equal signs of the simplified selection index equations; for example, the coefficient of b_2 in the first equation would be $(a_{12}h^2 + c_{12}^2)\sigma_P^2$ and $(p_i - 1)c_{ii'}^2$ would be added to the numerator of d_i. The coefficients on the right of the equals sign would not change.

How to set up the equations when these assumptions are not true is discussed in detail in Van Vleck (1983). A numerical example for a case with more than three kinds of records is given in Example 12-6.

Selection for producing ability

If selection is for $G_i + PE_i$ (producing ability), the simplified equations can be modified easily:

1. If the animal being evaluated has no records, the equations are exactly the same.
2. If the animal being evaluated has records which are included in the average X_i, then the right-hand side of the ith equation becomes $r\sigma_P^2$ rather than $h^2\sigma_P^2$. Only the right-hand side of one equation, the equation associated with the animal being evaluated, is modified.

The accuracy of predicting producing ability is calculated as the square root of the sum of products of the weights and the corresponding right-hand sides, but is divided by $r\sigma_P^2$, the variance of producing ability, rather than by $h^2\sigma_P^2$, the additive genetic variance.

Example 12-6 Selection index equations to find weights for records of many relatives

Assume: $h^2 = .40$, $r = .60$, and the phenotypic variance, $\sigma_P^2 = 400$.

Selection for additive genetic value The records available are:

X_1, average of 2 records of animal (α) to be evaluated;
X_2, record of sire;
X_3, record of dam; and
X_4, average of 3 records on each of 40 paternal half-sib progeny.
Draw a diagram to describe relationships needed:

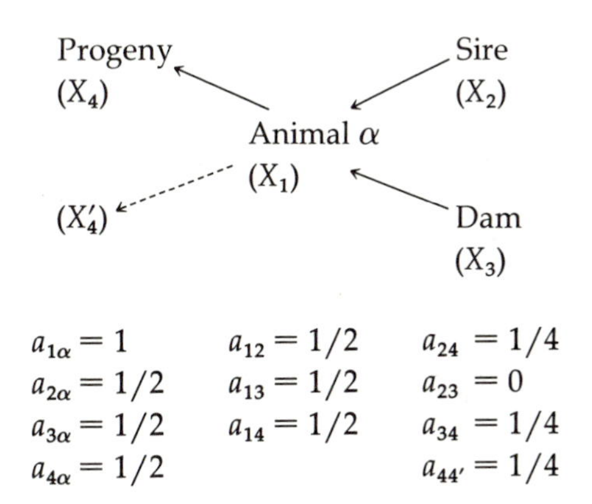

$a_{1\alpha} = 1$	$a_{12} = 1/2$	$a_{24} = 1/4$
$a_{2\alpha} = 1/2$	$a_{13} = 1/2$	$a_{23} = 0$
$a_{3\alpha} = 1/2$	$a_{14} = 1/2$	$a_{34} = 1/4$
$a_{4\alpha} = 1/2$		$a_{44'} = 1/4$

(X_4 and X_4' represent any pair of animals in group 4 for purposes of calculating $a_{44'}$.)

Symbolic selection index equations which determine the selection index weights are:

$$\begin{aligned}
(d_1)\sigma_P^2 b_1 + a_{12}h^2\sigma_P^2 b_2 + a_{13}h^2\sigma_P^2 b_3 + a_{14}h^2\sigma_P^2 b_4 &= a_{1\alpha}h^2\sigma_P^2 \\
a_{12}h^2\sigma_P^2 b_1 + \sigma_P^2 b_2 + a_{23}h^2\sigma_P^2 b_3 + a_{24}h^2\sigma_P^2 b_4 &= a_{2\alpha}h^2\sigma_P^2 \\
a_{13}h^2\sigma_P^2 b_1 + a_{23}h^2\sigma_P^2 b_2 + \sigma_P^2 b_3 + a_{34}h^2\sigma_P^2 b_4 &= a_{3\alpha}h^2\sigma_P^2 \\
a_{14}h^2\sigma_P^2 b_1 + a_{24}h^2\sigma_P^2 b_2 + a_{34}h^2\sigma_P^2 b_3 + (d_4)\sigma_P^2 b_4 &= a_{4\alpha}h^2\sigma_P^2
\end{aligned}$$

where

$$d_1 = [1 + (n - 1)r]/n_1$$

and

$$d_4 = \frac{[1 + (n_4 - 1)r]/n_4 + (p_4 - 1)a_{44'}h^2}{p_4}$$

After dividing by σ_P^2, the selection index equations can be written numerically once d_1 and d_4 are calculated:

$$d_1 = \frac{1 + (2 - 1)(.60)}{2} = .8$$

$$d_4 = \frac{[1 + (3 - 1)(.6)]/3 + (4 - 1)(1/4)(.4)}{40} = .02583$$

Then,

$$\begin{aligned}
(.80)b_1 + (1/2)(.4)b_2 + (1/2)(.4)b_3 + (1/2)(.4)b_4 &= (1.0)(.4) \\
(.20)b_1 + b_2 + (0)b_3 + (1/4)(.4)b_4 &= (1/2)(.4) \\
(.20)b_1 + (0)b_2 + b_3 + (1/4)(.4)b_4 &= (1/2)(.4) \\
(.20)b_1 + (.10)b_2 + (.10)b_3 + (.02583)b_4 &= (1/2)(.4)
\end{aligned}$$

The equations determine the weights for the selection index for genetic value, which is

$$I_\alpha = \hat{A}_\alpha = .1166X_1 + .0291X_2 + .0291X_3 + 1.4755X_4$$

The accuracy of evaluation is

$$r_{A,\hat{A}} = \sqrt{\frac{b_1 a_{1\alpha}h^2\sigma_P^2 + b_2 a_{2\alpha}h^2\sigma_P^2 + b_3 a_{3\alpha}h^2\sigma_P^2 + b_4 a_{4\alpha}h^2\sigma_P^2}{h^2\sigma_P^2}}$$

which is equivalent to

$$r_{A,\hat{A}} = \sqrt{b_1 a_{1\alpha} + b_2 a_{2\alpha} + b_3 a_{3\alpha} + b_4 a_{4\alpha}}$$

so that

$$r_{A,\hat{A}} = \sqrt{.1166(1) + .0291(1/2) + .0291(1/2) + 1.4755(1/2)} = .9399$$

Selection for producing ability Selection index equations are the same as for additive genetic value except that the first right-hand side is $r\sigma_P^2$ instead of $h^2\sigma_P^2$ (the covariance between $G_1 + PE_1$ and X_1 rather than between A_1 and X_1).

The index for producing ability is

$$PA_\alpha = .5538X_1 + .0146X_2 + .0146X_3 + .7377X_4$$

The accuracy of evaluation is

$$r_{PA,\widehat{PA}} = \sqrt{\frac{b_1 r\sigma_P^2 + b_2 a_{2\alpha} h^2\sigma_P^2 + b_3 a_{3\alpha} h^2\sigma_P^2 + b_4 a_{4\alpha} h^2\sigma_P^2}{r\sigma_P^2}}$$

(The denominator is the variance of producing ability, $r\sigma_P^2$, rather than the variance of additive genetic value.) This equation can be rewritten as

$$r_{PA,\widehat{PA}} = \sqrt{b_1 + (b_2 a_{2\alpha} + b_3 a_{3\alpha} + b_4 a_{4\alpha})(h^2/r)}$$

$$r_{PA,\widehat{PA}} = \sqrt{.5583 + [.0146(1/2) + .0146(1/2) + .7377(1/2)][(.4)/(.6)]}$$
$$= .9022$$

12-12 Summary

The selection of animals with the largest additive genetic values (breeding values) is usually the goal of animal breeders. The selection index procedure provides a way to rank animals using their own records and records of relatives. The properties of the selection index are:

1. average squared errors of prediction are *minimized,*
2. the accuracy of evaluation, $r_{A,\hat{A}}$, is *maximized,*
3. the probability of correctly ranking the animals for additive genetic value is *maximized,* and
4. the average additive genetic value of the selected animals is *maximized.*

The selection index to predict the additive genetic value of animal *i* has the general form

$$\hat{A}_i = b_1X_1 + \cdots + b_NX_N$$

where the X's are averages of adjusted records of the animal and/or its relatives, and the b's are the selection index weighting factors. Each record, P_i, is first adjusted for all identifiable nongenetic factors as

$$X_i = P_i - \mu$$

where μ represents the adjustment for nongenetic factors.

Once the weighting factors are determined the animals are ranked according to their numerical index values.

Selection is based on the ranking.

Associated with the index is an accuracy value which is not used in ranking or selection but which can be used to provide a confidence range for the true additive genetic value of the animal.

Weighting factors to predict the additive genetic value of an animal from records of relatives depend on heritability (h^2), repeatability (r), relationships among the animals, and the number of records and number of relatives.

Weighting factors are given in Table 12-2 for records for most combinations of up to three kinds of relatives.

The prediction of additive genetic value from the average, $\overline{X}_{i,n}$, of n records on animal i is

$$\hat{A}_i = b\overline{X}_{i,n}$$

where

$$b = nh^2/[1 + (n-1)r]$$

The accuracy of prediction is $r_{A,\hat{A}} = \sqrt{b}$.

The prediction of additive genetic value from the average, $\overline{X}_{i,p}$, of records on p half-sib progeny of sire i is

$$\hat{A}_i = b\overline{X}_{i,p}$$

where

$$b = 2p/[p + (4-h^2)/h^2]$$

The accuracy of prediction is $r_{A,\hat{A}} = \sqrt{b/2}$.

The confidence range for A_i in both cases is

$$\hat{A}_i \pm t\sqrt{(1 - r^2_{A,\hat{A}})\sigma^2_A}$$

where $r^2_{A,\hat{A}}$ is the square of accuracy of prediction, σ^2_A is the additive genetic variance, and t corresponds to the confidence percentage.

Estimated transmitting ability is $\text{ETA}_i = \hat{A}_i/2$. The accuracy of estimating transmitting ability is the same as for estimating additive genetic value, and the confidence range for transmitting ability is one-half the confidence range for additive genetic value.

Producing ability predicted from an animal's own average of n records is

$$\widehat{PA}_i = b\overline{X}_{i,n}$$

where

$$b = nr/[1 + (n-1)r]$$

A future record, R_i, is predicted by adding the predicted management level, M, to the additive genetic evaluation or to the prediction of producing ability if the animal has had previous records:

$$\hat{R}_i = M + \hat{A}_i$$

or

$$\hat{R}_i = M + \widehat{PA}_i$$

The predicted additive genetic value of an animal with no records is the average of the additive genetic evaluations of its sire and dam:

$$\hat{A}_{\text{progeny}} = \frac{\hat{A}_{\text{sire}} + \hat{A}_{\text{dam}}}{2}$$

Inbreeding will cause slight modifications of the equations which determine the weighting factors and accuracy of prediction.

Nonzero environmental correlations between records of relatives will reduce the weighting factors given to those relatives as well as reduce the accuracy of prediction.

Further Readings

Falconer, D. S. 1981. *Introduction to quantitative genetics.* 2d ed. Longman, Inc. New York.

Lush, J. L. 1945. *Animal breeding plans.* 3d ed. Iowa State College Press, Ames.

Pirchner, F. 1969. *Population genetics in animal breeding.* W. H. Freeman and Company, New York.

Turner, H. N., and S. S. Y. Young. 1969. *Quantitative genetics in sheep breeding.* Cornell University Press, Ithaca, N.Y.

Van Vleck, L. D. 1983. *Notes on the theory and application of selection principles for the genetic improvement of animals.* Department of Animal Science, Cornell University, Ithaca, N.Y.

CHAPTER 13

Prediction of Genetic Progress and Comparison of Selection Programs

Designing selection programs to maximize genetic improvement for specific situations and traits obviously is of great importance in animal breeding. The breeder can use the properties of the selection index, which was introduced in Chapter 12 as the best way to evaluate animals, to compare methods of genetic selection with the goal of optimizing genetic gain. Another important duty of the animal breeder is to monitor genetic progress to determine whether gain that has been made can be increased. This chapter discusses both topics—first, planning and comparing potential selection programs and second, monitoring genetic progress from implemented selection programs. Basically, genetic gain depends on

1. how well animals are evaluated, or accuracy of prediction;
2. the amount of selection, or selection intensity;
3. the magnitude of genetic differences among animals, or standard deviation of additive genetic values; and
4. how rapidly better younger animals replace their parents, known as generation interval.

These factors make up the key equation for predicting genetic progress.

13-1 The Key Equation

The superiority in average breeding value of selected animals over those available for selection depends on three factors—*genetic variation, selection intensity,* and *accuracy of predicting genetic value.* Genetic improvement per year also depends on the *generation interval,* or the average time in years between when an animal is born and when its replacement is born. These factors can be calculated for various selection plans to determine which plan would be expected to result in the largest genetic gain per year. For most traits, additive genetic value (also called breeding value) makes up most of the genetic value and is the part which is transmitted directly to progeny. Thus, in this chapter, selection will be assumed to be for additive genetic value.

Calculation of expected genetic superiority or genetic gain per year depends on two assumptions:

1. the records and additive genetic values have joint normal distributions, which appears approximately true for most quantitative traits; and
2. only animals ranked highest by the genetic evaluation procedure are selected to be parents.

The square root of the additive genetic variance, the additive genetic standard deviation, σ_A, is the measure of genetic variation used in the key equation.

The fraction of animals selected corresponds to a *selection intensity factor,* i, which also is an integral part of the key equation. A subscript on i may be used to indicate the fraction selected; for example, $i_{.05}$ would be the selection intensity factor when the top 5 percent of animals are selected.

The accuracy of predicting genetic value depends on the method of prediction and the number and kinds of records on the animal and its relatives. The symbol, $r_{A,\hat{A}}$, denotes accuracy of prediction, which is the correlation between the true additive genetic value, A, and the prediction of additive genetic value, $\hat{A}$. Accuracies of prediction using records of various kinds were briefly discussed in Chapter 12 and listed in Table 12-2 for many combinations of records used to predict genetic value. A more complete discussion of accuracy of prediction follows in a later section.

The symbol for expected additive genetic superiority due to selection of animals from the group of animals available for selection is ΔG and for genetic progress per year is Δg.

Thus, for any generation of animals,

expected additive genetic superiority due to selection	=	accuracy of prediction	×	selection intensity factor	×	additive genetic standard deviation

Symbolically, the equation for expected superiority is

$$\Delta G = (r_{A,\hat{A}})(i_p)(\sigma_A)$$

where p is the top fraction selected based on $\hat{A}$.

A more important equation is that for expected additive genetic gain per year which considers not only the selection superiority for a generation but also the generation interval. In words,

$$\text{expected additive genetic gain per year} = \frac{\text{expected additive genetic superiority due to selection}}{\text{generation interval in years}}$$

The symbol for the length of the generation interval is L.

Thus, in symbols, the key equation for expected gain per year is

$$\Delta g = \Delta G / L$$

or, equivalently,

$$\Delta g = \frac{(r_{A,\hat{A}})(i_p)(\sigma_A)}{L}$$

Genetic progress is expected to be as rapid as possible if the numerator terms are as large as possible and the generation interval as small as possible. Although that is essentially the goal, there are some biological and practical limitations. A change in any of the three factors (accuracy, selection intensity, generation interval) which are under control of the animal breeder may cause an undesirable change in one or both of the other factors because there are practical interrelationships among the three factors (σ_A is essentially a constant for a population). Thus the goal is to find the proper balance among the key factors in order to increase genetic progress to an optimum for a particular breeding situation. Costs of the expected gain must also be compared to the economic return of the expected gain.

13-2 Standard Deviation of Additive Genetic Value

Genetic variation is essential if selection is to be effective. If animals were all alike genetically, there would be no genetic variation, and there would be no genetic progress no matter how intense selection were to be. Unfortunately, genetic variation is the one factor in the equation for genetic progress which the breeder cannot easily change. The genetic variance of a trait is essentially a constant. Some traits have little, and others considerable, genetic variation. The additive genetic standard deviation for a trait is related to heritability and the total or phenotypic variation. Because

$$h^2 = \sigma_A^2/\sigma_P^2$$

then

$$\sigma_A^2 = h^2\sigma_P^2$$

The additive genetic standard deviation is $\sigma_A = \sqrt{\sigma_A^2}$. Either $\sqrt{\sigma_A^2}$ or $\sqrt{h^2\sigma_P^2}$ can be used in the key equation.

13-3 Accuracy of Predicting Additive Genetic Value

The accuracy of predicting genetic value will depend on the type of evaluation, for example, whether selection is based on an animal's own record or on records of many progeny.

The correlation (also called accuracy) between predicted and true additive genetic value, $r_{A,\hat{A}}$, can range from .00, the value when breeding value is guessed, to 1.00 when known exactly. Accuracy depends on the heritability of the trait and on the number of records on the animal and its relatives used in the evaluation. Records on close relatives increase accuracy more than records on distant relatives because more genes and their effects are alike; for example, twice as many gene effects are alike between a parent and a progeny as between an animal and its grandparent.

The most important type of selection is *mass selection,* in which all animals are available for selection, as contrasted to *family selection,* where whole families, for example, groups of full sibs, are selected. Progress from mass selection will be discussed in this chapter. Evaluations used for mass selection may be from the animal's own records or records on the animal and its relatives, including progeny.

Accuracy of evaluation from individual performance

Heritability of 1.00 means there are no random environmental influences on the trait; that is, a phenotypic measurement is a perfect measure of the genetic value of the animal. For such a trait, the accuracy of prediction would be 1.00 based on one record of the animal. In general, accuracy of prediction from one record on an animal is the square root of heritability.

$$r_{A,\hat{A}} = \sqrt{h^2}$$

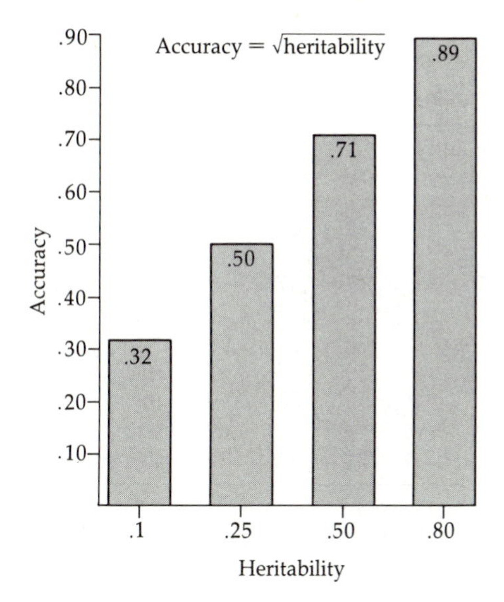

Figure 13-1 Accuracy of predicting genetic value depends on heritability of the trait. The graph shows the accuracy that results from evaluation on the basis of a single record of the animal for traits with different heritabilities.

Genetic values of traits with low heritability are predicted with less accuracy than traits with high heritability, as shown in Figure 13-1. Most quantitative traits have a heritability between .10 and .50. Within that range, accuracy of prediction is from $\sqrt{.10} = .32$ to $\sqrt{.50} = .71$.

If the animal has more than one record, the accuracy of predicting its genetic value increases:

$$r_{A,\hat{A}} = \sqrt{\frac{nh^2}{1 + (n - 1)r}}$$

where n is the number of records of the animal and r is repeatability of the trait. As seen in Table 13-1 for various values of n, h^2, and r, each additional record adds to the accuracy of prediction, but there is a diminishing increase in accuracy with each added record. The increase in accuracy depends on the ratio of h^2 to r, but the increase in relative accuracy is greater for low heritability than for high heritability as more records are added. As the number of records becomes large, accuracy becomes $\sqrt{h^2/r}$; this shows that when $r > h^2$ the accuracy of predicting additive genetic value is reduced as compared to the accuracy when $r = h^2$ (r cannot be less than h^2).

Table 13-1 Accuracy of predicting additive genetic value from an animal's records*

Heritability (h^2)	Repeatability (r)	Number of records on the animal (n)							
		1	2	3	5	...	10	...	∞†
.10	.10	.32	.43	.50	.60		.73		1.00
	.25	.32	.40	.45	.50		.55		.63
	.50	.32	.37	.39	.41		.43		.45
	.80	.32	.33	.34	.35		.35		.35
.25	.25	.50	.63	.71	.79		.88		1.00
	.50	.50	.58	.61	.65		.67		.71
	.80	.50	.53	.54	.55		.55		.56
.50	.50	.71	.82	.87	.91		.95		1.00
	.80	.71	.75	.76	.77		.78		.79

* $r_{A,\hat{A}} = \sqrt{nh^2/[1+(n-1)r]}$
† limit when $n \to \infty$ is $r_{A,\hat{A}} = \sqrt{h^2/r}$

Accuracy of evaluation from progeny records

A common method of predicting genetic values for mass selection of males is from records of their progeny. Evaluation from progeny records is especially common for traits such as milk production which cannot be measured on the male, or traits such as carcass quality which cannot be measured on breeding animals. If the progeny are all from different dams (the progeny are paternal half-sibs), accuracy of prediction will approach 1.00 as the number of progeny, p, becomes large:

$$r_{A,\hat{A}} = \sqrt{\frac{ph^2}{4+(p-1)h^2}}$$

which is algebraically equal to the more easily used form,

$$r_{A,\hat{A}} = \sqrt{\frac{p}{p+\dfrac{4-h^2}{h^2}}}$$

For example, when $h^2 = .25$, the ratio $(4-h^2)/h^2 = 15$ and $r_{A,\hat{A}} = \sqrt{p/(p+15)}$.

The increase in accuracy as the number of progeny increases depends on heritability and whether, for example, the added progeny is the fifth or seventieth. Going from zero to five progeny increases accuracy much more than

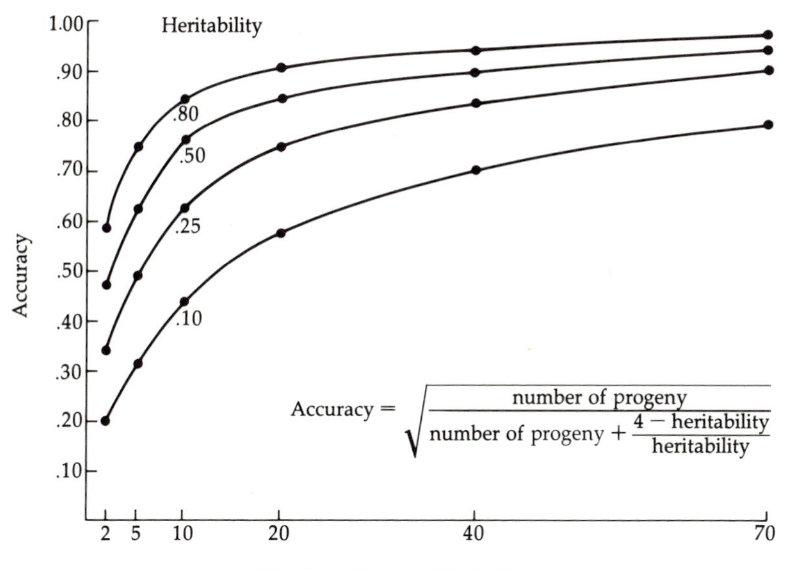

Figure 13-2 Accuracy of predicting additive genetic value from half-sib progeny depends on heritability of the trait and the number of progeny.

going from 65 to 70 progeny, as illustrated in Figure 13-2 for different heritabilities and numbers of progeny.

Figures 13-1 and 13-2 can be used to compare the accuracy of prediction from own records and from progeny records. Traits with high heritability have relatively high accuracy of prediction with own records so that progeny records generally can increase accuracy only slightly. With low heritability, however, the increase in accuracy with progeny records can be substantial. For example, with a heritability of .10 the accuracy of prediction is only $\sqrt{.10} = .32$ when based on the animal's own record, which is equivalent to prediction from four or five progeny records. The accuracy of prediction of genetic value from records on 30 progeny would be twice as large, that is $r_{A,\hat{A}} = \sqrt{30/(30 + 39)} = .66$, where $39 = (4 - h^2)/h^2 = (4 - .1)/.1$, as compared to $r_{A,\hat{A}} = .32$ for prediction from an animal's own record.

If heritability is high, records of relatives in addition to an animal's own record add relatively little to accuracy of prediction except for records of many progeny. The increase in accuracy with records of relatives is greater when heritability is low than when heritability is high. Accuracy of prediction from ancestor records decreases with each generation between the animal and the relatives with records, as illustrated numerically in Table 13-2.

Table 13-2 Accuracy of predicting genetic value from own and ancestor or progeny records

Records used	Heritability .10	.25	.50
Own	.32	.50	.71
Own + 1 parent	.35	.53	.73
Own + 2 parents	.38	.57	.76
Own + 1 grandparent	.32	.51	.71
Own + 4 grandparents	.35	.53	.73
Own	.32	.50	.71
Only 1 parent	.16	.25	.35
Only 2 parents	.23	.35	.50
Only 1 grandparent	.08	.12	.18
Only 4 grandparents	.16	.25	.35
Own	.32	.50	.71
Own + 1 progeny	.35	.53	.73
Own + 5 progeny	.44	.63	.79
Own + 10 progeny	.52	.71	.84
Own + 20 progeny	.62	.79	.89
Own + 40 progeny	.73	.87	.93
Own	.32	.50	.71
Only 1 progeny	.16	.25	.35
Only 5 progeny	.34	.50	.65
Only 10 progeny	.45	.63	.77
Only 20 progeny	.58	.76	.86
Only 40 progeny	.71	.85	.92

Accuracy of evaluation from sib performance

If a trait such as carcass quality is evaluated in the slaughterhouse, obviously the animal that supplied the carcass cannot produce offspring later. A common practice with such traits is to evaluate animals based on records of their sacrificed brothers and sisters (sibs). These are usually full sibs, but may be half-sibs.

The accuracy of evaluation from p sibs is

$$r_{A,\hat{A}} = (a_{\text{sibs}})\sqrt{\frac{ph^2}{1 + (p-1)(a_{\text{sibs}}h^2 + c^2_{\text{sibs}})}}$$

The additive relationship, $a_{\text{sibs}} = \frac{1}{2}$ if full sibs, and $a_{\text{sibs}} = \frac{1}{4}$ if half-sibs. The term c^2_{sibs} represents the nongenetic likeness between sibs where c^2_{sibs} is the correlation between environmental effects on the records of the sibs. For example, littermates have a common mother whose milk supply and mothering ability contribute to all of her progeny in that litter. The environmental correlation probably is different for full sibs and half-sibs and may be different for different situations such as when half-sibs are in the same or in different management systems. The effect of the environmental likeness is to reduce the accuracy of evaluation as can be seen by the positive term c^2_{sibs} in the denominator of $r_{A,\hat{A}}$.

For example, suppose evaluation is from records of six full sibs for a trait such as growth to weaning in swine with a heritability of about .20. The accuracy of predicting additive genetic value drops from

$$r_{A,\hat{A}} = (1/2)\sqrt{\frac{6(.2)}{1 + (6-1)[(1/2)(.2) + .00]}} = .45$$

when $c^2 = .00$, to

$$r_{A,\hat{A}} = (1/2)\sqrt{\frac{6(.2)}{1 + (6-1)[(1/2)(.2) + .05]}} = .41$$

when $c^2 = .05$

A nonzero environmental correlation will also reduce the accuracy of predicting the additive genetic value of a sire from records of his paternal half-sib (PHS) progeny. The accuracy of prediction for evaluation of a sire from his progeny with no environmental correlation is from Table 12-2:

$$r_{A,\hat{A}} = \sqrt{\frac{ph^2}{4 + (p-1)h^2}} = \sqrt{\frac{p}{p + (4 - h^2)/h^2}}$$

When the environmental correlation is not zero,

$$r_{A,\hat{A}} = \sqrt{\frac{ph^2/4}{1 + (p-1)(a_{\text{PHS}}h^2 + c^2_{\text{PHS}})}}$$

If $h^2 = .25$ and $c^2_{\text{PHS}} = .125$ then

$$r_{A,\hat{A}} = \sqrt{\frac{p(.25)(.25)}{1 + (p-1)[(.25)(.25) + (.125)]}}$$

which, after some algebra, can be shown to be

$$r_{A,\hat{A}} = \sqrt{\frac{p}{3p + 13}} = .58\sqrt{\frac{p}{p + 4.33}}$$

This form shows that, as the number of progeny becomes large, $r_{A,\hat{A}}$ becomes close to, but cannot exceed .58, whereas with no environmental correlation, $c^2_{PHS} = .00$, $r_{A,\hat{A}} = \sqrt{p/(p + 15)}$, and as p becomes large, $r_{A,\hat{A}}$ approaches 1.00.

In general, the limit for accuracy of prediction as the number of sibs or progeny becomes large depends on h^2, c^2, the additive relationship among the animals with records (a_{sibs} or a_{PHS}), a_{rec}, and the additive relationship between the animals with records and the animal being evaluated (a_{sibs} or $a_{father\text{-}progeny}$), denoted a_{eval}. Then

$$\text{limit of } (r_{A,\hat{A}}) \text{ as } p \rightarrow \infty = (a_{eval})\sqrt{\frac{h^2}{a_{rec}h^2 + c^2_{rec}}}$$

Accuracy of prediction using other relatives

Table 12-2 gives, in symbolic form, formulas to compute accuracy values for various combinations of relatives as well as weighting factors for their records. The procedures to derive the weights and accuracy values were described in Section 12-11.

In some situations, efforts made to increase accuracy of prediction result in a decrease in selection intensity as well as an increase in generation interval. For example, if more records on the animal are obtained, evaluation takes longer and there is an increased chance the animal will have died or have become unavailable for selection before selection decisions are made.

13-4 Selection Intensity Factor

A high selection intensity is important for rapid genetic gain. Selection of males is particularly important because fewer males than females are usually needed as parents of the next generation. The selection of mothers of dairy bulls to be tested through artificial insemination is also important because such a small fraction of cows is needed to produce sons.

The numerical factor of intensity of selection in the key equation depends on, but is not exactly proportional to, the fraction selected. Selection intensity factors for a range of selection percentages are given in Table 13-3. The table

Table 13-3 Selection intensity factors for variables with a normal distribution

Fraction selected	Selection intensity factor	Fraction selected	Selection intensity factor	Fraction selected	Selection intensity factor	Fraction selected	Selection intensity factor
.001	3.400	.01	2.660	.10	1.755	.55	.720
.002	3.200	.02	2.420	.15	1.554	.60	.644
.003	3.033	.03	2.270	.20	1.400	.65	.570
.004	2.975	.04	2.153	.25	1.271	.70	.497
.005	2.900	.05	2.064	.30	1.159	.75	.424
.006	2.850	.06	1.985	.35	1.058	.80	.350
.007	2.800	.07	1.919	.40	.966	.85	.274
.008	2.738	.08	1.858	.45	.880	.90	.195
.009	2.706	.09	1.806	.50	.798	.95	.109

values are based on a normal distribution. The selection intensity factor is a relative measure of how much the average of the selected group will exceed the average of the group from which the selected group was chosen. Multiplication by the standard deviation for the trait under selection converts the selection intensity factor to the units of measure appropriate for that trait. For example, the phenotypic standard deviations for fat percentage and protein percentage in milk of Holstein cows are about .30 percent and .19 percent. Thus, if the top 10 percent of cows were selected ($i_{.10} = 1.755$) on the basis of fat percentage, their phenotypic fat percentage would average 1.755(.30 percent) = .5265 percent above the average fat percentage of all Holsteins. However, if the top 10 percent were selected based on protein percentage, their phenotypic protein percentage would average 1.755(.19 percent) = .333 percent above the average protein percentage for all Holstein cows.

Selection must be based only on predicted genetic value in order for the intensity factors to be applicable for predicting genetic superiority. For example, selection of a random 20 out of the top 50 of 100 animals is not 20 percent selection but is equivalent to selection of the best 50 percent.

Derivation of selection intensity factors

The selection intensity factors in Table 13-3 are derived from the expected value (average) of the fraction of records, p, selected from a standardized normal distribution (average = 0, and variance of 1). As illustrated in Figure 13-3, the average of the selected group is $i_p = z/p$, where z is the height of the curve at the truncation point corresponding to selection of the top fraction, p, of records. This result for a standardized trait can be easily converted to that for a

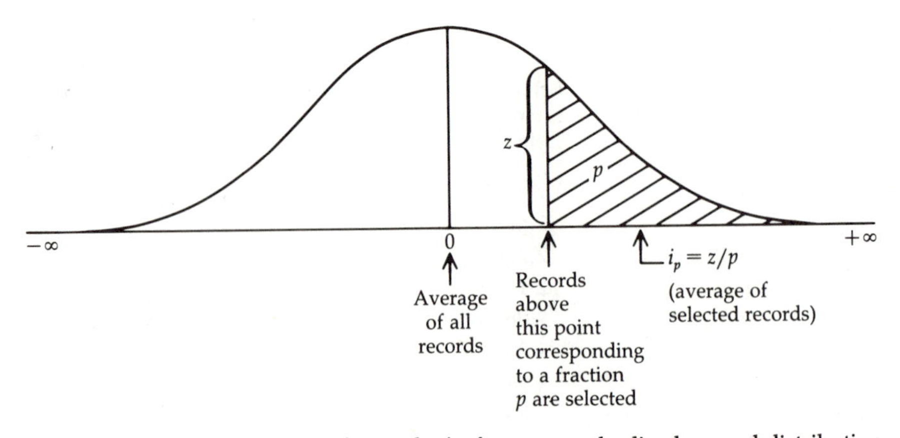

Figure 13-3 Average of selected records, i_p, from a standardized normal distribution (average $= 0$, and variance of 1) when a fraction p of the best records are selected.

distribution with standard deviation σ_P by multiplying by σ_P; that is, the average of the best fraction p of the records is $i_p\sigma_P$.

13-5 Generation Interval

Factors such as poor fertility, delayed sexual maturity, or the requirement of many records before selection decisions are made all tend to increase the generation interval and thus decrease genetic gain per year. Generation interval can be shortest for species with early sexual maturity and short gestation periods and for species having a high reproductive rate which allows replacements to be selected from the first offspring. Males often have shorter generation intervals than females because they can produce many offspring in a short period of time, whereas the female is limited to few offspring. Cows and mares, for example, usually can manage only one offspring per year. Because one-half of their progeny are males and also because some of the female progeny die before sexual maturity, about three parities are needed to produce a replacement cow or mare. Table 13-4 gives the near minimums for some domestic livestock.

Selection for traits which are not manifested until late in life will extend the generation interval. Methods of selection which increase accuracy of evaluation, such as progeny testing or requiring females to have several records before they are eligible for selection, also increase the generation interval. If a breed or species is being expanded in numbers the generation interval is likely to be decreased, and if a breed or species is declining in numbers the generation interval is often increased. Breeding programs for most species do not attain

Table 13-4 Generation intervals which are near the minimum length

Livestock	Generation interval, in years	
	Males	Females
Cattle	2	5
Swine	1	1
Sheep	1	$1\frac{1}{2}$
Goats	1	$1\frac{1}{2}$
Poultry	1	1
Horses	3	5
Dogs	2	3

the physiological minimum generation interval. Once the minimum is approached there is little chance of making large favorable changes in the generation interval for most classes of animals.

13-6 Optimizing Genetic Progress

Changes in some of the key factors for genetic progress may affect other factors. Decreasing generation interval, for example, would probably decrease accuracy of prediction because fewer records would be available for genetic evaluation. In many cases, accuracy needs to be balanced against intensity of selection, particularly if progeny testing of sires is being considered.

Obviously the key factors may be different for male and female selection —accuracy, selection intensity, and generation interval may all be different. In such cases, genetic progress per year can be predicted from the average superiority of selected males and selected females divided by their average generation interval. The key equation with two paths for genetic gain is

$$\text{genetic progress per year} = \frac{(\text{selected male superiority} + \text{selected female superiority})/2}{(\text{generation interval for males} + \text{generation interval for females})/2}$$

where

$$\text{selected male superiority} = \text{accuracy for males} \times \text{selection intensity factor for males} \times \text{genetic standard deviation}$$

and

selected female superiority
= accuracy for females × selection intensity factor for females × genetic standard deviation

The 2's in the formula cancel out, and the equation becomes

$$\Delta g = \frac{r_{A_m,\hat{A}_m} i_m \sigma_A + r_{A_f,\hat{A}_f} i_f \sigma_A}{L_m + L_f}$$

Considerable research has been devoted to finding how to maximize Δg. Some approaches have used the two-path equation (males and females as presented here) although most researchers dealing with traits that cannot be measured on males have divided selection into four paths—sires to produce future sires (*SS*), sires to produce future dams (*SD*), dams to produce future sires (*DS*) and dams to produce future dams (*DD*). Then.

$$\Delta g = \frac{\Delta SS + \Delta SD + \Delta DS + \Delta DD}{L_{SS} + L_{SD} + L_{DS} + L_{DD}}$$

which is the average superiority due to selection for each of the four paths (*SS, SD, DS, DD*) divided by the average generation interval. The divisor of 4 in both the numerator and denominator cancels out so it is not included in the above equation. The selection intensity factors for the four paths are especially likely to be different. The accuracy values and generation intervals for both male paths are likely to be similar, as are accuracies and generation intervals for both female paths. A derivation of the result is shown at the end of the chapter in Box 13-10. The calculations are illustrated in Example 13-1.

Example 13-1

Potential genetic gains from sex control (obtaining a bull calf or a heifer calf when desired) and from embryo transfer (superovulation of cows thought to be superior with up to 10 fertilized ova transferred to recipient cows) are to be compared for a dairy cattle breeding program using artificial insemination (AI) with programs using sex control (SC) and embryo transfer (ET) in addition to artificial insemination.

The table describes the characteristics which are assumed for the breeding programs. The phenotypic standard deviation is 2500 pounds of milk, and heritability is .25, so that the additive genetic standard deviation is 1250 pounds. Accuracies of prediction and generation intervals are assumed equal for all breeding programs.

Breeding Program	Fraction selected for path			
	SS	*SD*	*DS*	*DD*
Regular AI	.04	.20	.06	.90
AI with sex control	.04	.20	.03	.45
AI with embryo transfer	.04	.20	.01	.10
AI with sex control and embryo transfer	.04	.20	.005	.05
Accuracy of prediction	.79	.79	.65	.65
Generation interval	5	5	7	7

The selection intensity factors can be found in Table 13-3, for example, $i_{.005} = 2.90$, $i_{.03} = 2.27$, and $i_{.06} = 1.98$.

The formula for expected genetic gain per year is

$$\Delta g = \frac{\Delta SS + \Delta SD + \Delta DS + \Delta DD}{L_{SS} + L_{SD} + L_{DS} + L_{DD}}$$

where, for example,

$$\Delta SS = r_{A_{SS},\hat{A}_{SS}} i_{SS} \sigma_A$$

Then

$$\Delta g(\text{AI}) = \frac{.79(2.15)(1250) + .79(1.40)(1250) + .65(1.98)(1250) + .65(.20)(1250)}{5 + 5 + 7 + 7}$$

$= 220$ pounds

$$\Delta g\begin{pmatrix}\text{AI}\\ \text{SC}\end{pmatrix} = \frac{.79(2.15)(1250) + .79(1.40)(1250) + .65(2.27)(1250) + .65(.88)(1250)}{24}$$

$= 253$ pounds

$$\Delta g\begin{pmatrix}\text{AI}\\ \text{ET}\end{pmatrix} = \frac{.79(2.15)(1250) + .79(1.40)(1250) + .65(2.66)(1250) + .65(1.76)(1250)}{24}$$

$= 296$ pounds

$$\Delta g\begin{pmatrix}\text{AI}\\ \text{SC}\\ \text{ET}\end{pmatrix}$$

$$= \frac{.79(2.15)(1250) + .79(1.40)(1250) + .65(2.90)(1250) + .65(2.06)(1250)}{24}$$

$$= 314 \text{ pounds}$$

The expected genetic gains per year can then be used to determine what the dairy farmer can afford to pay for sex control and for embryo transfer. At a cost of $300 per embryo transfer and a net value of $.07 per pound of milk, more than 50 years would be required before the cumulative gain in excess of the regular AI program of 76 pounds per year (296 − 220) would have a value equal to the yearly embryo transfer cost.

The appropriate key equation (one path, two paths, or four paths) for predicting genetic gain should be used with each situation (heritability, types of available records, and generation intervals) to construct breeding programs.

The usual approach to maximizing Δg is to calculate the expected genetic gain for different combinations of records and fractions of animals selected which influence each other and also influence generation interval for a defined population of animals, and choose the combination with the largest Δg. Costs of, and returns from, the breeding program can then be balanced to find the program which is optimum under anticipated economic conditions.

Table 13-5 suggests which general type of breeding program is optimum for various situations. With high heritability ($\geqq .40$) an animal's own record (performance test) provides a high accuracy of evaluation and a short generation interval, thus maximizing genetic gain. An animal's own records, however, are easily affected by deliberate preferential treatment which would make the records useless for selection. Performance tests must also be adjusted for management levels to avoid the apparent phenotypic superiorities arising from better treatment rather than greater genetic value.

Maternal traits such as milk yield and mothering ability cannot be measured on males. For selection of males, progeny tests may be necessary to increase the accuracy of evaluation, especially for traits with low heritability, even though the generation interval will increase. As the intensity of selection for females is generally low, their own records are usually suitable for selection because of the small influence of the female path on genetic gain. Records of ancestors (pedigree information) may provide added accuracy for female evaluations, especially for traits with low heritability. Evaluation from progeny records is usually not recommended for female selection, although for some traits progeny records actually measure the female's performance. For example, weaning weight of her calf is an indication of a beef cow's milk production.

Traits which require slaughter for measurement often have high heritability. Progeny records are usually necessary for selection of males for carcass traits; however when heritability is moderate to high, sib records for litter-bearing species may provide sufficient accuracy and a shorter generation interval as compared to progeny records. A breeder should usually base selection of females on sib records or records of collateral relatives.

Sib records, especially records of full sibs in the same litter for traits measured early in life, may be less accurate than expected from the high relationship with the animal being evaluated. The reason is the sibs' common environment which causes their records to be more alike than they would be because of gene effects in common. The net result is that these environmentally correlated records may lose some or, in special cases, all of their usefulness for predicting genetic value as was discussed earlier in the chapter.

Table 13-5 is only a guide. Calculations to find the optimum selection program must be performed for each situation. All available records of rela-

Table 13-5 Likely choices of records for optimum breeding programs depending on heritability and restrictions on available records

	Heritability	
Restriction on records	**Low ($h^2 < .20$)**	**High ($h^2 > .40$)**
	Selection of males	
None	Progeny*	Own†
Females only (example, milk yield)	Progeny	Progeny, maternal relatives
Relatives only (example, carcass traits)	Progeny	Sibs‡, progeny
	Selection of females	
None	Own, pedigree	Own
Males only (example, semen production)	Sibs, pedigree	Sibs, pedigree
Relatives only	Sibs, pedigree	Sibs

* Progeny proofs as compared to own records: (1) usually increase generation interval, (2) usually decrease selection intensity, and (3) increase accuracy of evaluation.
† Own records: (1) minimum generation interval, (2) maximum selection intensity, and (3) may be subject to management bias.
‡ Sib records: May be environmentally correlated which decreases accuracy of evaluation.

tives, within the limits of computing capability, except those affected by deliberate management distortions should be used to evaluate each animal. Small increases in accuracy can be economically important.

13-7 Negative Selection— Selection for Small Values

Breeders usually select for animals with the largest genetic evaluation because generally it is high phenotypic values that are desired. For some traits, however, the smallest scores are most desirable: for example, amount of feed eaten per unit of weight gained, or time to run a race. Thus, selection for such traits will be for a fraction p of animals with the smallest genetic evaluations—which might be called negative selection. Predicted response is calculated just as for positive selection except the sign is negative:

$$\Delta G = -r_{A,\hat{A}} i_p \sigma_A$$

13-8 Measuring Genetic Gain

The appropriate key equation from sections 13-1 or 13-6 for expected response to selection can be used to predict response for a given selection program and, thus, can be used to design optimum breeding programs on a theoretical basis.

Often the question is asked whether any genetic gain has resulted from selection decisions which have previously been made. The question is applied to the situation where relatively complete knowledge of how the decisions were made exists and to field data in order to monitor what is happening in the population. In the situation where the selection procedure is known—usually an experimental situation—the question is modified to whether the genetic response matches what would be expected based on assumed heritability, accuracy of evaluation, and selection intensity practiced.

When selection decisions are based on a single record of the animal, an estimate of genetic superiority can be obtained from the difference between the phenotypic averages of the animals selected to be parents, $\overline{P}_s$, and of all animals available for selection, $\overline{P}$, at the time the selection decision is made. The difference, $\overline{P}_s - \overline{P}$, is called the *phenotypic selection differential.* The equation for the estimate of genetic superiority of progeny of the selected parents is

$$\text{estimated genetic superiority} = h^2(\overline{P}_s - \overline{P})$$

If a selection experiment is repeated (or other selection decisions made under similar situations) the phenotypic selection differential probably will be different each time unless a very large number of animals is involved. If a very large number of animals, N, is used or if the experiment is repeated many times, the average phenotypic selection differential will approach the numerical value of the theoretical phenotypic selection differential, $i_p\sigma_P$, for a fraction p selected; that is,

$$\text{as } N \to \infty,\ \overline{P}_s - \overline{P} \to i_p\sigma_P$$

Thus the average genetic superiority becomes:

$$\text{as } N \to \infty,\ \Delta G = h^2(\overline{P}_s - \overline{P}) \to h^2 i_p \sigma_P$$

which is the same as the expected genetic superiority for selection on an animal's own record.

The difference between the average of progeny of selected parents and the average of progeny of random mating of the animals available when selection was made is a measure of genetic superiority which does not depend on an assumed value of heritability; that is,

$$\widehat{\Delta G} = \overline{P}_{\text{progeny selected parents}} - \overline{P}_{\text{progeny random parents}}$$

This estimate of genetic superiority can be compared to what is expected from the phenotypic selection differential times heritability:

$$h^2(\overline{P}_s - \overline{P})$$

The progeny difference for the average of repeated experiments approaches the true genetic superiority, that is,

$$\text{as } N \to \infty,\ \widehat{\Delta G} = \overline{P}_{\text{progeny selected parents}} - \overline{P}_{\text{progeny random parents}} \to \Delta G = h^2 i_p \sigma_P$$

as theory predicts.

For any one experiment, or the average of many experiments, an estimate of *realized heritability* can be obtained from the ratio

$$h^2_{\text{realized}} = \frac{\overline{P}_{\text{progeny selected parents}} - \overline{P}_{\text{progeny random parents}}}{\overline{P}_s - \overline{P}}$$

based on what ratio is expected as N becomes very large:

$$\text{as } N \to \infty, \quad \frac{\overline{P}_{\text{progeny selected}} - \overline{P}_{\text{progeny random}}}{\overline{P}_s - \overline{P}} \to \frac{h^2 i_p \sigma_P}{i_p \sigma_P} = h^2$$

If progeny are not obtained from random parents, by substituting $\overline{P}$ for $\overline{P}_{\text{progeny random parents}}$, the same procedure can be used if the assumption is made that the nongenetic factors affecting records do not change from the parent generation to the progeny generation so that

$$h^2_{\text{realized}} = \frac{\overline{P}_{\text{progeny selected parents}} - \overline{P}}{\overline{P}_s - \overline{P}}$$

Estimates of additive genetic value based on all records of each animal and its relatives can be used to monitor genetic gain in a population. The estimated average genetic superiority over the previous generation is the difference between average estimated genetic value of animals selected to be parents of that generation and the average estimated genetic value of all animals born in the previous generation; that is,

$$\text{average genetic superiority} = \text{average } \hat{A} \text{ for selected parents} - \text{average } \hat{A} \text{ for animals born}$$

Genetic trend can be estimated by averaging estimates of additive genetic value for all animals born in each year.

13-9 Expected Change in Other Traits

This chapter has described prediction of expected genetic change and how to estimate genetic change—for a single trait only. The next chapter discusses principles of selection for more than one trait and expected change in any trait when selection is for several traits. Even when one trait is considerably more important than all other traits, some care must be taken to insure that the levels of some of the other traits do not change so much that the economic value of the population of animals selected for one trait decreases. Expected change in any trait from selection based on records for another trait can be calculated. The formula depends primarily on the genetic correlation between the traits, heritabilities of the traits, and the phenotypic standard deviations. Discussion of selection response when more than one trait is considered will be postponed until the next chapter.

13-10 Equations for Genetic Gain per Year

Expected genetic gain per year is the average expected superiority of the selected males and females for each generation divided by the average generation intervals. Let S and D be the average breeding values of sires and dams selected to produce the next generation. Also, let

$$\Delta S = r_{A_S,\hat{A}_S} i_S \sigma_A$$

where ΔS is the genetic superiority of selected sires,
$r_{A_S,\hat{A}_S}$ is the accuracy of evaluations for sires, and
i_S is the selection intensity factor for sire selection.

Similarly, let

$$\Delta D = r_{A_D,\hat{A}_D} i_D \sigma_A$$

the genetic superiority of selected dams.

The key steps in the proof are that:

1. the average genetic value of the progeny, $\bar{A}$, is the average of the parents, because progeny receive a sample half of the genetic value of each parent:

$$\bar{A} = (S + D)/2$$

2. and gain in previous years has been at a steady rate of Δg.

Genetic improvement per year—two selection paths

If Δg is genetic improvement per year, L_S is the generation interval in years for sires, and L_D is the generation interval for dams, then

$$\Delta g = (\Delta S + \Delta D)/(L_S + L_D)$$

which is not $[\Delta S/L_S) + (\Delta D/L_D)]/2$, as might be suspected. The proof is as follows and is illustrated for an example in Figure 13-4.

The selected sires with genetic value S are born L_S years before they produce replacement progeny with genetic value $\bar{A}$. The genetic average of males born L_S years ago is $\bar{A} - L_S\,\Delta g$. The superiority of the selected sires over that average is ΔS.
Thus,

$$S = \bar{A} - L_S\,\Delta g + \Delta S$$

and similarly,

$$D = \bar{A} - L_D\,\Delta g + \Delta D$$

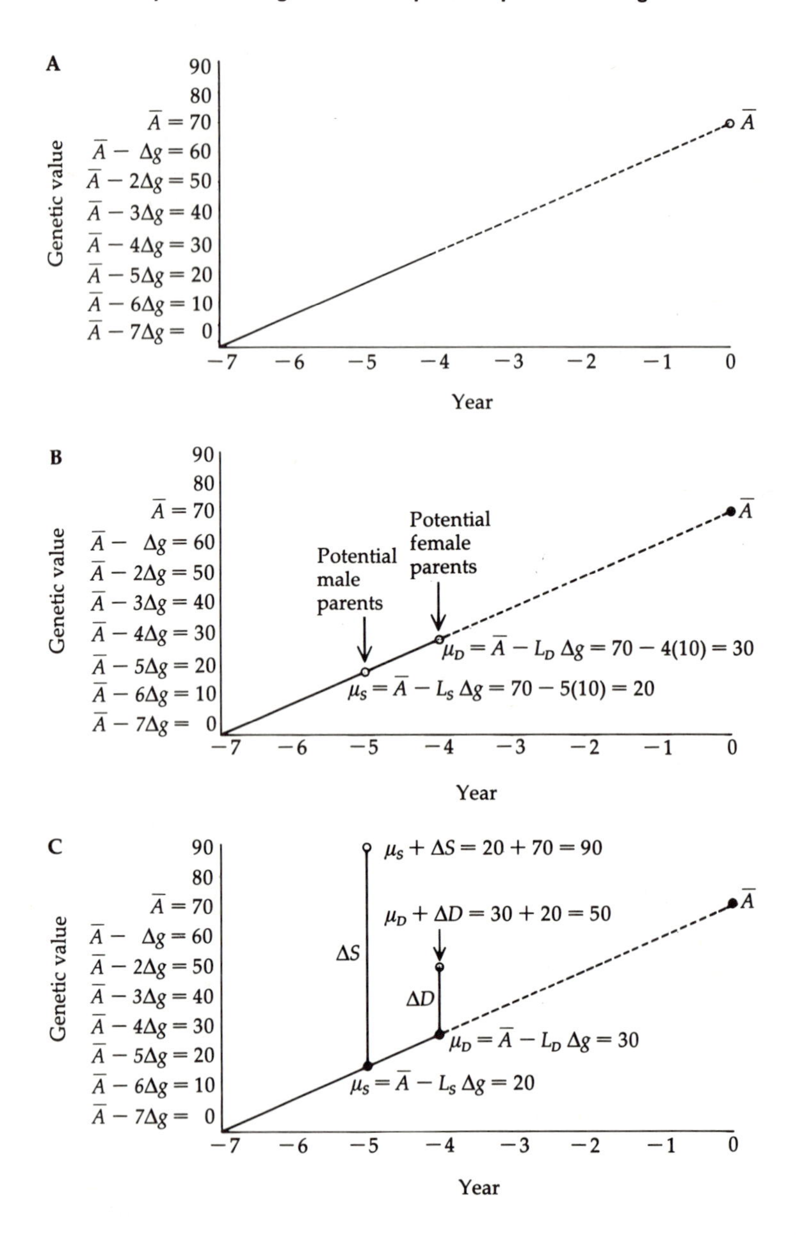

Figure 13-4 Example of how generation intervals, L_S and L_D, and selection superiority of sires, ΔS, and dams, ΔD, determine genetic gain per year. Assume $\Delta g = 10$, $\Delta S = 70$, $\Delta D = 20$, $L_S = 5$ and $L_D = 4$. (*A*) The average of progeny born in year zero is $\bar{A} = 70$. (*B*) The average genetic value of males born L_S years ago was $\bar{A} - L_S\Delta g$ (from which male parents were selected). Similarly, the average genetic value of females born L_D years ago was $\bar{A} - L_D\Delta g$ (from which female parents were selected).

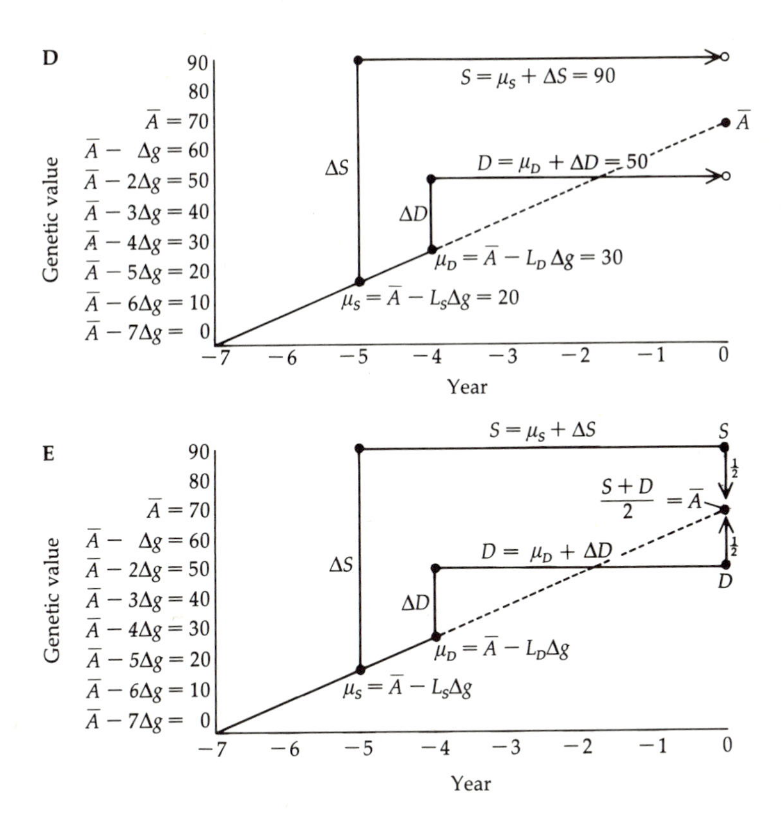

$$\bar{A} = \frac{S + D}{2}$$

$$\bar{A} = \frac{(\mu_S + \Delta S) + (\mu_D + \Delta D)}{2}$$

$$\bar{A} = \frac{(\bar{A} - L_S \Delta g + \Delta S) + (\bar{A} - L_D \Delta g + \Delta D)}{2}$$

Thus, $\Delta g = \dfrac{\Delta S + \Delta D}{L_S + L_D}$

(C) The superiority of selected sires to the average of males born L_S years ago was ΔS, and the superiority of selected dams to the average of females born L_D years ago was ΔD. (D) The animals born in year zero have parents with genetic value $S = \mu_S + \Delta S$ and $D = \mu_D + \Delta D$. (E) The genetic value of animals born in year zero is $(S + D)/2$ and by definition is $\bar{A}$.

Because

$$\bar{A} = (S + D)/2$$

by substitution,

$$\bar{A} = (\bar{A} - L_S\,\Delta g + \Delta S + \bar{A} - L_D\,\Delta g + \Delta D)/2$$

After subtracting $\bar{A}$ from both sides,

$$0 = (-L_S\,\Delta g - L_D\,\Delta g + \Delta S + \Delta D)/2$$

Rearranging gives

$$\Delta g(L_S + L_D)/2 = (\Delta S + \Delta D)/2$$

and finally,

$$\Delta g = (\Delta S + \Delta D)/(L_S + L_D)$$

Genetic improvement per year—four selection paths

This procedure can be extended to consider genetic values of sires of sires (*SS*), dams of sires (*DS*), sires of dams (*SD*), and dams of dams (*DD*), selected as grandparents each with different generation intervals (L_{SS}, L_{DS}, L_{SD}, and L_{DD}, respectively).

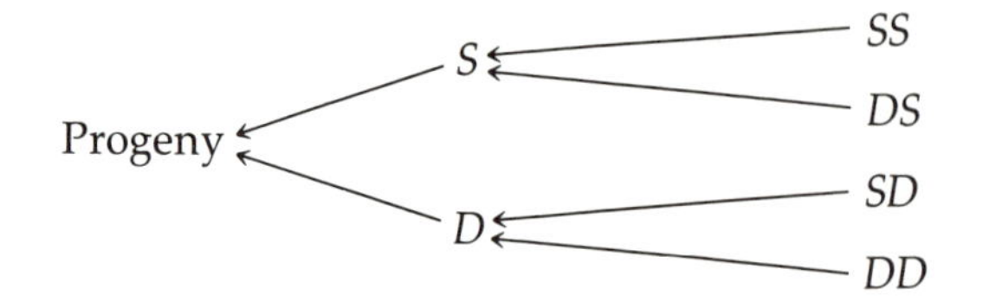

Let ΔSS, ΔDS, ΔSD, and ΔDD be the respective genetic superiorities of the selected grandparents above their generation average. For example,

$$\Delta SS = r_{A_{SS},\hat{A}_{SS}} i_{SS} \sigma_A$$

By reasoning similar to that used previously,

$$SS = S - L_{SS}\,\Delta g + \Delta SS$$
$$DS = S - L_{DS}\,\Delta g + \Delta DS$$
$$SD = D - L_{SD}\,\Delta g + \Delta SD$$
$$DD = D - L_{DD}\,\Delta g + \Delta DD$$

Because $S = (SS + DS)/2$, $D = (SD + DD)/2$, and $\overline{A} = (S + D)/2 = (SS + DS + SD + DD)/4$, then by substitution,

$$(S + D)/2 = (S - L_{SS}\,\Delta g + \Delta SS + S - L_{DS}\,\Delta g + \Delta DS + D - L_{SD}\,\Delta g + \Delta SD + D - L_{DD}\,\Delta g + \Delta DD)/4$$

After rearranging and subtracting $(S + D)/2$ from both sides,

$$\Delta g(L_{SS} + L_{DS} + L_{SD} + L_{DD}) = \Delta SS + \Delta DS + \Delta SD + \Delta DD$$

and

$$\Delta g = (\Delta SS + \Delta DS + \Delta SD + \Delta DD)/(L_{SS} + L_{DS} + L_{SD} + L_{DD})$$

Genetic progress per year, then, is equivalent to the average superiority of the selected grandparents divided by the average generation interval of the different grandparents. These are the paths by which selection is actually practiced. For example, the selected sires of sires produce sons, some of which may then become, after further selection, sires of dams.

This expression, or the one preceding involving just sires and dams, can be used to compare expected genetic progress for different selection programs which consider differences in generation intervals, selection intensities, and accuracies of prediction.

13-11 Summary

Genetic superiority, ΔG, of a selected group of animals depends on accuracy of genetic evaluation, $r_{A,\hat{A}}$; the selection intensity factor, i_p, associated with the fraction selected, p; and the additive genetic standard deviation of the trait, σ_A:

$$\Delta G = r_{A,\hat{A}} i_p \sigma_A$$

The additive genetic standard deviation, $\sigma_A = \sqrt{h^2\sigma_P^2}$, where h^2 is heritability and σ_P^2 is the phenotypic variance.

Genetic gain per year, Δg, is predicted by dividing ΔG by the generation interval in years, L:

$$\Delta g = \Delta G/L$$

Generation interval, L, is the average time between birth of an animal and birth of its replacement.

If accuracy of prediction, selection intensity, and generation interval are different for the four paths for selection, then Δg is the average genetic superiority of sires of sires, ΔSS, sires of dams, ΔSD, dams of sires, ΔDS, and dams of dams, ΔDD, divided by average generation interval:

$$\Delta g = \frac{\Delta SS + \Delta DS + \Delta SD + \Delta DD}{L_{SS} + L_{DS} + L_{SD} + L_{DD}}$$

Accuracy of predicting genetic value, $r_{A,\hat{A}}$, depends on h^2 and the number and kinds of relatives with records.

The fraction of animals selected, p, determines the standardized selection intensity factor, i_p.

Table 13-5 provides a guide to whether performance testing, progeny testing, or sib selection is likely to optimize genetic gain. The optimum choice depends on heritability, whether selection is for males or females, and whether records are available on the animal being selected. The accuracy of predicting genetic value, selection intensity, and generation interval all are interrelated and must be balanced to make genetic gain as rapid as possible.

Progeny testing is usually indicated for selection of males when heritability is low and particularly when the trait is not measured on males. Increased accuracy of evaluation must be balanced against decreased selection intensity and increased generation interval. Progeny testing is not ordinarily advised for evaluation of females.

Performance testing (selection on own record) is usually optimum for traits with high heritability when records are available on all animals, because generation interval is as short as possible and selection intensity is as high as possible.

Selection on records of sibs (half-sibs or full sibs) may be advised for such traits as carcass measurements which cannot be obtained on the live animal. Generation interval may be as short as with performance testing, but selection intensity will be less.

Actual gain from selection can be estimated as

$$\widehat{\Delta G} = \overline{P}_{\text{progeny of selected parents}} - \overline{P}$$

where $\overline{P}_{\text{progeny of selected parents}}$ is the phenotypic average of progeny of selected parents, and $\overline{P}$ is the phenotypic average of animals available for selection when the parents were selected, under the assumption that the environmental conditions are the same in both generations. When environmental conditions may change an estimate is

$$\widehat{\Delta G} = h^2(\overline{P}_s - \overline{P})$$

where $\overline{P}_s$ is the phenotypic average of selected parents, and $\overline{P}_s - \overline{P}$ is the phenotypic selection differential.

On the average, the selection differential is expected to equal the product of the selection intensity factor, i_p, and the phenotypic standard deviation, σ_P, that is

$$\text{as } N \to \infty,\ \overline{P}_s - \overline{P} \to i_p \sigma_P$$

Similarly, on the average,

$$\text{as } N \to \infty,\ \overline{P}_{\text{progeny of selected parents}} - \overline{P}_{\text{progeny of random parents}} \to h^2 i_p \sigma_P$$

Thus, realized heritability is defined as

$$h^2_{\text{realized}} = (\overline{P}_{\text{progeny of selected parents}} - \overline{P}_{\text{progeny of random parents}})/(\overline{P}_s - \overline{P})$$

Further Readings

Falconer, D. S. 1981. *Introduction to quanitative genetics.* 2d ed. Longman, Inc. New York.

Lush, J. L. 1945. *Animal breeding plans.* 3d ed. Iowa State College Press, Ames.

Pirchner, F. 1969. *Population genetics in animal breeding.* W. H. Freeman and Company, New York.

Turner, H. N., and S. S. Y. Young. 1969. *Quantitative genetics in sheep breeding.* Cornell University Press, Ithaca, N.Y.

Van Vleck, L. D. 1983. *Notes on the theory and application of selection principles for the genetic improvement of animals.* Department of Animal Science, Cornell University, Ithaca, N.Y.

CHAPTER 14

Consideration of More than One Trait in Selection

Selection for a single trait using records on the animal and its relatives to index each animal was discussed in Chapters 12 and 13. Selection for a single trait, however, is not usually adequate for most classes of livestock because other traits of economic value would be ignored. In fact, selection for just one trait implies that other traits have no economic value. Most breeders must select to improve more than one trait because several traits usually contribute to total economic value of the animal.

This chapter will (1) describe methods of selection when more than one trait is considered — the selection index, independent culling levels, and tandem selection, and (2) show the expected change in other traits when selection is for a single trait. Genetic and phenotypic correlations are often needed for selection for more than one trait. Genetic correlations also are required for calculating expected response from selection. The next section reviews these correlations.

14-1 Genetic, Environmental, and Phenotypic Correlations

Efficient multiple-trait selection requires that correlations be known. Just as two traits such as weight and height can be correlated with each other — the *phenotypic correlation* — the genetic values for weight and height can be correlated — the *genetic correlation* — and the environmental effects on the two traits can also be correlated — the *environmental correlation.* Figure 14-1 sym-

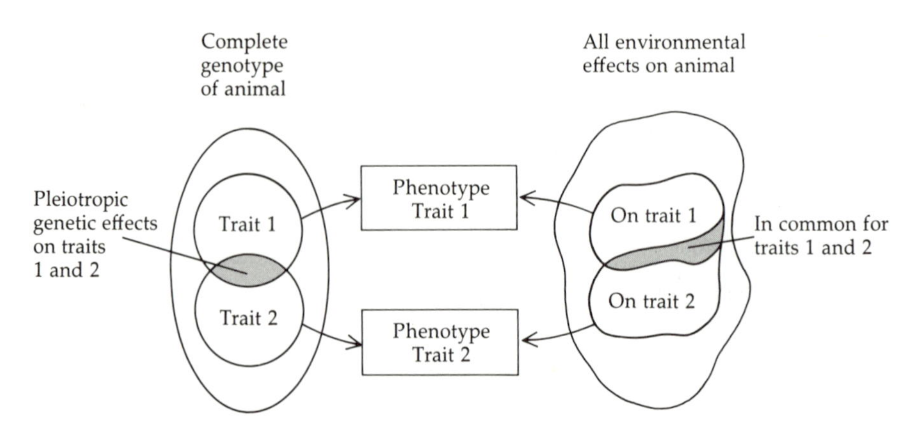

Figure 14-1 Diagram showing origin of genetic and environmental correlations between two traits. Shaded areas show genetic effects and environmental effects influencing both traits and correspond to genetic and environmental correlations.

bolically describes these correlations. Of all the genes of the animal, only relatively few directly affect differences in each trait. If some of the genes affect two traits, then the traits are correlated genetically. This phenomenon for the single locus model was called *pleiotropy*. Similarly, out of all the possible environmental influences, only some operate on each trait. When some influence two traits, the traits are said to be environmentally correlated. The correlations that are needed for multiple-trait selection, however, are the genetic and phenotypic correlations.

Model for genetic and phenotypic correlations

Another approach to explaining phenotypic, genetic, and environmental correlations is to examine models for the two traits.

Let

$$X_1 = P_1 - \mu_1 = G_1 + E_1$$

be the phenotype for trait 1 which expresses the genetic and environmental effects for trait 1. Also, let

$$X_2 = P_2 - \mu_2 = G_2 + E_2$$

be the phenotype for trait 2 which expresses the genetic and environmental effects for trait 2.

From the rules in Section 11-3, the variances of $X_1 = G_1 + E_1$ and $X_2 = G_2 + E_2$ are

$$\sigma^2_{P_1} = \sigma^2_{G_1} + \sigma^2_{E_1}$$

and

$$\sigma^2_{P_2} = \sigma^2_{G_2} + \sigma^2_{E_2}$$

(with subscripts now needed to identify the traits). Similarly, the phenotypic covariance between the two traits is

$$\sigma_{P_1P_2} = \sigma_{G_1G_2} + \sigma_{E_1E_2}$$

The phenotypic covariance is the genetic covariance plus the environmental covariance between the traits. The genetic covariance can, in turn, be partitioned into parts due to additive, dominance, and epistatic genetic effects. Because selection is ordinarily for additive genetic value, the additive genetic covariance, $\sigma_{A_1A_2}$, between pairs of traits such as traits 1 and 2 will often be the only part of the genetic covariance used in selection index procedures and in calculating expected change in additive genetic value. In fact, genetic covariances are often estimated with the assumption that only additive genetic effects contribute to the total genetic covariance.

The definition of a correlation is the covariance between the two traits divided by the square root of the product of the variances of the two traits. The traits have been partitioned into genetic and environmental parts with the parts having corresponding covariances and correlations. Thus,

$$\text{phenotypic correlation} = r_{P_1,P_2} = \frac{\sigma_{P_1P_2}}{\sqrt{\sigma^2_{P_1}\sigma^2_{P_2}}}$$

$$\text{genetic correlation} = r_{G_1,G_2} = \frac{\sigma_{G_1G_2}}{\sqrt{\sigma^2_{G_1}\sigma^2_{G_2}}}$$

$$\text{environmental correlation} = r_{E_1,E_2} = \frac{\sigma_{E_1E_2}}{\sqrt{\sigma^2_{E_1}\sigma^2_{E_2}}}$$

and

$$\text{additive genetic correlation} = r_{A_1,A_2} = \frac{\sigma_{A_1A_2}}{\sqrt{\sigma^2_{A_1}\sigma^2_{A_2}}}$$

An assumption necessary for this representation of the phenotypic correlation is that there is no correlation between genetic and environmental effects.

In many situations phenotypic and additive genetic covariances are needed to calculate correlated responses or selection index weights. These covariances can be determined from the definitional form if the phenotypic and additive genetic correlations, the phenotypic variances (or standard deviations), and heritabilities of the traits are known. For example, because

$$r_{P_1,P_2} = \frac{\sigma_{P_1P_2}}{\sqrt{\sigma^2_{P_1}\sigma^2_{P_2}}}$$

then

$$\sigma_{P_1P_2} = r_{P_1,P_2}\sqrt{\sigma^2_{P_1}\sigma^2_{P_2}}$$

The equation for the additive genetic covariance is more complicated but is derived similarly by remembering that $\sigma^2_{A_1} = h^2_1\sigma^2_{P_1}$ and $\sigma^2_{A_2} = h^2_2\sigma^2_{P_2}$. Thus, because

$$r_{A_1,A_2} = \frac{\sigma_{A_1A_2}}{\sqrt{\sigma^2_{A_1}\sigma^2_{A_2}}}$$

then

$$\sigma_{A_1A_2} = r_{A_1,A_2}\sqrt{\sigma^2_{A_1}\sigma^2_{A_2}}$$

or equivalently,

$$\sigma_{A_1A_2} = r_{A_1,A_2}\sqrt{h^2_1h^2_2}\,\sigma_{P_1}\sigma_{P_2}$$

These formulas allow going from correlations to covariances or from covariances to correlations.

The next section will make use of genetic covariances to calculate expected genetic change in one trait if selection is based on another trait.

14-2 Correlated Responses If Selection Is on a Single Trait

Response in selected trait

The simplest case of correlated response is when selection is for one trait based on a record for that trait. The expected additive genetic superiority from selection for a trait, say trait 1, was shown in Section 13-1 to be

$$\Delta A_1 = r_{A_1,\hat{A}_1} i_p \sigma_{A_1}$$

When selection is based on a single record of the animal $r_{A_1,\hat{A}_1} = \sqrt{h_1^2}$, and because $\sigma_{A_1} = \sqrt{h_1^2 \sigma_{P_1}^2}$, the expected genetic superiority also can be written as

$$\Delta A_1 = h_1^2 i_p \sigma_{P_1}$$

If selection is based on phenotypic records, the *expected phenotypic superiority* of the selected group is $i_p \sigma_{P_1}$. Sometimes this term, $i_p \sigma_{P_1}$, is called the *selection differential* or, more precisely, the *expected phenotypic selection differential* when a fraction, p, is selected. The *actual phenotypic selection differential* obtained from a set of records is the average phenotypic record of the selected animals minus the average of the group from which selection occurred (see Section 13-8).

Response in correlated trait

Expected correlated genetic superiority in another trait, for example, trait 2, when selection is on trait 1 depends on the genetic correlation between the traits and their heritabilities:

$$\Delta A_2 = r_{A_1,A_2} \sqrt{h_1^2 h_2^2} (i_p \sigma_{P_2})$$

This formula can be used to monitor what is expected to happen to other traits when selection is for a single trait. The sign of the genetic correlation, r_{A_1,A_2}, is most important because it determines the direction of correlated change. The selection intensity factor, i_p, and the factors $\sqrt{h_1^2 h_2^2}$ and σ_{P_2} also determine the magnitude of the expected change.

Indirect selection

The formulas for expected genetic superiority and correlated response can be used to determine whether indirect selection would be more effective than direct selection.

If selection is directly for trait 2, expected genetic superiority is

$$\Delta A_2 = h_2^2 (i_p \sigma_{P_2})$$

If selection is indirectly for trait 2 by selecting for trait 1, the expected correlated superiority can be found from the genetic regression of trait 2 on trait 1, that is

${}_c\Delta A_2 = \{[\sigma_{A_1A_2}]/\sigma^2_{A_1}\}\Delta A_1$. This equation can be written, after simplification, as follows for the situation where selection is based on a single record of trait 1:

$${}_c\Delta A_2 = r_{A_1,A_2}\sqrt{h_1^2 h_2^2}\,(i_p\sigma_{P_2})$$

Thus, the relative selection progress for trait 2 by selection for trait 1 as compared to progress from direct selection for trait 2 is

$$\frac{{}_c\Delta A_2}{\Delta A_2} = \frac{r_{A_1,A_2}\sqrt{h_1^2}}{\sqrt{h_2^2}}$$

This equation shows that more genetic gain in trait 2 is expected by indirect selection than by direct selection if $r_{A_1,A_2}\sqrt{h_1^2}$ is greater than $\sqrt{h_2^2}$.

Indirect selection is often considered for traits that are expensive and difficult to measure or cannot be measured until the animal is quite old. For example, survival (s) to five years of age in Holstein cattle has a heritability of .04, but the genetic correlation with first lactation milk production (available by about three years of age) is .60, and the heritability of production is .30. Then the ratio of expected gain in survival by indirect selection for production to expected gain in survival by direct selection is:

$$\frac{{}_c\Delta A_s}{\Delta A_s} = \frac{.6\sqrt{.30}}{\sqrt{.04}} = 1.64$$

Thus, genetic gain in survival by indirect selection for survival by selecting for production early in life would be 64 percent faster than by selection directly for survival if generation intervals were the same. If the generation interval were smaller with indirect selection, as might be the situation, then the gain per year by indirect selection would be more than 64 percent faster than by direct selection.

Response from progeny testing

The preceding formulas apply to expected response, correlated responses, and indirect response for mass selection when selection is based on the performance of an animal. The major part of selection response in most species, however, comes from selection of sires which is often based on progeny records.

The corresponding formulas for selection responses when selection of sires is based on records of p half-sib progeny are as follows.

Expected genetic superiority of selected males by direct selection for trait 1 from p_1 progeny:

$$\Delta A_1 = r_{A_1,\hat{A}_1}\, i_p \sqrt{h_1^2 \sigma_{P_1}^2} = \left(\sqrt{\frac{p_1}{p_1 + \frac{4 - h_1^2}{h_1^2}}} \right) (i_p)(\sqrt{h_1^2})\sigma_{P1}$$

Expected genetic superiority of selected males by direct selection for trait 2 from p_2 progeny:

$$\Delta A_2 = \left(\sqrt{\frac{p_2}{p_2 + \frac{4 - h_2^2}{h_2^2}}} \right) (i_p)(\sqrt{h_2^2})\sigma_{P_2}$$

Expected correlated genetic superiority of selected males for trait 2 from selection for trait 1 (indirect for trait 2):

$$\Delta_{c} A_2 = \left(\sqrt{\frac{p_1}{p_1 + \frac{4 - h_1^2}{h_1^2}}} \right) (r_{A_1,A_2})(i_p)(\sqrt{h_2^2})\sigma_{P_2}$$

Ratio of indirect to direct superiority from selection of males is:

$$\frac{\Delta_{c} A_2}{\Delta A_2} = \frac{r_{A_1,A_2} \sqrt{\frac{p_1}{p_1 + \frac{4 - h_1^2}{h_1^2}}}}{\sqrt{\frac{p_2}{p_2 + \frac{4 - h_2^2}{h_2^2}}}}$$

Relative response in general

In general, when selection is from records of a single trait, the ratio of indirect to direct genetic response depends on the accuracies of evaluating each trait and the genetic correlation:

$$\frac{\Delta_{c} A_2}{\Delta A_2} = \frac{r_{A_1,A_2} r_{A_1,\hat{A}_1}}{r_{A_2,\hat{A}_2}}$$

Example of relative selection response

For an example of correlated response if sires are selected from evaluation of progeny, consider the previous example for survival (s) and milk production (m) in dairy cattle: with $h_{\mathrm{m}}^2 = .30$, $(4 - h_m^2)/h_m^2 = 12.3$, and with $h_s^2 = .04$, $(4 - h_s^2)/h_s^2 = 99.0$. The genetic correlation is .60. If all sires have 50 progeny with both production records and records on survival to 60 months, then the ratio of indirect to direct response to selection is

$$\frac{{}_C\Delta A_s}{\Delta A_s} = \frac{.6\sqrt{\dfrac{50}{50 + 12.3}}}{\sqrt{\dfrac{50}{50 + 99}}} = .93$$

Thus, with sire selection, indirect selection for survival is only 93 percent as effective as direct selection, whereas, with individual cow selection indirect selection was more effective (164 percent) than direct selection. Direct selection is more effective with sire selection because the ratio of the accuracies of evaluating sires for production (.90) and for survival (.58) is 1.55, but the ratio of the accuracies of cow evaluation (.55 and .20) is 2.75. If the extra two years needed to obtain the survival data are considered, genetic gain per year from indirect selection based on daughter production will be found to be greater than from direct selection of males for survival based on their daughters' survival. Situations in which indirect selection is more effective, however, are rare.

Many selection decisions are best made by considering more than one trait and by predicting total economic value.

14-3 Total Economic Value

The problem of selecting for total economic value is complicated for several reasons. First, genetic values, even for individual traits, can be estimated only with less than perfect accuracy. Second, proper economic values are difficult to determine for many traits, particularly those without any direct market value such as temperament or straightness of leg. Third, genetic correlations that are needed for multiple-trait selection are difficult to estimate accurately. Estimates of genetic correlations which are not close to, or are different in sign from, the true correlations (for example, a large and positive estimate when the true correlation is large and negative) would increase the number of incorrect selection decisions.

Economic genetic value

The theory of selecting for *economic genetic value* (also called *aggregate* or *overall* genetic value) is straightforward when economic genetic value is defined as the sum of the product of the additive genetic value for each trait and the trait's net economic value:

$$H_i = v_1 A_{1i} + v_2 A_{2i} + \cdots + v_m A_{mi}$$

where H_i is the overall genetic value in money units for animal i,

v_j's $(j = 1, \ldots, m)$ are net economic values per unit for the m economic traits (net value = market value − cost of production), and

A_{ji}'s $(j = 1, \ldots, m)$ are the additive genetic values of animal i for the m economic traits.

If a trait has no economic value, the corresponding v will be zero and genetic value for that trait drops out of the equation for overall genetic value.

Determination of net economic value for a trait is often difficult. Conformation traits, for example, may not have a direct value but may contribute to sale price for breeding stock or may contribute to longevity. Traits which have a selling price are easier to assign an economic value. For example, an added pound of milk per lactation and an added egg per year have a market price. The added costs to produce the extra pound of milk or the extra egg could be calculated reasonably well.

One approach to assigning economic values is first to assign economic values to traits that have direct values and then to assign relative values to other traits. Such a procedure seems, and is, imprecise, but expected responses to selection can be calculated for different relative values of the other traits to allow a basis for choosing appropriate economic values. As an example, weaning weight of beef calves usually has a direct value while the conformation score does not. Calculation of economic response in weaning weight could be done for different amounts of relative emphasis on weaning weight and conformation score. The results may allow a decision to be made as to how much selection emphasis to put on conformation score. The procedure for these complicated calculations will be illustrated in the next section.

Relative economic emphasis

Different traits are measured in different units, for example, pounds, inches, points; therefore the value of each trait is expressed as value per unit of that trait. Variability also is not the same for all traits. For example, milk yield per lactation has a standard deviation of 2500 pounds, and type score (conformation score), a standard deviation of 4 points. About 16 percent of all milk records would be 2500 pounds or more above the milk average, and about 16 percent of all type scores would be 4 points or more above the type average. Thus, the opportunity to select animals with milk records more than 2500 pounds above average is the same as the opportunity to select animals with type records more than 4 points above average. Relative economic emphasis accounts for this opportunity for selection and is the ratio of the economic value of an increase of one standard deviation in one trait to the economic value of an increase of one standard deviation in another trait. For example, if the net value of milk is $.08 per pound and the net value of a one point increase in type score is $100, the relative economic values would be

$$\text{for milk:}\quad (\$.08/\text{pound})(2500\ \text{pound}) = \$200$$

and

$$\text{for type:}\quad (\$100/\text{point})(4\ \text{points}) = \$400$$

The relative economic emphasis on milk and type for this example is the proportionality $200:$400, which is equivalent to a relative economic emphasis of 1 : 2.

Illustration of relative economic emphasis

Figure 14-2 illustrates the relative economic emphasis for milk production and type score. In this example, considering value per unit and opportunity for selection, the relative economic emphasis is 1 : 1; that is, 2500 pounds of milk has the same value as 4 points of type score.

The value of milk can be determined more easily than that of type. Therefore, one approach to calculating relative economic emphasis is to calculate expected genetic response in milk and type for various proportions of relative emphasis assigning as the base value the economic value for a 2500 pound (one standard deviation) increase in milk yield. Suppose the desired emphasis is 8 : 1 for milk : type, and that a standard deviation of milk is worth ($.08)(2500) = $200. The value of a standard deviation of milk—$200—is eight times the desired value of a standard deviation (4 points) of type, so that the value of a standard deviation of type must be $200/8 = $25. Then

$$(\text{value per point of type}) \times (\text{standard deviation of type}) = \$25$$

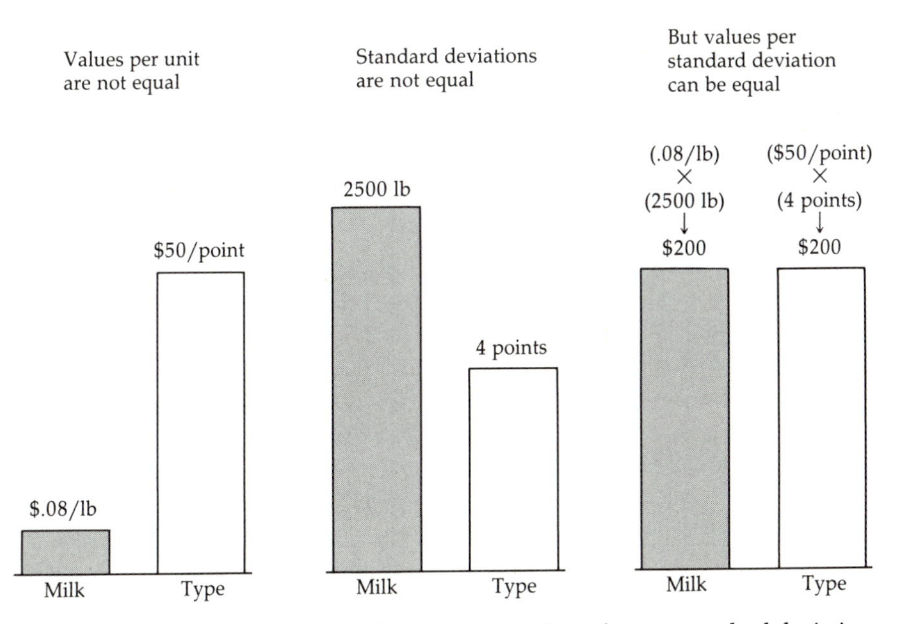

Figure 14-2 Relative economic emphasis considers the value per standard deviation of each trait so that the value, for example, of more milk and of a higher type score can be compared. The example is for the case of equal (1 : 1) relative selection emphasis.

Because the standard deviation of type is 4 points, the equation to find the corresponding value per point of type is:

$$(\text{value per point of type}) \times (4 \text{ points}) = \$25$$

and

$$\begin{aligned}(\text{value per point of type}) &= \$25/4 \text{ points}\\ &= \$6.25/\text{point}\end{aligned}$$

The results of calculating expected responses for different relative economic emphasis for this example are shown in Figure 14-3. The ratio, 6 : 0, means that all emphasis is on milk and none is on type. The ratio 0 : 6 (or 0 : N, 0 : any positive number) means all emphasis is on type and none on milk. When all economic emphasis is on one trait, any positive economic value assigned to that trait will cause the animals to be ranked in the same way as if any other positive economic value has been used. For example, if all predictions of genetic value for milk are multiplied by 5, then those predictions are ranked the same as when all predictions are multiplied by 2.

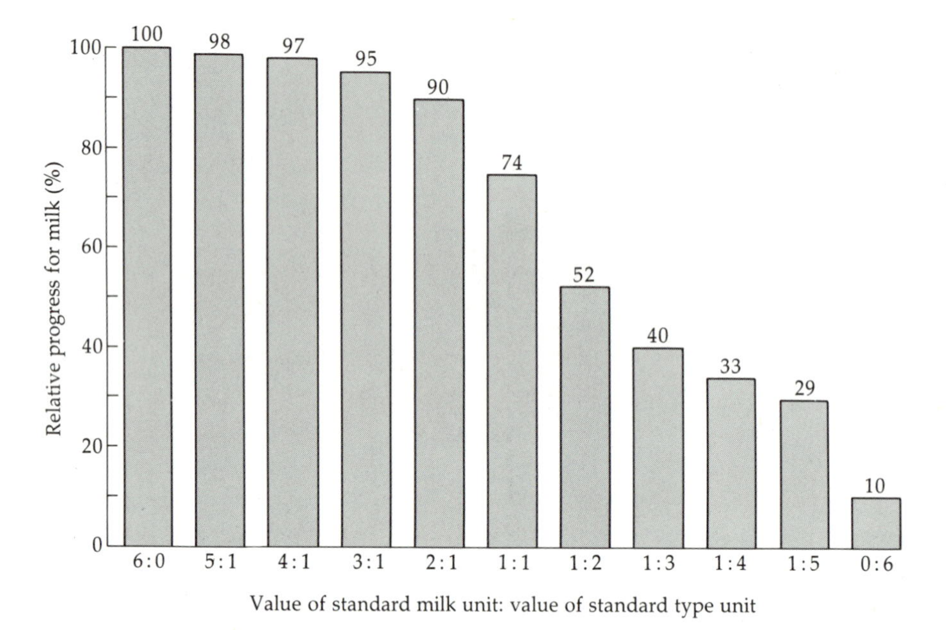

Figure 14-3 Comparison of progress in milk yield when selecting for both milk yield and type and when selecting for milk alone. (From G. H. Schmidt and L. D. Van Vleck, 1974. *Principles of Dairy Science,* W. H. Freeman and Company, New York, 1974, p. 273.)

Examination of the expected responses for milk, as shown in Figure 14-3, would give the breeder information for deciding how much emphasis to put on type relative to milk in a breeding program.

The example illustrated in Figure 14-3 shows that a relative economic emphasis of 3 : 1 gives essentially the same gain in milk yield as more emphasis on milk. The figure also shows that when type receives more emphasis, progress for milk yield is drastically reduced. The best method, however, of determining relative economic emphasis is to determine, if possible, the economic value for each trait based on an analysis of relationships to economic traits.

14-4 Prediction of Overall Economic Value

The following example illustrates, using only two traits, how to calculate the selection index to predict overall economic value and how to calculate expected genetic response to selection for a trait included in the index or for a trait not included in the index. Extending the procedures to the general case of many traits will become obvious.

In this section overall economic value for animal i is defined as:

$$H_i = v_1 A_{1i} + v_2 A_{2i}$$

where

H_i is the overall (holistic) economic genetic value,

A_{1i} is the additive genetic value for trait 1,

A_{2i} is the additive genetic value for trait 2, and

v_1 and v_2 are the net economic values per unit of traits 1 and 2.

If X_{1i} and X_{2i} are the phenotypic measures (adjusted for identifiable non-genetic factors) on traits 1 and 2 then there are two equivalent ways to predict H_i. Both ways require calculation of weighting factors, B_1 and B_2, for X_{1i} and X_{2i} for the index of overall genetic value $\hat{H}_i = B_1 X_{1i} + B_2 X_{2i}$.

One-step calculation of weighting factors

The general procedure for finding the selection index weights is to solve the following equations:

$$B_1 \sigma^2_{P_1} + B_2 \sigma_{P_1P_2} = \mathrm{Cov}(H_i, X_{1i})$$
$$B_1 \sigma_{P_2P_1} + B_2 \sigma^2_{P_2} = \mathrm{Cov}(H_i, X_{2i})$$

The selection index for animal i will be

$$\hat{H}_i = B_1 X_{1i} + B_2 X_{2i}$$

where B_1 and B_2 are the selection index weights for traits 1 and 2,

$\sigma^2_{P_1}$ is the phenotypic variance for trait 1,

$\sigma^2_{P_2}$ is the phenotypic variance for trait 2, and

$\sigma_{P_1P_2}$ is the phenotypic covariance between traits 1 and 2. (Note that $\sigma_{P_1P_2} = \sigma_{P_2P_1}$.)

The first right-hand side, $\mathrm{Cov}(H_i, X_{1i})$, is the covariance between overall genetic value, H_i, and trait 1, X_{1i}. The second right-hand side is the covariance between H_i and X_{2i}. The rules for covariances of sums can be used to determine these covariances. For example,

$$\begin{aligned}\text{Cov}(H_i, X_{1i}) &= \text{Cov}(v_1 A_{1i} + v_2 A_{2i}, A_{1i} + D_{1i} + I_{1i} + E_{1i}) \\ &= v_1 \sigma^2_{A_1} + v_2 \sigma_{A_1 A_2}\end{aligned}$$

and similarly

$$\text{Cov}(H_i, X_{2i}) = v_1 \sigma_{A_2 A_1} + v_2 \sigma^2_{A_2}$$

where $\sigma^2_{A_1}$ is the additive genetic variance for trait 1,
$\sigma^2_{A_2}$ is the additive genetic variance for trait 2, and
$\sigma_{A_1 A_2}$ is the additive genetic covariance between traits 1 and 2. (Note that $\sigma_{A_1 A_2} = \sigma_{A_2 A_1}$.)

The additive, dominance, and epistatic components of G_{1i}, that is, $A_{1i} + D_{1i} + I_{1i}$, were substituted for G_{1i} to show that the only effects contributing to the covariance with H_i are additive genetic values. The right-hand sides will be the same for any animal being evaluated from its own records. Thus the equations can be written as

$$\begin{aligned}B_1 \sigma^2_{P_1} + B_2 \sigma_{P_1 P_2} &= v_1 \sigma^2_{A_1} + v_2 \sigma_{A_1 A_2} \\ B_1 \sigma_{P_2 P_1} + B_2 \sigma^2_{P_2} &= v_1 \sigma_{A_1 A_2} + v_2 \sigma^2_{A_2}\end{aligned}$$

Although these equations appear quite formidable, all of the terms except B_1 and B_2 are numerical values once traits 1 and 2 are defined. Before considering a numerical example, the other method of predicting overall genetic value will be described.

Two-step calculation of weighting factors

The alternative method is (1) to calculate an index for each trait separately using the records for all traits and (2) to combine the separate indexes into a single index. The index for trait 1 will be

$$\hat{A}_{1i} = b_{11} X_{1i} + b_{21} X_{2i}$$

The weights, b_{11} and b_{21}, for the index for trait 1 are obtained from the equations

$$\begin{aligned}b_{11} \sigma^2_{P_1} + b_{21} \sigma_{P_1 P_2} &= \text{Cov}(A_{1i}, X_{1i}) \\ b_{11} \sigma_{P_2 P_1} + b_{21} \sigma^2_{P_2} &= \text{Cov}(A_{1i}, X_{2i})\end{aligned}$$

By the rules for covariances,

$$\mathrm{Cov}(A_{1i}, X_{1i}) = \sigma^2_{A1}$$

and

$$\mathrm{Cov}(A_{1i}, X_{2i}) = \sigma_{A_1A_2}$$

The notation for the weights is that the second subscript refers to the trait being predicted and the first subscript refers to the trait being weighted; that is, b_{21} is the weight for X_{2i} when predicting A_{1i}.

Similarly the index for trait 2 will be

$$\hat{A}_{2i} = b_{12}X_{1i} + b_{22}X_{2i}$$

with b_{12} and b_{22} being calculated from

$$b_{12}\sigma^2_{P_1} + b_{22}\sigma_{P_1P_2} = \sigma_{A_2A_1}$$
$$b_{12}\sigma_{P_2P_1} + b_{22}\sigma^2_{P_2} = \sigma^2_{A_2}$$

A comparison of the equations which define the weights shows that (B_1, B_2), (b_{11}, b_{21}), and (b_{12}, b_{22}) have the same coefficients $\sigma^2_{P_1}$, $\sigma_{P_1P_2}$ (or, equivalently, $\sigma_{P_2P_1}$), and $\sigma^2_{P_2}$. Only the right-hand sides of the equations are different for predicting $H_i = v_1A_{1i} + v_2A_{2i}$ or A_{1i} or A_{2i}. The different right-hand sides for different predictions force the weights to be different for the three indexes.

The second step is to combine the separate indexes. The way to remember how the indexes are combined is to remember the definition of overall economic value, so that

$$\hat{H}_i = v_1\hat{A}_{1i} + v_2\hat{A}_{2i}$$

The equation can be rewritten in terms of the separate indexes as

$$\hat{H}_i = v_1(b_{11}X_{1i} + b_{21}X_{2i}) + v_2(b_{12}X_{1i} + b_{22}X_{2i})$$

By combining terms, $\hat{H}_i$ also can be rewritten as

$$\hat{H}_i = (v_1b_{11} + v_2b_{12})X_{1i} + (v_1b_{21} + v_2b_{22})X_{2i}$$

Some rather difficult algebra can be used to show that

$$B_1 = v_1b_{11} + v_2b_{12}$$

and

$$B_2 = v_1 b_{21} + v_2 b_{22}$$

and therefore that the two procedures are algebraically identical. The advantage of the two-step procedure is that, if different economic values become appropriate, the overall index can be changed without setting up and resolving the equations for the weights. For example, if the economic values should be v_1^* and v_2^*, then

$$\hat{H}_i^* = v_1^*(b_{11}X_{1i} + b_{21}X_{2i}) + v_2^*(b_{12}X_{1i} + b_{22}X_{2i})$$

The extension of the procedures described for two traits to many traits should be obvious from the pattern used to set up the equations for two traits. The coefficients of the weights (B's or b's) are the appropriate phenotypic variances and covariances, and the right-hand sides are determined from the rules for covariances of sums and follow the pattern developed with two traits. Example 14-1 illustrates the calculation of an index for two traits.

Example 14-1

Let trait 1 be milk yield and trait 2 be protein yield. The net values are \$.09 for a pound of milk and \$.80 for a pound of protein. Use the correlations and phenotypic variances in Table 14-1 to find the weights to predict overall genetic value for milk and protein from a milk record, X_{1i}, and a protein record, X_{2i}, for animal i. Convert the correlations to covariances using the formulas of Section 14-1.

$$\begin{aligned}
\sigma_{P_1P_2} &= r_{P_1,P_2}\sqrt{\sigma^2_{P_1}\sigma^2_{P_2}} \\
&= (.95)\sqrt{(6{,}250{,}000)(6{,}600)} = 192{,}946 \\
\sigma_{A_1A_2} &= r_{A_1,A_2}\sqrt{h_1^2h_2^2}\,\sigma_{P_1}\sigma_{P_2} \\
&= (.80)\sqrt{(.25)(.20)}(\sqrt{6{,}250{,}000})(\sqrt{6{,}600}) = 36{,}332
\end{aligned}$$

Convert the heritabilities to additive genetic variances:

$$\begin{aligned}
\sigma^2_{A_1} &= h_1^2\sigma^2_{P_1} \\
&= .25(6{,}250{,}000) = 1{,}562{,}500 \\
\sigma^2_{A_2} &= h_2^2\sigma^2_{P_2} \\
&= .20(6{,}600) = 1{,}320
\end{aligned}$$

Table 14-1 Phenotypic* and genetic† correlations between milk yield, protein yield, and protein content and their heritabilities and phenotypic variances.

	Milk (lb)	Protein (lb)	Protein (%)	Variance
Milk (lb)	—	.80	−.45	6,250,000 lb²
Protein (lb)	.95	—	.15	6,600 lb²
Protein (%)	−.35	.05	—	.036 %²

* Phenotypic correlations above the diagonal.
† Genetic correlations below the diagonal.
Source: Adapted from K. R. Butcher, et al., *Journal of Dairy Science* **50:**185–193, 1967.

Two-step method The equations to find the weights to predict additive genetic value for milk are

$$b_{11}(6{,}250{,}000) + b_{21}(192{,}946) = (1{,}562{,}500)$$
$$b_{11}(192{,}946) + b_{21}(6{,}600) \quad = (36{,}332)$$

Then,

$$\hat{A}_{1i} = .8211X_{1i} - 18.5000X_{2i}$$

The equations to find the weights to predict additive genetic value for protein are

$$b_{12}(6{,}250{,}000) + b_{22}(192{,}946) = 36{,}332$$
$$b_{12}(192{,}946) + b_{22}(6{,}600) \quad = 1{,}320$$

Then,

$$\hat{A}_{2i} = -.0037X_{1i} + .3083X_{2i}$$

The overall index, $H_i = v_1\hat{A}_{1i} + v_2\hat{A}_{2i}$, is

$$\hat{H}_i = .09(.8211X_{1i} - 18.5000X_{2i}) + .80(-.0037X_{1i} + .3083X_{2i})$$

or, after combining like terms,

$$\hat{H}_i = .0709X_{1i} - 1.4184X_{2i}$$

The weights reflect three factors—the phenotypic and genetic variability of the traits, the economic values of the traits, and the phenotypic and genetic correlations between the traits.

One-step method The right-hand sides are

$$\text{Cov}(.09A_{1i} + .80A_{2i}, X_{1i}) = .09\sigma^2_{A_1} + .80\sigma_{A_1A_2} = 169{,}691$$
$$\text{Cov}(.09A_{1i} + .80A_{2i}, X_{2i}) = .09\sigma_{A_1A_2} + .80\sigma^2_{A_2} = 4{,}326$$

The equations to find the weights are

$$B_1(6{,}250{,}000) + B_2(192{,}946) = 169{,}691$$
$$B_1(192{,}946) \quad + B_2(6{,}600) \quad = 4{,}326$$

Solutions for B_1 and B_2 provide the overall index,

$$\hat{H}_i = .0709X_{1i} - 1.4182X_{2i}$$

which, except for rounding, is the same as from the two-step method.

14-5 Correlated Responses If Selection Is for Overall Genetic Value

Calculation of the expected superiority of the selected group for overall genetic value requires simple arithmetic and is not difficult. Similarly, the expected correlated response for any trait can be calculated if the additive genetic covariances between that trait and the traits used in the index are known.

Overall response

The expected superiority in overall genetic value, ΔH, is in economic units. The calculation requires:

1. the selection intensity factor for the fraction p selected, i_p;
2. the standard deviation of $\hat{H}_i$, the prediction of H, which will be denoted as $\sigma_{\hat{H}}$; and
3. the multiplication of i_p by $\sigma_{\hat{H}}$; that is,

$$\Delta H = (i_p)(\sigma_{\hat{H}})$$

The standard deviation of $\hat{H}_i$ is the square root of the variance of $\hat{H}_i$ which can be found from the rules listed in section 11-3. For two traits in the index such that the index is $\hat{H}_i = B_1X_{1i} + B_2X_{2i}$, then

$$\sigma^2_{\hat{H}} = B_1^2\sigma^2_{P_1} + B_2^2\sigma^2_{P_2} + 2B_1B_2\sigma_{P_1P_2}$$

The standard deviation is

$$\sigma_{\hat{H}} = \sqrt{\sigma^2_{\hat{H}}}$$

The numerical values of B_1 and B_2 are the weights for X_{1i} and X_{2i} for prediction of overall economic value, H_i. $\sigma^2_{P_1}$, $\sigma^2_{P_2}$, $\sigma_{P_1P_2}$ are numerical values of the phenotypic variances and covariances for X_{1i} and X_{2i} which are the coefficients of B_1 and B_2 in the equations that were solved to determine the numerical values of B_1 and B_2. Thus the calculation of $\sigma_{\hat{H}}$ simply requires substituting the numerical values for the symbols.

Correlated response for any trait

The expected correlated change for any trait, for example trait c, can be calculated easily from $\sigma_{\hat{H}}$, the index weights B_1 and B_2, and the additive genetic covariances between trait c and traits 1 and 2.

In general, the expected change in any trait is obtained from the regression of the additive genetic value of trait c on the index, $\hat{H}$. Thus

$$\Delta A_c = [\text{Cov}(A_{ci}, \hat{H}_i)/\sigma^2_{\hat{H}}](\Delta H)$$

where the regression coefficient is the covariance between A_{ci} and the index $\hat{H}_i$, divided by the variance of the index. The regression coefficient, as usual, determines the number of units of change in A_c that are expected for each unit change in $\hat{H}$. The average $\hat{H}$ of the selected group is expected to be $\Delta H = (i_p)(\sigma_{\hat{H}})$. Thus ΔA_c is the expected (average) additive genetic value for trait c when the overall genetic value is expected to be ΔH.

The equation for correlated response can be rewritten in several ways:

$$\Delta A_c = [\text{Cov}(A_{ci}, \hat{H}_i)(i_p)(\sigma_{\hat{H}})]/\sigma^2_{\hat{H}}$$
$$\Delta A_c = [\text{Cov}(A_{ci}, \hat{H}_i)/\sigma_{\hat{H}}](i_p)$$

Only $\text{Cov}(A_{ci}, \hat{H}_i)$ has not yet been determined. From the rules for covariances of sums;

$$\begin{aligned}\text{Cov}(A_{ci}, \hat{H}_i) &= \text{Cov}(A_{ci}, B_1X_{1i} + B_2X_{2i})\\ &= B_1\text{Cov}(A_{ci}, X_{1i}) + B_2\text{Cov}(A_{ci}X_{2i})\\ &= B_1\text{Cov}(A_c, A_1) + B_2\text{Cov}(A_c, A_2)\\ &= B_1\sigma_{A_cA_1} + B_2\sigma_{A_cA_2}\end{aligned}$$

The covariance is the sum of the products of the selection index weight for each trait and the additive genetic covariance between that trait and the correlated trait.

The equation for correlated response in trait c can now be written in its usual form:

$$\Delta A_c = [(B_1\sigma_{A_cA_1} + B_2\sigma_{A_cA_2})/\sigma_{\hat{H}}](i_p)$$

The formula also can be used if trait c is included in the index, that is, if $c = 1$ or $c = 2$, in which case one of the covariance terms in the numerator will be an additive genetic variance. For example, if $c = 2$, then $\sigma_{A_cA_2} = \sigma_{A_2A_2} = \sigma^2_{A_2}$.

When more than two traits are included in the index, calculation of $\sigma^2_{\hat{H}}$ and $\text{Cov}(A_{ci}, \hat{H}_i)$ follows the same pattern as with two traits. The variance of the index is the sum of all the phenotypic variances, each multiplied by the square of the corresponding weighting factor, and of all the different phenotypic covariances, each multiplied by twice the product of the two corresponding weighting factors. The covariance of A_{ci} with $\hat{H}_i$ for m traits in the index is

$$\text{Cov}(A_{ci}, \hat{H}_i) = B_1\sigma_{A_cA_1} + B_2\sigma_{A_cA_2} + \cdots + B_m\sigma_{A_cA_m}$$

Example 14-2 will demonstrate the calculation of expected superiority in overall genetic value and of expected correlated responses (1) for a trait not in the overall index and (2) for a trait included in the index for an index with two traits.

Example 14-2

Use the index of overall genetic value developed in Example 14-1 and the variances and correlations in Table 14-1 to calculate the following:

1. expected superiority from selection for overall genetic value,
2. expected correlated superiority for protein content,
3. expected correlated superiority for milk yield, and
4. expected correlated superiority for protein yield.

Assume selection of the top 20 percent based on the overall index with economic values of \$.09 for milk yield and \$.80 for protein yield.

The overall index was determined to be

$$\hat{H}_i = .0709\ X_{1i} - 1.4182\ X_{2i}$$

1. *Expected superiority for overall genetic value,* H
   a. the selection intensity factor, $i_p = i_{.20} = 1.400$, from Table 13-3.
   b. $\sigma^2_{\hat{H}} = (.0709)^2(6{,}250{,}000) + (-1.4182)^2(6{,}600)$
   $+ 2(.0709)(-1.4182)(192{,}946)$
   $= 5890; \sigma_{\hat{H}} = 76.75$
   c. $\Delta H = (1.400)(76.75)$
   $= \$107.45$
2. *Expected superiority for protein content*
   Let protein content be trait 3. The values for the additive genetic covariances between protein content and milk yield and between protein content and protein yield, $\sigma_{A_3A_1}$ and $\sigma_{A_3A_2}$ are needed.

   $$\sigma_{A_3A_1} = r_{A_3,A_1} \sqrt{h_3^2 h_1^2}\, \sigma_{P_3}\sigma_{P_1}$$
   $$= (-.35) \sqrt{(.50)(.25)} \sqrt{.036} \sqrt{6{,}250{,}000}$$
   $$= -58.6968$$

   $$\sigma_{A_3A_2} = (.05) \sqrt{(.50)(.20)} \sqrt{.036} \sqrt{6{,}600}$$
   $$= .2437$$

   a. $\text{Cov}(A_{3i}, \hat{H}_i) = B_1\sigma_{A_3A_1} + B_2\sigma_{A_3A_2}$
   $= (.0709)(-58.6968) + (-1.4182)(.2437)$
   $= -4.5072$
   b. $\Delta A_3 = [-4.5072/76.75](1.400) = -.082$ percent protein
3. *Expected superiority for milk yield*
   Values needed are:
   $\sigma_{A_1A_1} = \sigma^2_{A_1} = h_1^2\sigma^2_{P_1} = 1{,}562{,}500$
   $\sigma_{A_1A_2} = 36{,}332$ (from Example 14-1)
   a. $\text{Cov}(A_{1i}, \hat{H}_i) = B_1\sigma^2_{A_1} + B_2\sigma_{A_1A_2}$
   $= (.0709)(1{,}562{,}500) + (-1.4182)(36{,}322)$
   $= 59{,}255$
   b. $\Delta A_1 = [59{,}255/76.75](1.400) = 1081$ pounds milk
4. *Expected superiority for protein yield*
   Values needed are:
   $\sigma_{A_2A_1} = 36{,}332$
   $\sigma_{A_2A_2} = \sigma^2_{A_2} = h_2^2\sigma^2_{P_2} = 1{,}320$
   a. $\text{Cov}(A_{2i}, \hat{H}_i) = B_1\sigma_{A_2A_1} + B_2\sigma^2_{A_2}$
   $= (.0709)(36{,}332) + (-1.4182)(1{,}320)$
   $= 703.9148$
   b. $\Delta A_2 = [703.9148/76.75](1.400) = 12.84$ pounds protein

What can be seen from this example is that joint selection for milk yield and protein yield is expected to result in a decrease in protein content while substantially increasing yield of both milk and protein. The reasons for the expected decrease are the negative correlation of protein content with milk yield and the predominance of milk yield in the overall genetic value. The relative economic emphasis of milk yield to protein yield is

$$(.09)(\sqrt{6{,}250{,}000}):(.80)(\sqrt{6{,}600}) = 225:65$$

or approximately 3.5:1.

What can also be seen is that if the expected responses in the two traits having economic value are each multiplied by their economic values and summed, the sum equals the expected superiority in overall genetic value; that is.

$$\begin{aligned}\Delta H &= v_1\,\Delta A_1 + v_2\,\Delta A_2\\ &= \$.09(1081) + \$.80(12.84)\\ &= \$107.56\end{aligned}$$

which is the same result as in part 1 except for the rounding error.

14-6 Approximate Procedure for Predicting Total Genetic Value

The accuracy of predicting total genetic value is maximized when all traits on all relatives are included in the index for an animal. This index is the best index of overall genetic value. In some cases an approximate index may be nearly as accurate, and, in other cases, more accurate than the "best" index calculated from incorrect covariances. The approximate method is also computationally easy.

An approximate method of selecting for overall genetic value is to use methods from Chapter 12 to predict the additive genetic value for *each trait* separately using only records of *that* trait from the animal and its relatives. For example, records on weaning weight would be used to predict the genetic value for weaning weight, and records on yearling weight to predict genetic value for yearling weight, and so on. Records on weaning weight would not be used to predict genetic value for yearling weight. These predictions would be weighted by their net economic values as before:

$$\tilde{H}_i = v_1\tilde{A}_{1i} + v_2\tilde{A}_{2i} + \cdots + v_m\tilde{A}_{mi}$$

where the ~ indicates that the approximate procedure is used rather than the best selection index procedure.

Some research has suggested the approximate procedure may be better than the "best" procedure when the genetic correlations needed for the "best" procedure are estimated from small sets of records. Genetic correlations estimated from small data sets are notoriously unreliable. Weights for different traits in the overall index are very sensitive to the genetic correlations (covariances); therefore "best" indexes constructed using correlations estimated from small data sets may provide less accurate predictions than the approximate index which assumes all phenotypic and genetic correlations are zero.

In some cases the multiple-trait procedures may not be needed. If each trait is accurately predicted from records on that trait (for example, $r_{A_j,\hat{A}_j}$ becomes close to 1.0 for a sire from a progeny evaluation for trait j as the number of progeny becomes large), then records on other traits cannot increase the accuracy of evaluation of that trait much, if at all.

The approximate procedure is the same as the exact procedure when all of the genetic and phenotypic correlations are zero. Thus, if the correlations are close to zero, the approximation is quite good.

If traits are available only on the animal being evaluated, the approximate index becomes what is called the *base index*. Each trait is weighted by the product of its net economic value and heritability:

$$\tilde{H}_i = v_1\tilde{A}_{1i} + v_2\tilde{A}_{2i} + \cdots + v_m\tilde{A}_{mi}$$

and equivalently

$$\tilde{H}_i = v_1h_1^2X_{1i} + v_2h_2^2X_{2i} + \cdots + v_mh_m^2X_{mi}$$

where now $\tilde{A}_{ji} = h_j^2X_{ji}$.

If all traits are assumed to have equal heritabilities (h^2), the base index simplifies to weighting phenotypic records by their economic values:

$$\tilde{H}_i = h^2(v_1X_{1i} + v_2X_{2i} + \cdots + v_mX_{mi})$$

A very useful approximate index is one for males whereby each trait is evaluated from records of progeny for that trait. For example, for trait j,

$$\tilde{A}_{j,\text{ sire}} = \left(\frac{2p_j}{p_j + \dfrac{4 - h_j^2}{h_j^2}}\right)\overline{X}_{j,\text{progeny}}$$

where $\overline{X}_{j,\text{progeny}}$ is the average of records for trait j on p_j progeny and h_j^2 is the heritability of trait j. Then

$$\tilde{H}_{\text{sire}} = v_1\tilde{A}_{1,\text{sire}} + v_2\tilde{A}_{2,\text{sire}} + \cdots + v_m\tilde{A}_{m,\text{sire}}$$

In some cases the male may have performance (own) records on some traits and progeny evaluations on other traits. Then, if

$$\tilde{A}_{j,\text{sire}} = \left(\frac{2p_j}{p_j + \dfrac{4 - h_j^2}{h_j^2}} \right) \overline{X}_{j,\text{progeny}}$$

for the traits measured on progeny and $\tilde{A}_{k,\text{sire}} = h_k^2 X_{k,\text{sire}}$, where $X_{k,\text{sire}}$ represents a record of a trait measured on the male, the approximate index is a mixture of the base index and an approximate index based on progeny records.

14-7 Independent Culling Levels and Tandem Selection

In addition to the selection index for multiple trait selection, two other methods, tandem selection and independent culling levels, have been studied. If economic values of the traits are linear (each added unit of a trait has the same economic value as the first added unit of that trait) then the selection index is expected to result in the greatest economic gain from selection. The method of independent culling levels is never better than the selection index and never worse than tandem selection in expected linear economic response to selection, although the expected responses may be equal in some cases. These methods were first compared by Hazel and Lush (1942). Turner and Young (1969) have included a comprehensive discussion in their book.

Index selection, tandem selection, and independent culling levels are illustrated in Figure 14-4 for an equal fraction of animals selected, p, when only two traits are of interest.

Index selection weights each trait for economic value so that an animal with a high record for one trait may be selected even if the record for the other trait is very low. Thus, animals A and B in Figure 14-4*A* and animals X and W in Figure 14-4D would be selected by index selection although their records are low for either trait 1 or trait 2.

Tandem selection in any generation is selection on only one trait. If trait 1 is being selected then the fraction p of animals with high records for trait 1 are selected no matter how low their records are for trait 2. Thus, animal B in Figure 14-4*B* and animal X in Figure 14-4*E* are culled even though their records for trait 2 are relatively high. On the other hand, animals A and W are selected even though their records for trait 2 are low.

With independent culling levels, the levels for selection are set for each trait separately with the restriction that the overall fraction selected is p. As

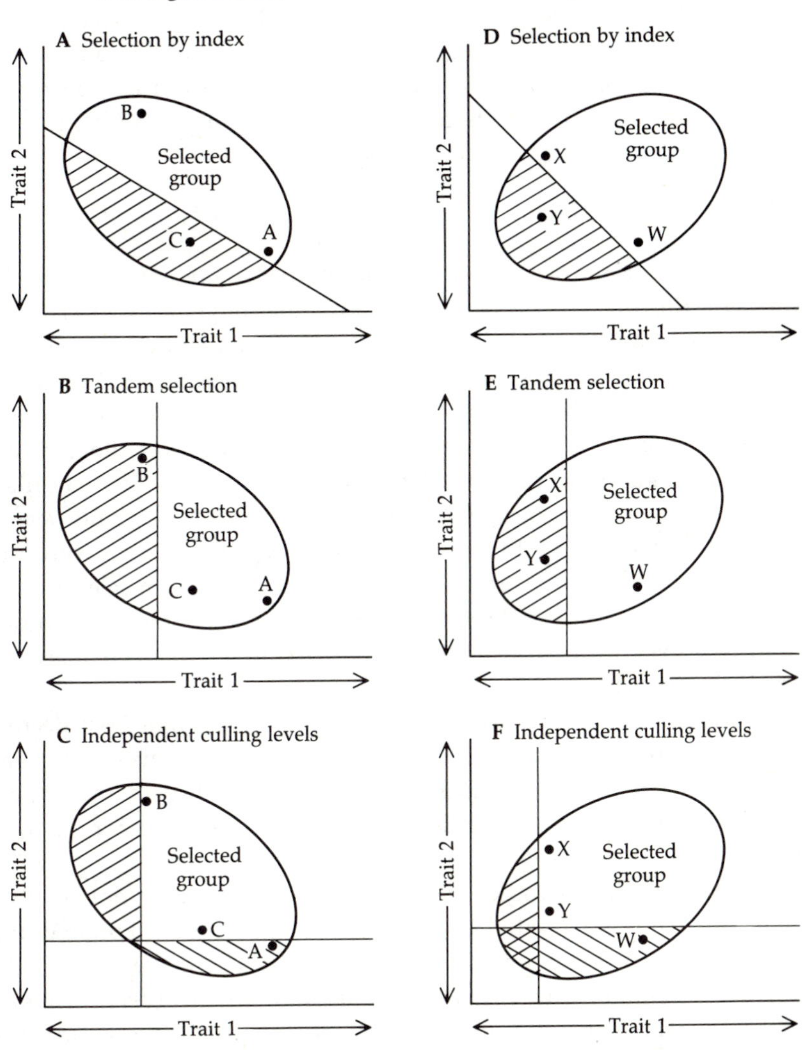

Figure 14-4 Plots of joint occurrences of traits 1 and 2. (*A* and *D*) Selection by index considers records of all traits depending on economic values, heritabilities, correlations and variances. (*B* and *E*) Tandem selection considers record of only one trait at a time no matter how good other traits are. (*C* and *F*) Selection by independent culling levels considers records of each trait independently so that a poor record on any trait will lead to culling.

with tandem selection, animals with high records for trait 2 may be culled if their records for trait 1 are low, but, with independent culling levels, animals with high records for trait 1 also may be culled if their records for trait 2 are low. In Figures 14-4*C* and 14-4*F*, animals A and W are culled because of low records for trait 2 even though their records for trait 1 are high enough to allow selection by index selection or tandem selection.

Tandem selection

Tandem selection is the ordering of traits according to importance followed by selection for the most important trait for a number of generations. Then the next most important trait is the basis of selection for a number of generations and so on. Thus, selection is for one trait at a time, ignoring all other traits. A practical advantage of the method is that only one trait needs to be measured at a time. In practice, however, a breeder may want to measure other traits to monitor changes in those traits in order to avoid difficulties which may arise if the trait currently under selection is negatively correlated with traits near the borderline of acceptable performance. The effect of the sign of the correlation of trait 1 with trait 2 is illustrated in Figure 14-5 for tandem selection with selection on trait 1. If the genetic correlation is positive then trait 2 will improve as trait 1 improves. If the correlation is zero then trait 2 will not change except for random drift as trait 1 improves. The troublesome case is that in which the genetic correlation is negative so that the average merit for trait 2 will decrease as trait 1 is improved. Determining the number of generations to select each trait is also difficult, although a high frequency of animals with unacceptable performance for another trait would be an indication to select for that other trait.

Determination of the order of importance of the traits is not simple. If tandem selection is practiced, the traits should be considered in order of their economic value. The major question is what criterion should be used to determine a trait's economic value. Values per unit for traits with different units and standard deviations are not comparable. Value per phenotypic standard deviation might be considered. Progress from selection, however, is proportional to the genetic standard deviation. Thus, the value per genetic standard deviation seems to be a more appropriate economic value for ranking traits. An even more accurate determination of economic value depends on correlations among the traits.

Section 14-2 outlined a procedure to calculate the correlated response in one trait when selection was for another trait. When selection is for trait 1 (based on an animal's own record) the expected responses are:

$$\Delta A_1 = i_p h_1^2 \sigma_{P_1}$$

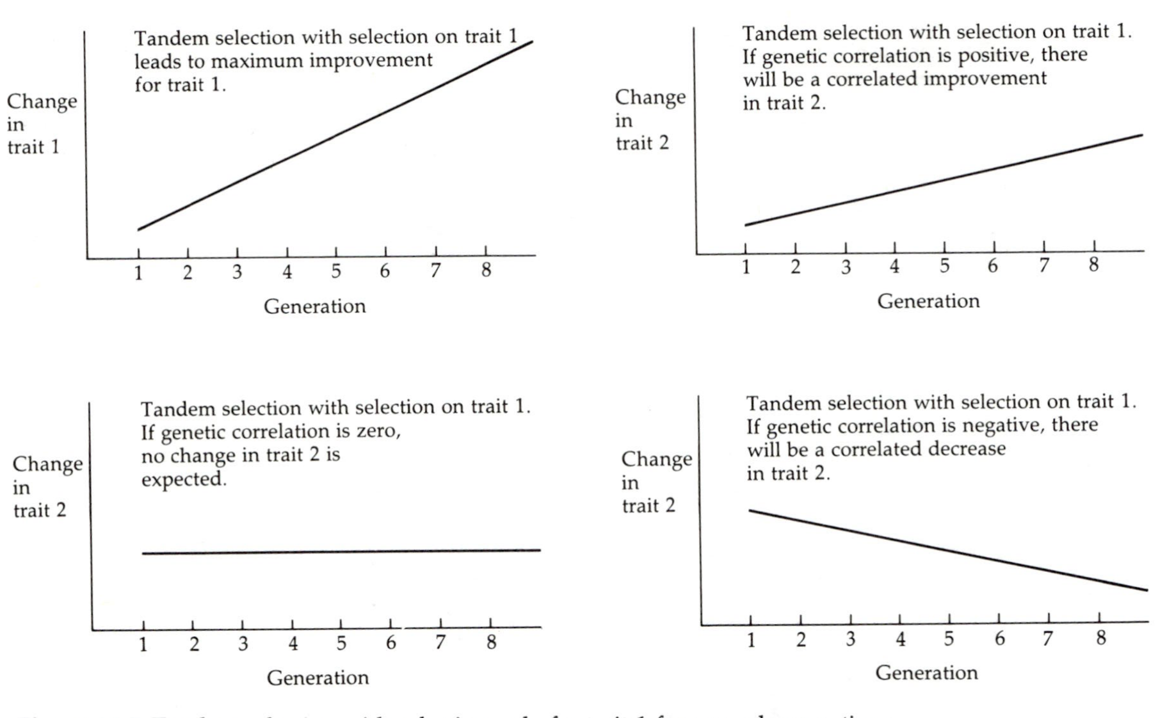

Figure 14-5 Tandem selection with selection only for trait 1 for several generations will provide maximum improvement in trait 1, but other traits will improve, stay the same, or become worse depending on the genetic correlation with trait 1.

for trait 1, and

$$ {}_{c}\Delta A_j = i_p r_{A_1A_j}\sqrt{h_1^2h_j^2}\,\sigma_{P_j} $$

for other traits ($j = 2, \ldots, m$). If the economic values are known and are linear, then the total economic response from selection for trait 1 is:

$$ {}_{1}\Delta H = v_1\,\Delta A_1 + v_2\,{}_{c}\Delta A_2 + \cdots + v_m\,{}_{c}\Delta A_m $$

The expected total economic response from selection for each trait could be calculated similarly. These expected economic responses (${}_{1}\Delta H$, ${}_{2}\Delta H$, . . . , ${}_{m}\Delta H$) could then be ranked to determine the order of importance of the traits (Turner and Young, 1969). Some difficulties with this approach are that

1. genetic correlations must be known,
2. correlations and heritabilities must remain the same over several generations of intense selection on single traits,
3. economic values are linear, and
4. economic values do not change over time.

Tandem selection may be useful when some economic values are nonlinear, as, for example, when the level of a trait falls below acceptable market standards. In marketing of milk, an increase or decrease in fat content from near the average has relatively little value. If, however, the fat content falls below the legal minimum required for sale, then fat content becomes very important. A similar example is crimp count of Merino wool, which determines price per pound of wool. Turner and Young (1969) suggested that tandem selection could be used occasionally to keep such traits above threshold levels.

Independent culling levels

The method of independent culling levels is relatively simple. For each trait, a certain fraction of remaining animals will be selected based on that trait alone, that is, for trait 1, a fraction p_1; for trait 2, a fraction, p_2; . . . , for trait m, a fraction p_m. Because selection is done independently for each trait, the total fraction selected is the product of the fractions selected for each trait: $p = p_1p_2 \cdots p_m$. For example, if, on a beef cattle ranch, 80 percent of heifer calves are selected at weaning based on weaning weight, $p_1 = .8$, and if 60 percent of the remainder are selected at one year of age on yearling weight, $p_2 = .6$, then $p = (.8)(.6) = .48$, the fraction surviving to enter the breeding pasture. In practice, p, the final fraction selected, is predetermined by the breeding requirements. Any number of combinations of the fractions saved for each trait can be found to give the same total p. For example, $p_1 = .6$ and $p_2 = .8$ would also give $p = .48$. The difficulty lies in determining how much selection pressure should be devoted to each trait, that is, the optimum fraction selected for each trait.

If the traits are uncorrelated and normally distributed, with selection for each using only the animal's record, the expected response to selection for each trait is

$$\Delta A_j = i_{p_j} h_j^2 \sigma_{P_j}$$

where i_{p_j} is the selection intensity factor for the fraction, p_j, selected for trait j. If v_j is the economic value of trait j, the total economic response with independent culling levels for m traits is:

$$\Delta H = v_1 i_{p_1} h_1^2 \sigma_{P_1} + v_2 i_{p_2} h_2^2 \sigma_{P_2} + \cdots + v_m i_{p_m} h_m^2 \sigma_{P_m}$$

Even with uncorrelated traits, the problem of finding the optimum combination of fractions selected for all traits is complicated. With rapid computational facilities, different combinations of $p_1, \ldots, p_m$ could be tried, within the limit that their product equals p, to determine the largest ΔH.

If the traits are correlated, the problem of calculating expected responses is much more difficult (see, for example, Turner and Young, 1969).

The principle advantage of the method of independent culling levels for many species is that it can follow the biological development of the animal; that is, selection will occur in stages corresponding to level of maturity. For example, a fraction of calves could be culled at weaning, others as yearlings, and others at breeding age. This selection sequence would reduce the number of animals that must be maintained for the whole time period. Maintenance and growth costs as well as sale prices at different ages would become important considerations.

In many situations, some form of independent culling is almost always used. Animals which fall below an acceptable threshold for certain traits such as fertility, disease resistance, or temperament are likely to be culled more or less independently of their potential genetic values for traits with more clearly defined economic values. Such threshold traits do not have a linear economic value; above the threshold for acceptability, an increase in the trait may have only a relatively small value per unit, whereas falling below the threshold is reason for culling despite the redeeming qualities of other traits. The animals that survive the threshold culling levels then can be indexed for the other economic traits.

14-8 Summary

The additive genetic correlation between traits 1 and 2 is the additive genetic covariance between traits 1 and 2 divided by the square root of the product of the additive genetic variances:

$$r_{A_1,A_2} = \sigma_{A_1A_2}/\sqrt{\sigma^2_{A_1}\sigma^2_{A_2}}$$

Similarly the phenotypic correlation is

$$r_{P_1P_2} = \sigma_{P_1P_2}/\sqrt{\sigma^2_{P_1}\sigma^2_{P_2}}$$

Genetic and phenotypic correlations or covariances are needed when all traits are used to predict the genetic value of one trait or to predict total genetic value.

Overall (aggregate or economic) genetic value of animal i is

$$H_i = v_1A_{1i} + v_2A_{2i} + \cdots + v_m\mathrm{A}_{mi}$$

where the v's are the net economic values for the m economic traits and the A_{ji}'s are the additive genetic values of animal i for the $j = 1, \ldots, m$ economic traits.

The selection index prediction of H_i is to replace the A_{ji}'s with their predictions using an index of records on all traits for each:

$$\hat{H}_i = v_1\hat{A}_{1i} + v_2\hat{A}_{2i} + \cdots + v_m\hat{A}_{mi}$$

where $\hat{A}_{ji} = b_{1j}X_{1i} + b_{2j}X_{2i} + \cdots + b_{mj}X_{mi}$ with $b_{1j}, \ldots, b_{mj}$ being the weights to predict A_{ji} from records $X_{1i}, \ldots, X_{mi}$.

An equivalent index can be obtained directly from the selection index equations that have right-hand sides determined by economic values and additive genetic variances and covariances among the traits:

$$\hat{H}_i = B_1X_{1i} + B_2X_{2i} + \cdots + B_mX_{mi}$$

An approximate index includes only records on a trait to predict the additive genetic value for that trait. When only a single record on each trait of animal i (X_{ji}; $j = 1, \ldots, m$) is available, the approximation is the base index:

$$\tilde{H}_i = v_1h^2_1X_{1i} + v_2h^2_2X_{2i} + \cdots + v_mh^2_mX_{mi}$$

Relative economic emphasis is the economic value of an increase of one phenotypic standard deviation (σ_{P_1}) in one trait as compared to the value of an increase of one phenotypic standard deviation (σ_{P_2}) in another trait. Thus, relative economic emphasis for traits 1 and 2 is the ratio

$$v_1\sigma_{P_1} : v_2\sigma_{P_2}$$

When selection for trait 1 is based only on records of trait 1, the expected genetic superiority for trait 1 is

$$\Delta A_1 = i_p r_{A_1\hat{A}_1}\sigma_{A_1}$$

The expected correlated response in trait 2 is obtained from the genetic regression of trait 2 on trait 1:

$$C\Delta A_2 = [\sigma_{A_1A_2}/\sigma^2_{A_1}]\Delta A_1$$

Expected genetic superiority of selected animals for trait 1 when selection is based on a single record of each animal for trait 1 is

$$\Delta A_1 = h_1^2 i_p \sigma_{P_1}$$

where p is the fraction selected and i_p is the corresponding selection intensity factor. Expected indirect correlated response in trait 2 when selection is on trait 1 is

$$C\Delta A_2 = r_{A_1A_2}\sqrt{h_1^2 h_2^2}\, i_p \sigma_{P_2}$$

The ratio of indirect response in trait 2 from selection on trait 1 to direct response from selection on trait 2 is called relative selection efficiency. Relative selection efficiency depends on the genetic correlation between the traits and the ratio of accuracy of predicting genetic value of trait 1 from records on trait 1 to the accuracy of predicting genetic value of trait 2 from records on trait 2:

$$\frac{C\Delta A_2}{\Delta A_2} = \frac{r_{A_1A_2} r_{A_1\hat{A}_1}}{r_{A_2\hat{A}_2}}$$

If selection is based on a single record on the animal being evaluated, relative selection efficiency is

$$\frac{C\Delta A_2}{\Delta A_2} = \frac{r_{A_1A_2}\sqrt{h_1^2}}{\sqrt{h_2^2}}$$

If selection is based on progeny records, relative selection efficiency is

$$\frac{\Delta_c A_2}{\Delta A_2} = \frac{r_{A_1A_2}\sqrt{\dfrac{p_1}{p_1 + \dfrac{4 - h_1^2}{h_1^2}}}}{\sqrt{\dfrac{p_2}{p_2 + \dfrac{4 - h_2^2}{h_2^2}}}}$$

where p_1 and p_2 are the number of progeny with records for traits 1 and 2.

Further Readings

Falconer, D. S. 1981. *Introduction to Quantitative Genetics.* 2d ed. Longman. New York.

Hazel, L. N. 1943. The genetic basis for constructing selection indexes. *Genetics* **28**:476.

Hazel, L. N., and J. L. Lush. 1942. The efficiency of three methods of selection. *J. Heredity* **33**:393.

Pirchner, F. 1969. *Population Genetics in Animal Breeding.* W. H. Freeman and Company. New York.

Smith H. F. 1936. A discriminant function for plant selection. *Ann. Eugen.* **7**:240.

Turner, H. N., and S. S. Young. 1969. *Quantitative Genetics in Sheep Breeding.* Cornell University Press. Ithaca, New York.

Van Vleck, L. D. 1983. *Notes on the Theory and Application of Selection Principles for the Genetic Improvement of Animals.* Department of Animal Science, Cornell University. Ithaca, New York.

CHAPTER 15

Mating Systems

Previous chapters dealt in depth with the principles of selection for quantitative traits. Selection of animals with superior additive genetic values to be parents of the next generation results in increasing the frequency of "beneficial" genes at loci influencing the traits of interest. This chapter concentrates on alternative strategies for the mating of animals selected to be parents. These strategies, called mating systems, influence the gene combinations received by progeny.

The mating systems discussed in this chapter are

1. assortative mating,
2. inbreeding and crossing of inbred lines,
3. linebreeding,
4. crossbreeding, and
5. grading up.

Each of these systems influences the degree of homozygosity in the population. It is important to understand the consequences of either increasing or decreasing homozygosity through each system.

15-1 Assortative Mating

Assortative mating is a strategy based on the phenotypic expression of some characteristic. Assortative mating can be either positive or negative. Positive assortative mating is the mating of individuals of like phenotypes, and negative assortative mating is the mating of those of unlike phenotypes. With size as an example, mating large to large is positive assortative mating (as is small to small), and mating large to small is negative assortative mating. Positive and negative assortative mating change the phenotypic and genetic variation in the resulting progeny from that observed in the parents.

Positive assortative mating causes more variation in the progeny than does random mating of parents. To demonstrate this, assume a trait that is normally distributed in the parent population. The parents with above average phenotypes are inter se mated, as are the parents with below average phenotypes. The expected progeny distribution can be examined using the principles of selection response. The expected genetic average of progeny from inter se matings of above-average parents can be found by calculating ΔG for selection of the top 50 percent of the animals:

$$\Delta G_H = r_{G_H,\hat{G}} i_p \sigma_G$$

Assume the trait is measured once on each parent and has an h^2 of .36. ΔG for the high group becomes

$$\Delta G_H = .48\sigma_G$$

since $r_{G,\hat{G}} = \sqrt{h^2} = .6$ and i_p for 50 percent saved is .8 in large populations. The inter se mating of parents with below average phenotypes gives the negative result:

$$\Delta G_L = r_{G_L,\hat{G}}(-i_p)\sigma_G$$
$$= -.48\sigma_G$$

Hence, 50 percent of the progeny are expected to show a positive genetic change and the other half a negative change. Note the total genetic change, ΔG, is

$$\Delta G = (1/2)(\Delta G_H + \Delta G_L) = 0$$

The progeny have diverged while there is no expected change in the mean of the total population. The result is an increase in σ_G^2 and also in σ_P^2. Figure 15-1 shows the distribution of progeny breeding values for this example.

Negative assortative mating tends to decrease the variance in progeny as compared to random mating. The breeding value of a parent estimated from its own record is

$$\text{EBV}_i = h^2(P_i - \mu)$$

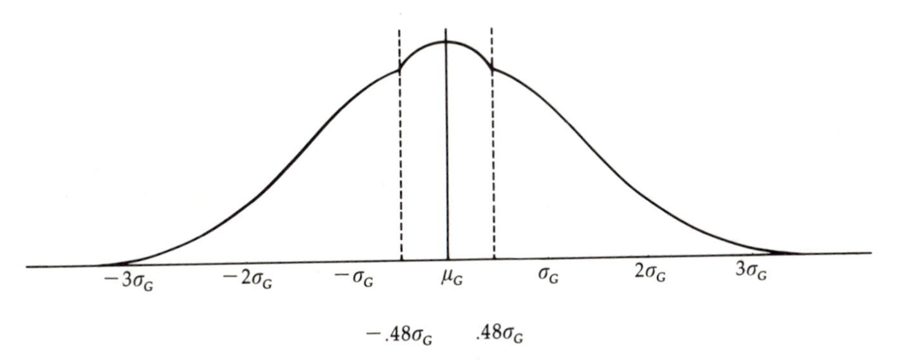

Figure 15-1 Distribution of breeding values after one generation of positive assortative mating.

The estimated breeding value of a progeny is half the sum of the estimated breeding values of the parents:

$$\begin{aligned} \mathrm{EBV}_{\text{progeny}} &= .5[h^2(\mathrm{P}_{\text{dam}} - \mu) + h^2(P_{\text{sire}} - \mu)] \\ &= .5h^2[(P_{\text{dam}} - \mu) + (P_{\text{sire}} - \mu)] \end{aligned}$$

If above average animals are mated to below average animals, there is, as a rule, the tendency for the positive phenotypic deviation to cancel the negative deviation. More progeny are thus expected to have breeding values close to the mean of zero than occurs with random mating, therefore σ_G^2 is reduced.

15-2 Inbreeding

As discussed in Chapter 10, inbreeding results from the mating of related animals. The intensity of inbreeding is a function of how closely the parents are related. For example, the mating of full sibs is a more intense form of inbreeding than mating half-sibs. This section examines inbreeding in regular systems of inbreeding and in small closed populations.

A regular system of inbreeding is a system in which, in each generation, matings are between designated types of relatives. For example, in each generation matings might be between half-sibs. Table 15-1 shows the average inbreeding coefficient of progeny by generation from matings of different types of relatives. Selfing, or the mating of an individual to itself, can occur in many plant species and is the most intense form of inbreeding.

Inbreeding can also occur in small closed populations because related animals will be mated by chance. A closed population is one into which no

Table 15-1 Expected average inbreeding by generation from systematic matings among relatives

Generation	Type of relationship		
	Selfing	Full sibs	Half-sibs
1	.5	.250	.125
2	.75	.375	.219
3	.875	.500	.305
4	.938	.594	.381
5	.969	.672	.449

outside (hence, unrelated) breeding animals are introduced. The increase in inbreeding in such populations is related to the number of males and females used as breeding animals. The decrease in the fraction of heterozygotes from one generation to the next in a closed population will be denoted as ΔF. An approximate formula for ΔF is

$$\Delta F = 1/(2N_e)$$

where N_e is the *effective number* of animals in the population and

$$1/N_e = 1/(4N_m) + 1/(4N_f)$$

where N_m and N_f are the number of males and females used as parents, respectively. Thus,

$$\Delta F = 1/(8N_m) + 1/(8N_f)$$

For most livestock species, the number of males is the major factor in determining the increase in inbreeding because far fewer males than females usually are used.

The average inbreeding coefficient at some generation, say t, can be written as a function of ΔF and t as

$$F_t = 1 - (1 - \Delta F)^t$$

which reflects the loss of a fraction, ΔF, of the heterozygotes in each generation and a corresponding increase in the fraction of homozygotes. The loss of heterozygosity (or gain in homozygosity) has several implications. First, inbreeding tends to "uncover" undesirable recessive genes, that is, lethals or semilethals. If an undesirable gene is in the population at a very low frequency, the probability of its expression is small. Mating of relatives that have an ancestor carrying the gene, however, will dramatically increase the frequency of expression. As a simple example, assume a sire carries a unique recessive gene not otherwise present in the population. Half of his progeny are expected to be carriers; hence, although the frequency of the gene may be infinitely small in the population, in his progeny the frequency is .25. Random mating of his progeny would produce the homozygous recessive genotype in one-eighth of his grandprogeny as compared to none if his progeny were mated to unrelated animals. In Chapter 8, inbreeding was used as one method of detecting lethal genes, specifically, the mating of a sire with his daughters to screen genes at all loci.

Second, inbreeding reduces genetic variation within an inbred line. If σ^2_{G0} is the genetic variation in the base population, the genetic variance among animals within the line in a subsequent generation, say the nth generation, is

$$\sigma^2_{Gn} = (1 - F_n)\sigma^2_{G0}$$

where F_n is the average inbreeding coefficient of animals in the nth generation. As F approaches 1, the genetic variation in the line approaches zero. Theoretically, all animals in the line have identical genotypes when $F_n = 1$ so that there can be no genetic variation. Highly inbred lines should not be as responsive to selection as noninbred lines. However, as individuals within a line become genetically more similar, the genetic difference between lines increases as a function of F_n. Figure 15-2 shows the division of a population into n inbred lines. The genetic variation between lines is $2F_n\sigma^2_{G0}$. Hence, the total variation in the population at generation n is $\sigma^2_{Gn} = (1 + F_n)\sigma^2_{G0}$ because the total genetic variance is the sum of the within and between line variances. In summary, at

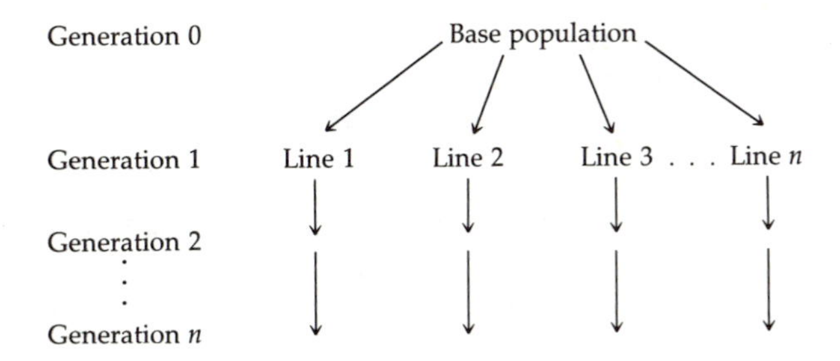

Figure 15-2 Division of base population into n inbred lines.

the nth generation of inbreeding,

$$\text{Within line genetic variance} = (1 - F_n)\sigma^2_{G0}$$
$$\text{Between line variance} = 2F_n\sigma^2_{G0}$$
$$\text{Total variance} = (1 + F_n)\sigma^2_{G0}$$

A third consequence of inbreeding is a phenomenon called inbreeding depression. Inbreeding depression is the decrease in the average performance of individuals in the population resulting from increased homozygosity of genes with sublethal and deleterious effects. This will be demonstrated using the single locus model. Assume a population where

Genotype	Frequency	Value
AA	p^2	a
Aa	$2pq$	d
aa	q^2	$-a$

From Chapter 9, the mean genotypic value is

$$m_0 = a(p - q) + 2pqd$$

If all individuals are selfed, then one-fourth of the progeny of the heterozygotes are AA, one-half are Aa, and one-fourth are aa. Hence, in the progeny after one generation of selfing,

Genotype	Frequency	Value
AA	$p^2 + .5pq$	a
Aa	pq	d
aa	$q^2 + .5pq$	$-a$

The new mean genotypic value is

$$m_1 = a(p - q) + pqd$$

The difference in generation means, $m_1 - m_0$, is $-pqd$, a decrease due to the loss of heterozygotes. For quantitative traits, the summation of these losses over all loci is defined as inbreeding depression. Note that no depression occurs if $d = 0$ because $m_0 = m_1 = a(p - q)$. Hence, for quantitative traits where the dominance variance, σ^2_D, is small relative to the additive genetic variance, σ^2_A, inbreeding depression is small.

15-3 Linecrossing

Figure 15-2 shows a population split into distinct breeding lines in which inbreeding was performed for n generations. As stated previously, variation among the animals within a line decreases as F increases, but the variation between the lines increases. One strategy for producing progeny in the next generation is to mate individuals from different inbred lines. This strategy, called linecrossing, takes advantage of both the increased homozygosity within a line and the differences between lines.

Highly inbred animals are homozygous at a large proportion of their loci which leads to less variation in the gametes they produce. Within a highly inbred line there is less variation among genotypes of animals; hence, the gametes produced by a line are genetically more uniform than gametes from a noninbred population. Progeny resulting from crossing two highly inbred lines are therefore generally uniform in genotype and consequently more uniform in their performance than are those from random mating populations. This uniformity is useful in intensive production systems where large individual variation may be a disadvantage. For example, chicken producers want broiler chickens to reach market size at the same time so that the group of birds can be efficiently processed as a unit. It is a costly process to repeatedly check the group for birds that have reached the desired weight.

A second reason for linecrossing is to take advantage of variation between lines. This variation results from differences between lines due to different alleles becoming "fixed" within lines. Assume the genotypes of individuals in two lines can be compared after n generations of inbreeding. Some loci in both lines would still be segregating (not necessarily the same ones). For loci that are fixed in both lines, the same allele may be fixed in both lines at some loci and different alleles fixed at other loci. Progeny of crosses between these lines will be heterozygous at these later loci. If dominance effects exist for a trait, the progeny will exceed the average performance of the parent lines. This phenomenon is termed *hybrid vigor* or *heterosis.* As an example, assume a trait controlled by genes at three loci with the following values:

Locus	Genotype	Value
1	*AA*	2
	Aa	2
	aa	0
2	*BB*	2
	Bb	2
	bb	0
3	*CC*	2
	Cc	2
	cc	0

Assume two lines exist with the following genotypes:

Line 1 *AA bb cc* Line 2 *aa BB CC*

A linecross progeny has the genotype *Aa Bb Cc*. The value of each genotype is obtained by summing genotypic values over all loci for individuals in each line. Assuming no epistatic effects, the values for lines 1 and 2 and their cross are

	Line 1	**Line 2**	**Cross**
Locus 1	$AA = 2$	$aa = 0$	$Aa = 2$
Locus 2	$bb = 0$	$BB = 2$	$Bb = 2$
Locus 3	$cc = 0$	$CC = 2$	$Cc = 2$
Total	2	4	6

With a completely additive inheritance model the mean predicted for the progeny would be the average of the two lines, in this example, three. However, the progeny value from summing over loci is 6 due to dominance. Note that the magnitude of heterosis depends on how different the lines are. For example, a third line may have the genotype *AA BB cc*, that has a total value of 4 (the same as line 2). When crossed with line 1, the progeny genotype is *AA Bb cc* with value 4 because two of the three loci in each line are identical. It is also important to note that if the value of the heterozygous genotype *at each locus* is equal to the average of homozygous values, as is true with no dominance, no heterosis is observed and the linecross progeny performance would equal the average performance of each line. Heterosis is observed only when gene frequencies are different between lines at loci where dominance exists.

Linecrossing can be systematically studied using a diallel cross. Assume nine lines are selected for crossing. Average performance of progeny of these crosses may be represented in a table as follows:

	Females from line					
Males from line	**1**	**2**	**3**	. . .	**9**	**Mean for row (male line)**
1		$\bar{X}_{1,2}$	$\bar{X}_{1,3}$	. . .	$\bar{X}_{1,9}$	$\bar{X}_{1.}$
2	$\bar{X}_{2,1}$		$\bar{X}_{2,3}$	. . .	$\bar{X}_{2,9}$	$\bar{X}_{2.}$
3	$\bar{X}_{3,1}$	$\bar{X}_{3,2}$		. . .	$\bar{X}_{3,9}$	$\bar{X}_{3.}$
.	.	.	.		.	.
.	.	.	.		.	.
.	.	.	.		.	.
9	$\bar{X}_{9,1}$	$\bar{X}_{9,2}$	$\bar{X}_{9,3}$	. . .		$\bar{X}_{9.}$
Mean for column (female line)	$\bar{X}_{.1}$	$\bar{X}_{.2}$	$\bar{X}_{.3}$	. . .	$\bar{X}_{.9}$	$\bar{X}_{..}$

The first subscript identifies the sire line, the second represents the dam line. The information in this table of crosses has three uses. First, the lines to breed to all other lines to produce the highest average performance can be determined by averaging the marginal means [e.g., $\frac{1}{2}(\overline{X}_{1.} + \overline{X}_{.1})$] for a line. Average performance is useful when deciding which lines to use in a general crossing program. The term used to describe average performance is *general combining ability.* Differences between lines for general combining ability reflect additive genetic variation. Second, the "best" cross for producing hybrid progeny for commercial use can be determined by comparing the averages for each specific cross:

$$(\overline{X}_{i,j} + \overline{X}_{j,i})/2$$

The performance of progeny of a specific cross is referred to as *specific combining ability* and is a function of dominance effects as well as additive genetic effects. Finally, information in this type of table shows, if differences between reciprocal crosses exist, whether it is the line which furnishes the female parent or male parent that is important. The difference in reciprocal crosses between two lines is

$$X_{i,j} - X_{j,i}$$

This represents the difference between progeny means from the two ways the progeny of two lines can be generated; in other words, the female comes from either line j or line i, with the corresponding effects of the maternal genotypes on her progeny. The importance of this concept will be developed when crossbreeding is discussed.

Several points with regard to linecrossing should be considered. First, success of developing inbred lines for linecrossing as a mechanism for producing commercial hybrids often depends on reproductive capability and generation interval of the species. Development of highly inbred lines in large domestic species, such as cattle, takes many years due to the long generation interval. Because inbreeding depression may cause some lines to be lost, the expense of developing many lines is large. For species with a high reproductive rate, selection can be effective in offsetting inbreeding depression in reproductive rate. Before initiating a program to develop inbred lines for crossing after n generations of inbreeding, the breeder must consider whether selection for the same number of generations would be as effective. Finally, evaluation of linecross progeny should be based on total productivity which is a composite of all productive traits. For example, line 1 may excel in one trait and be inferior in a second while the reverse may be true for line 2. The progeny obtained from a specific cross may not be superior to other crosses for either trait, but if both traits are important, the specific cross may be superior in total productivity to other crosses. This property of crossing is often termed *complementation.* Complementarity refers to combining breeds which excel in different characteristics. The crossbred progeny may not exceed the performance of the better breed for any trait but may exceed both parent breeds when the traits are combined in a total performance evaluation.

15-4 Linebreeding

Linebreeding is a system of mating which might be considered under inbreeding but which will be considered separately since the objective is slightly different and the amount of inbreeding may be kept low. Linebreeding is used to maintain a high degree of relationship of individuals in a given generation to some ancestor. The ancestor is usually some animal which has been deemed to be quite superior. Figure 15-3 shows an example of linebreeding. Assume for the pedigree in Figure 15-3 that animal A has been recognized as a superior individual. His progeny (animals in generation 1) are related to A by one-half and his grandprogeny (animals in generation 2) are related to him by one-fourth. The genotype of animal A is being diluted by one-half with each generation. By mating his grandprogeny, as exemplified by mating F and G, the relationship of H to A is one-fourth even though H is three generations removed.

Although the high degree of relationship has been maintained for essentially three generations, only a minimum amount of inbreeding has occurred. The inbreeding coefficient of both H and I is .03.

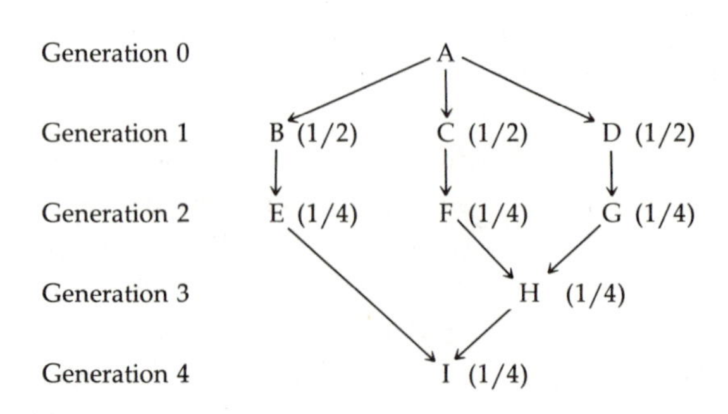

Figure 15-3 Example of linebreeding with the purpose of maintaining a high degree of relationship to animal A. (The fractions in parentheses represent the relationship of the animal to A.)

15-5 Crossbreeding

Linecrossing usually refers to crossing of inbred lines within a specific breed. Crossbreeding describes crossing individuals of different breeds, for example, Hereford and Angus cattle. Breeds, however, can be considered as lines within a species that differ in gene frequencies. These differences in frequency are usually a result of emphasis on different traits in selection, or of natural selection if the breeds originate from different geographical locations. An example of breed development for different traits is selection for milk yield versus beef production in cattle; an example of geographic selection is the development of tropical breeds of cattle as compared to temperate zone breeds.

Breeds represent tremendous resources of varying genetic material. The variation between breeds for most quantitative traits, such as size, represents opportunities to combine breeds to enhance productivity.

Crossbreeding often provides an opportunity to make progress in one generation that would require generations of selection to obtain. Assume, for example, a sheep breeder with Targhee ewes wishes to increase the frequency of multiple births. Selection, though effective, would take generations to increase substantially the fraction of twin lambings. However, crossing with a prolific breed such as the Finnish Landrace sheep would produce a crossbred population in one generation that would show a very large increase in frequency of twinning (see Table 15-2).

Complementarity is a second consideration in crossbreeding. Crossbreeding may be used to combine and improve the characteristics of two breeds. For example, the Angus is a beef breed notable for carcass quality; however, its performance in tropical environments may not match that of tropical breeds adapted to the area, such as Brahman cattle. Crossing may be used to combine

Table 15-2 Average number of lambs born to Targhee, Finnsheep, and Targhee-Finnsheep crossbred ewes

	Average number of lambs born	
Genotype of ewe	**Good environment**	**Poor environment**
Targhee	1.5	1.2
Finnsheep	2.6	No data
$\frac{1}{2}$ Targhee × $\frac{1}{2}$ Finnsheep	2.1	1.9
$\frac{3}{4}$ Targhee × $\frac{1}{4}$ Finnsheep	1.7	1.5

Data from G. E. Dickerson, 1977, *Crossbreeding Evaluation of Finnsheep and Some U.S. Breeds for Market Lamb Production,* North Central Regional Publication 246, Lincoln, Neb.: Agricultural Research Service, USDA, and University of Nebraska.

these characteristics and, in fact, such crosses have formed the basis for a breed called the Brangus.

As with linecrosses, crossbred progeny usually show heterosis in performance. The amount of heterosis again depends on the difference in gene frequencies at loci where dominance effects exist in the two breeds. Heterosis is defined as a function of the difference of the crossbred progeny from the average of the straightbred progeny. Assume, for example, that Angus and Hereford cattle are crossed:

		Female	
		Hereford	**Angus**
Male	**Hereford**	$\overline{HH}$	$\overline{HA}$
	Angus	$\overline{AH}$	$\overline{AA}$

The usual notation is to put the breed of sire first; hence, $\overline{AH}$ is the average of crossbreds with Angus sires and Hereford dams. $\overline{HA}$ is the average of the reciprocal cross, that is, Hereford sires on Angus dams. The heterosis for any trait is:

$$\text{heterosis} = \frac{\dfrac{\overline{AH} + \overline{HA}}{2} - \dfrac{\overline{HH} + \overline{AA}}{2}}{\dfrac{\overline{HH} + \overline{AA}}{2}}$$

This formula can be rearranged to predict the performance of crossbred animals; that is,

$$\frac{\overline{AH} + \overline{HA}}{2} = \frac{\overline{HH} + \overline{AA}}{2} + \text{heterosis}\left(\frac{\overline{HH} + \overline{AA}}{2}\right)$$

The first term, $(\overline{HH} + \overline{AA})/2$, represents the additive value of the breed combination and the second term, heterosis times $(\overline{HH} + \overline{AA})/2$, the difference from additive (the nonadditive contributions of the cross). For future reference the latter term will be represented by heterosis $(A \times B)$ where the terms in brackets represent the cross in question.

The difference in reciprocal crosses usually represents differences in maternal ability and is a second factor to consider. For the Hereford-Angus example, the reciprocal cross difference is $\overline{AH} - \overline{HA}$. The genotypes of the calves in each crossbred group are the same (with the exception that X chromosomes in males come from different breeds of dams). The major difference in this comparison is that the calves have a different breed of dam. For example, in considering weaning weight, milk producing ability of the dam can cause differences in weight even though the genotypes of the calves are the same.

Again a model can be used to demonstrate reciprocal differences. The performance of one cross, say Angus sires on Hereford cows, can be represented as:

$$\overline{AH} = \frac{\overline{HH} + \overline{AA}}{2} + \text{heterosis } (H \times A) + \text{maternal}(H)$$

The term maternal(H) refers to the maternal effect of the dam, the subscript indicating the breed of dam. The performance of the reciprocal cross, $\overline{HA}$ can be represented as

$$\overline{HA} = \frac{\overline{HH} + \overline{AA}}{2} + \text{heterosis } (H \times A) + \text{maternal}(A)$$

and the difference in reciprocal crosses,

$$\overline{HA} - \overline{AH} = \text{maternal}(A) - \text{maternal}(H),$$

simply reflects the maternal performance comparison of the two breeds.

15-6 Crossbreeding Programs

There are several types of programs which take advantage of the benefits of crossbreeding, such as terminal crosses and rotational crossbreeding.

Assume the original population consists of individuals of breed A. A terminal cross program consists of mating females in this population to males of breed B and using *all* crossbred progeny as market animals. The crossbred progeny never enter the breeding populations, and so the cross is called a *terminal cross* with the sires from breed B referred to as terminal cross sires (see Figure 15-4). This system is used to advantage when females are from a breed known for excellence in maternal characteristics and males are chosen to complement the maternal breed progeny performance; for example, in swine, boars from a breed known for fast growth are mated to sows from a breed with large litters. Two requirements of this system are (1) that a portion of the population of breed A females must be bred to breed A males to produce replacement females and (2) that a source of breed B males is available. The

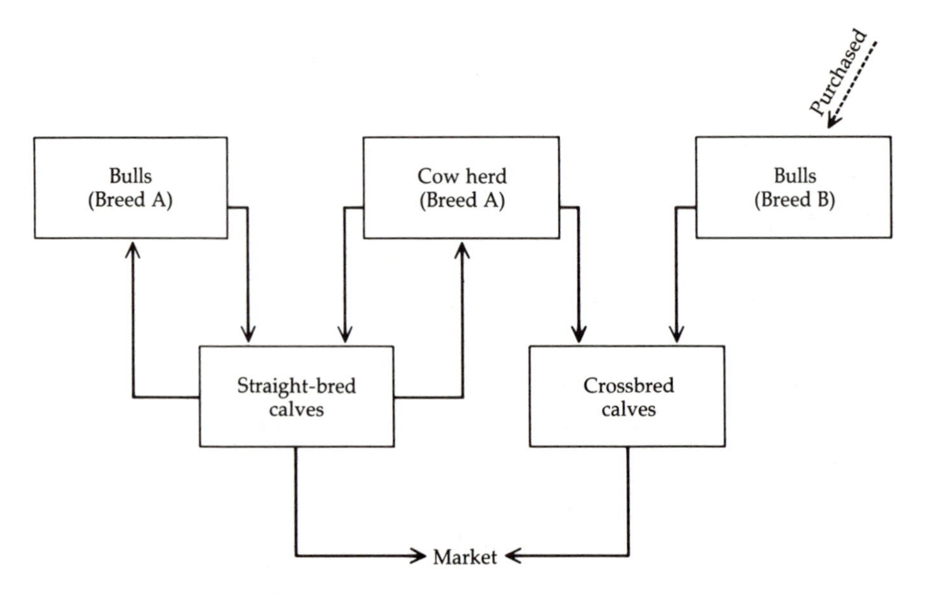

Figure 15-4 Diagram of terminal cross system.

former requirement may reduce drastically the proportion of females available for crossbreeding as shown by the following example.

Assume a population of breed A females which have only one progeny per year. Also assume that each year 30 percent of the A females need to be replaced either due to death, reproductive failure, poor health, or poor production. Even with 100 percent conception and survival from birth to replacement age, 60 percent of the breed A females must be bred to breed A males to produce enough replacement females; hence, only 40 percent are available for crossbreeding. Considering less than perfect reproductive performance in the population and progeny losses, the fraction of the population available for crossbreeding is reduced further. This system is more applicable for litter-bearing species as fewer females are required to produce straightbred replacements.

Other types of crossbreeding programs involve the use of crossbred females as replacements. Doing so takes advantage of any heterosis associated with reproduction, survival, and maternal ability and allows for combining breeds excelling in different maternal traits. The question arises, however, as to what breed to mate with these crossbred females. One option is a terminal cross to a third breed. One example of such a program is the so-called "AJEX" system (see Figure 15-5). This system uses the Angus breed (A) combined with Jerseys (J) as the maternal line. The AJ cows are then crossed (X) to sires from what were formerly exotic breeds (E) but which now includes any large mature

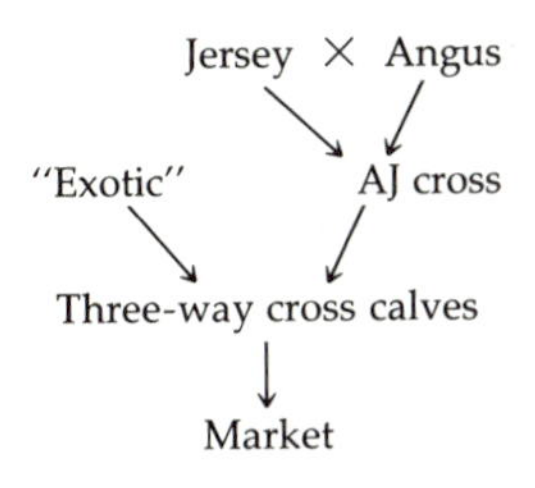

Figure 15-5 "AJEX" crossing system.

size breed. The AJ cow is early maturing, has high milk production, is small in mature size, and has desirable carcass qualities while the terminal sire from a large breed supplies the growth potential to the calf. This system requires a supply of crossbred heifers from herds maintaining straightbred animals as well as bulls from a third breed.

A second option to the use of crossbred females is to enter a rotational crossbreeding system. As suggested by the name, these systems involve rotating the use of sire breeds on crossbred cows. The next two sections consider two-and three-breed rotational cross programs.

15-7 Two-breed Rotational Cross Program

Assume crossbred females are created from breeds A and B. These females can be designated as *AB* or $\frac{1}{2}$A$\frac{1}{2}$B. The latter designation represents the proportion of genes coming from each breed. In a two-breed rotational program, the crossbred females are mated back to males from one of the original breeds which results in either

$$\begin{array}{ccc} \text{B mated to } \tfrac{1}{2}\text{A}\tfrac{1}{2}\text{B} & \text{or} & \text{A mated to } \tfrac{1}{2}\text{A}\tfrac{1}{2}\text{B} \\ \downarrow & & \downarrow \\ \tfrac{1}{4}\text{A}\tfrac{3}{4}\text{B} & & \tfrac{3}{4}\text{A}\tfrac{1}{4}\text{B} \end{array}$$

The genotype of the progeny has different proportions ($\frac{3}{4}$ or $\frac{1}{4}$) of the original breeds represented, depending on the sire breed. For all subsequent generations, the breed of sire is alternated (see Figure 15-6), giving the breed proportions across generations shown in Table 15-3. Note from Table 15-3 that the male used to mate with crossbred females is from the breed least represented in the female's genotype. After many generations the system reaches equilibrium (see generations n and $n + 1$ in Table 15-3) at two-thirds of genes from the most recent sire breed and one-third from the other breed.

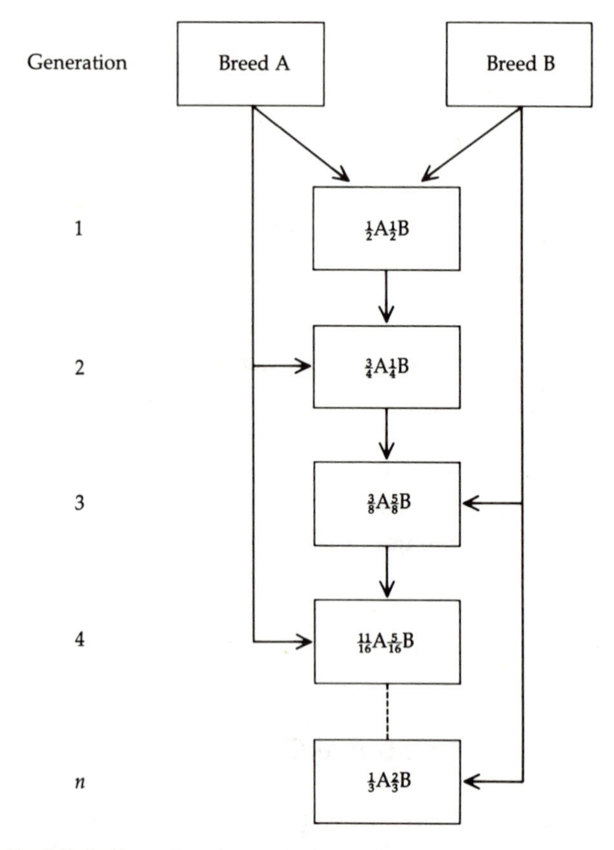

Figure 15-6 Two-breed rotational cross system.

A point of interest is that heterosis observed in each generation decreases. The predicted performance of the first cross from breed A males mated to breed B females can be represented as

$$\tfrac{1}{2}A\tfrac{1}{2}B = \frac{\overline{AA} + \overline{BB}}{2} + \text{heterosis } (A \times B) + \text{maternal}(B)$$

Crossing these progeny back to one of the original breeds, breed A, results in

$$\tfrac{3}{4}A\tfrac{1}{4}B = \tfrac{3}{4}AA + \tfrac{1}{4}BB + \text{heterosis } (A \times \tfrac{1}{2}A\tfrac{1}{2}B) + \text{maternal}(\tfrac{1}{2}A\tfrac{1}{2}B)$$

Note the fractions relating to the straightbred performance of the two breeds are $\frac{3}{4}$ and $\frac{1}{4}$, not $\frac{1}{2}$ for both as with the first cross. Also, the heterosis now is that which results from the mating $A \times \frac{1}{2}A\frac{1}{2}B$. To determine what heterosis to

Table 15-3 Expected proportion of genes from breeds represented in progeny by generation from a two-breed rotational cross program

Genotype				Breed %	
Sire	Dam	Generation	Progeny genotype	A	B
A	B	1	$\frac{1}{2}A\frac{1}{2}B$	50	50
A	$\frac{1}{2}A\frac{1}{2}B$	2	$\frac{3}{4}A\frac{1}{4}B$	75	25
B	$\frac{3}{4}A\frac{1}{4}B$	3	$\frac{3}{8}A\frac{5}{8}B$	38	62
A	$\frac{3}{8}A\frac{5}{8}B$	4	$\frac{11}{16}A\frac{5}{16}B$	69	31
B	$\frac{11}{16}A\frac{5}{16}B$	5	$\frac{11}{32}A\frac{21}{32}B$	34	66
.					
.					
.					
B	$\frac{2}{3}A\frac{1}{3}B$	n	$\frac{1}{3}A\frac{2}{3}B$	33	67
A	$\frac{1}{3}A\frac{2}{3}B$	$n+1$	$\frac{2}{3}A\frac{1}{3}B$	67	33

expect, the mating is carried through as ($\frac{1}{2}AA + \frac{1}{2}AB$), which shows the expected fraction of times an "A" gene from the sire combines with an "A" gene from the crossbred dam and the "A" gene combines with a "B" from the dam. Heterosis by definition is zero for the $\frac{1}{2}AA$, hence, what remains is one-half the heterosis from the original cross. Each subsequent generation can be examined similarly for the amount of the original heterosis expected. At equilibrium, two-thirds of the original heterosis is expected to be maintained.

15-8 Three-breed Rotational Cross Program

Assume crossbred females are obtained from crossing individuals from breeds A and B, that is, they are $\frac{1}{2}A\frac{1}{2}B$. These females are then mated to males from breed C, which gives

$$\begin{array}{c} \text{C mated to } \frac{1}{2}A\frac{1}{2}B \\ \downarrow \\ \frac{1}{2}C\frac{1}{4}A\frac{1}{4}B \end{array}$$

The expected performance of the $\frac{1}{2}C\frac{1}{4}A\frac{1}{4}B$ calves can be represented as:

$$\begin{aligned} \tfrac{1}{2}C\tfrac{1}{4}A\tfrac{1}{4}B = {} & \tfrac{1}{2}CC + \tfrac{1}{4}AA + \tfrac{1}{4}BB + \text{heterosis}(C \times \tfrac{1}{2}A\tfrac{1}{2}B) \\ & + \text{maternal}(\tfrac{1}{2}A\tfrac{1}{2}B) \end{aligned}$$

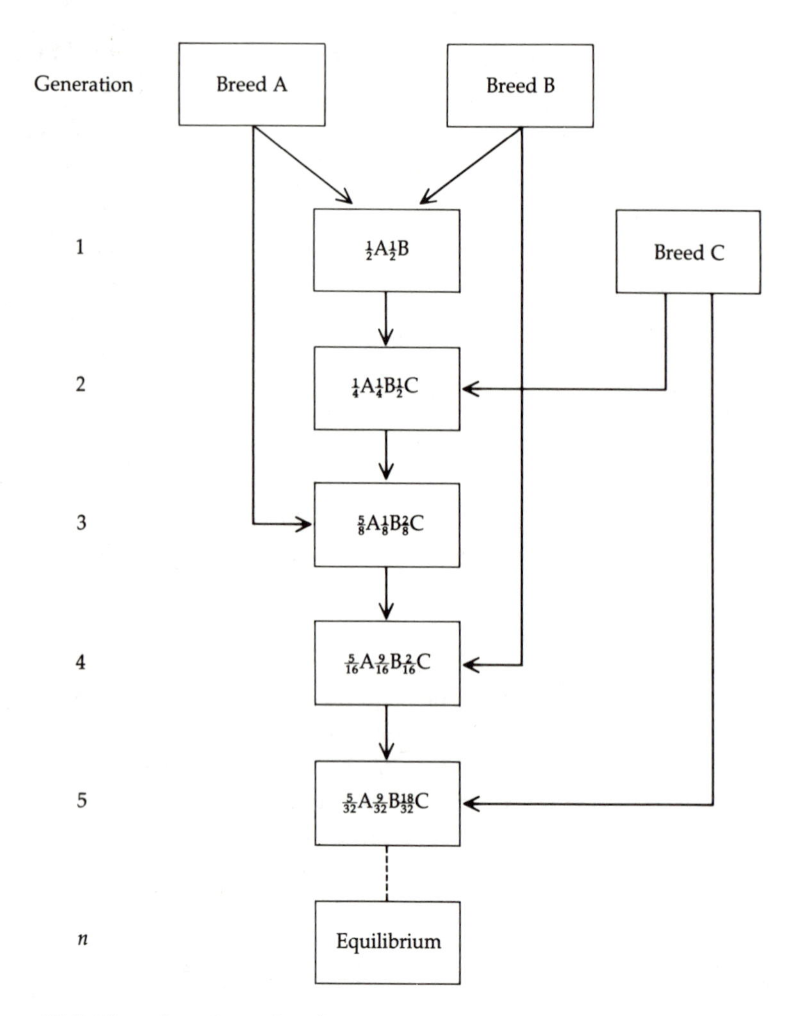

Figure 15-7 Three-breed rotational cross system.

The term for the contribution of heterosis can be written as

$$(1/2) \text{ heterosis } (C \times A) + (1/2) \text{ heterosis } (C \times B)$$

As with the two-breed system, there is a rotation through the breeds with each generation (see Figure 15-7) with the breed proportions shown in Table 15-4. Equilibrium is reached at proportions of approximately 57 percent of the breed of the sire, 14 percent of breed of the sire to be used next, and 29 percent of the other breed.

There are two points of interest relative to the rotational cross programs.

Table 15-4 Expected proportion of genes of breeds represented in progeny by generation from a three-breed rotational cross system

Genotype				Breed %		
Sire	Dam	Generation	Progeny genotype	A	B	C
A	B	1	$\frac{1}{2}A\frac{1}{2}B$	50	50	0
C	$\frac{1}{2}A\frac{1}{2}B$	2	$\frac{1}{4}A\frac{1}{4}B\frac{1}{2}C$	25	25	50
A	$\frac{1}{4}A\frac{1}{4}B\frac{1}{2}C$	3	$\frac{5}{8}A\frac{1}{8}B\frac{2}{8}C$	62	13	25
B	$\frac{5}{8}A\frac{1}{8}B\frac{2}{8}C$	4	$\frac{5}{16}A\frac{9}{16}B\frac{2}{16}C$	31	56	13
C	$\frac{5}{16}A\frac{9}{16}B\frac{2}{16}C$	5	$\frac{5}{32}A\frac{9}{32}B\frac{18}{32}C$	16	28	56

The first relates to the percentage of genes from each breed in each cross. In the first cross the progeny receive half of their genes from each parent; hence, the progeny genes are exactly $\frac{1}{2}$A and $\frac{1}{2}$B. However, in further crosses segregation can influence the progeny's genotype. For example, in the cross B $\times$ $\frac{1}{2}A\frac{1}{2}B$, the progeny receives half of its genes from the B male and half from the crossbred parent. The crossbred parent's chromosomes are segregating randomly; it can therefore pass on any fraction of genes between 0 and 1 that it received from either of its parents. The expectation of the progeny genotype is $\frac{1}{4}A\frac{3}{4}B$, but there is variation about this expectation as shown in Figure 15-8. At one extreme is the case in which the crossbred parent passes on all A genes so that the progeny is $\frac{1}{2}A\frac{1}{2}B$, and at the other extreme is when the crossbred parent passes all B genes so that the progeny is no A and all B. For species with many chromosomes, the extreme events are highly unlikely; however, variation from the expected $\frac{1}{4}A\frac{3}{4}B$ can exist as a result of random segregation. The second point is that selection can influence the proportion of a breed's genes represented in the animals selected to be parents for any crossbred progeny other than the first

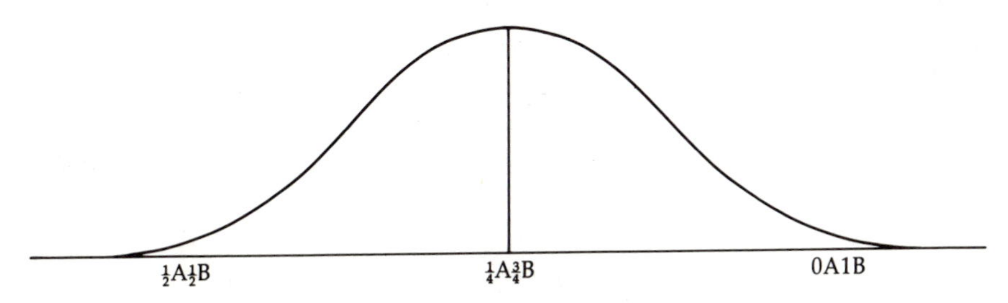

Figure 15-8 Mean and distribution of expected progeny genotypes from the backcross of the F_1 crossbred to the breed B parent.

cross. For example, if selection is for size and breed A is large and breed B is small, then, because genes for size are favored, selected individuals might average more than the expected $\frac{1}{4}$A of genes from breed A.

15-9 Grading Up

Assume a producer has animals of breed A and wants to change over to animals of breed B. A method of accomplishing this is to continually mate his females to males of breed B. In each generation, the proportion of the progeny genes traceable to breed B increases (see Table 15-5). After five generations, the progeny theoretically have 96.9 percent of their genes from breed B. The proportion of genes from breed A is halved with each generation to the point that there are essentially no genes of A remaining in the progeny after eight generations—$(\frac{1}{2})^8$. This procedure represents *grading up* from breed A to breed B and is essentially the procedure that was used in the United States to establish such beef breeds as Charolais and Simmental where initially only bulls were imported into the country.

The term grading up originally referred to successive matings of grade cattle to registered animals within the same breed. The results for each generation are exactly the same as shown in Table 15-5, with the proportion from B now meaning the proportion of genes received from registered parents.

15-10 Summary

Mating strategies influence gene combinations received by the progeny.

Positive assortative mating is the mating of individuals with like phenotypes. Positive assortative mating results in more variation in the progeny than would occur from random mating of parents. Negative assortative mating is the mating of individuals with unlike phenotypes. Negative assortative mating results in less variation in the progeny than would be observed from random mating. Mean performance of progeny in the next generation is expected to be the same as it was in the parent generation for both positive and negative assortative mating.

Inbreeding results from the mating of related animals. The intensity of inbreeding depends on how closely the parents are related. A regular system of inbreeding is a system in which matings in each generation are between desig-

Table 15-5 Expected progeny genotypes by generation for five generations of grading up to breed B

Genotype			Breed proportions	
Sire	Dam	Generation	A	B
B	A	1	.5	.5
B	$\frac{1}{2}$A$\frac{1}{2}$B	2	.25	.75
B	$\frac{1}{4}$A$\frac{3}{4}$B	3	.125	.875
B	$\frac{1}{8}$A$\frac{7}{8}$B	4	.063	.937
B	$\frac{1}{16}$A$\frac{15}{16}$B	5	.031	.969

nated types of relatives. Inbreeding can also occur in small, closed populations. A closed population is one into which no outside breeding animals are introduced. An approximate formula for the change in inbreeding coefficient from one generation to the next is

$$\Delta F = 1/(2N_e)$$

where N_e is the effective population size with

$$1/N_e = 1/(4N_m) + 1/(4N_f)$$

where N_m and N_f are the number of males and females used as parents in each generation. Inbreeding increases homozygosity and reduces genetic variation within an inbred line. However, genetic variation between inbred lines is increased so that total variation in a population of inbred lines is larger than in the base population.

Linecrossing is a system of crossing inbred lines that takes advantage of increased homozygosity within a line and differences between lines. General combining ability is defined as the mean performance of crossbred progeny for a line mated to all other lines. Specific combining ability is the mean performance of a cross between two specified lines.

Linebreeding maintains a high degree of relationship of a population to some ancestor sometimes with only a minimal amount of inbreeding.

Crossbreeding is a system of mating individuals from different breeds. Crossbreeding increases heterozygosity. Two considerations in the evaluation of a crossbreeding program are complementarity and heterosis. Heterosis, or hybrid vigor, is a result of dominance effects. Heterosis is measured as the deviation of crossbred progeny from what is expected from a completely additive genetic model. The fraction of heterosis is calculated as

$$\text{heterosis} = \left(\frac{\text{average of crossbred progeny} - \text{average of parent breeds}}{\text{average of parent breeds}} \right)$$

Grading up is the process of mating individuals of one breed to individuals of a second breed and back to resulting progeny in successive generations.

CHAPTER 16

The End or the Beginning?

Often when the last page of a textbook is closed the reader is finished with the subject. For many readers that will be true of this book—the end. For others the end of this book should be a signal to begin to apply the basic principles of genetics for the improvement of animals. The step through the threshold from principles to application is a long one, much too great to be covered in a "final" chapter. The goal of this chapter is to tie some of the basic principles to some general problems of application with the hope that those readers who are so inclined can begin to apply the principles to problems of their interest.

Two key factors for implementation of plans for genetic improvement are: (1) accurate identification of the animal, its ancestors, and its descendents; and (2) written records of performance for quantitative or Mendelian traits of possible importance. The lack of these factors will prevent, first, determination of the modes of inheritance, and, second, improvement from selection.

The basic principles of inheritance apply to all animals. The circumstances may be considerably different—what traits can be measured, when they can be measured, the generation interval, the reproductive rate, and the economic situation. The two main classes of principles are those governing: (1) population genetics, which deals primarily with simply inherited traits, and (2) quantitative traits.

16-1 Importance of Mendelian and Population Genetics

Mendelian genetics

The principles of Mendelian genetics can be used to determine the mode of inheritance, that is, the type of gene action which determines the phenotype. Few readers will ever have to determine, or will have the facilities to investigate, how a trait is inherited. Whether phenotypic expression is determined by dominant-recessive, incompletely dominant, epistatic, homozygous lethal, or other types of gene action will usually be listed in a reference book. After the mode of inheritance is determined, the question is how to apply that knowledge to improvement by selective mating in a population of animals.

Population genetics

Several chapters developed the principles which show how quickly desirable genes can be fixed in a population or, more importantly, how quickly undesirable genes can be eliminated from a population. The size of the population was shown to affect the probability that "good" or "bad" genes can become fixed or lost from a population by chance. Mutation does not generally have much effect on a short time basis. Two factors of great importance are (1) the mode of inheritance, that is, how difficult it is to discover the genotype of prospective parents, and (2) the intensity of selection practiced for or against certain phenotypes. Selection can be natural in that nature is responsible for which animals become parents, or selection can be artificial. Artificial selection by man may or may not lead to improvement. Animals selected for by man-made standards must carry the genes to improve traits of importance. In some cases artificial selection for one trait has led to deterioration in other traits. For example, selection of Collie dogs with long narrow heads has led to vision problems, as illustrated by the Collie eye syndrome.

Mendelian traits of importance

Many characteristics are simply inherited. Most of these are fixed genetically —the alleles which are present in the population determine acceptable phenotypes. Other such traits are of transitory importance due to breeder fads which act to create demand for unusual animals. For example, at one time, red Holsteins were of little value and could not be registered in the Holstein

Table 16-1 Some possibly important simply inherited traits

All species
Hair color
Coat pattern
Red blood cell antigens
Serum proteins
Milk proteins
Leukocyte antigens
Major histocompatibility complex
Various lethals and semilethals
Some species
Horns
Cattle
Dwarfism
Mulefoot (syndactylism)
Double muscling
Swine
Porcine stress syndrome (PSS)
Sheep
Multiple ovulations (Booroola gene)
Horse
Hemolytic disease
Combined immune deficiency (Arabians)
Dog
Hair length
Poultry
Skin color
Egg color
Slow and rapid feathering

registry; however, in the 1980s red Holsteins and carriers of the gene for red often command higher prices than their black and white herdmates.

Some simply inherited traits which in some situations may be economically important are listed in Table 16-1. Many other traits could be added, especially those involving lethal characteristics. The economic importance of such extreme as well as less extreme defects depends on the frequencies of the responsible genes.

Similarly, the economic importance of quantitative traits will vary depending on the marketing situation and heritability, among other factors, as will be discussed in the next section.

16-2 Important Quantitative Traits

Important classes of traits

If selection is to be successful, the trait or traits under selection must be carefully defined. There are four general groups of traits:

1. reproductive traits such as services per conception, calving interval in cattle and litter size in swine;
2. production traits such as age to market weight in swine, milk yield for 305 days in dairy cattle, and weaning weight for beef cattle;
3. quality traits such as carcass score for beef cattle, swine, and sheep; fraction of milk solids in dairy cattle; and backfat thickness in swine; and
4. aesthetic or personal preference traits such as shade of coat color or pattern, fine points of conformation, and head shape for most species.

Production and reproductive traits can usually be measured objectively. Although some quality traits can be measured objectively, many are given subjective ratings.

Most reproductive traits have low heritability (0 to 15 percent).

Many production traits have moderate heritability (20 to 40 percent).

Many quality traits have high heritability (50 to 70 percent).

The relative economic values of such traits depend on the situation. Unusual situations can result in unusual economic values depending on a combination of high production and high scores of aesthetic traits. For example, a dairy cow with an estimated transmitting ability for milk yield in the top two percent and with a type score of excellent would be in demand for superovulation. The many calves produced by embryo transfer would sell for several times the price of other calves. Most breeders, however, are affected little, if at all, by such unusual and unpredictable occurrences.

Generally, production traits have the largest relative economic value; however, for some classes of livestock, such as swine, litter size or survival to weaning may be most important. Quality traits such as milk composition or carcass score usually have relatively little economic value for a wide range of scores, but exceptional scores may provide an unexpected bonus or loss. The relative economic value of aesthetic traits is often a function of the "eye (or mind) of the beholder."

Correlations, particularly negative genetic correlations, are important in designing a selection program. A negative genetic correlation tends to lead to a decrease in desirability for one trait if the other trait is improved. For example, a classic dilemma in breeding dairy cattle is that high production is associated with deep udders which are more susceptible to disease. Thus, calculation of relative economic values and the use of selection index procedures become more and more important with large negative correlations between traits.

Fair comparisons

A problem encountered in real situations, but essentially ignored in this book, is that, for records to be compared fairly, they must be adjusted to a common basis for many identifiable and nongenetic effects. When establishing an evaluation procedure to use for selection, an important step is to list factors for which adjustment may be needed. Some typical examples are

1. management level (which can usually be associated with the herd, flock, ranch, time of year, and year),
2. age of animal,
3. age of dam,
4. sex of animal,
5. parity of animal, and
6. number of littermates

Estimation of the proper adjustment factors is not a trivial problem. Often adjustment factors can be found in the technical literature, but there is always a question of whether those factors apply to the population of animals the breeder is evaluating. No attempt will be made here to answer the questions of how to estimate the factors or how to determine whether previously estimated factors are valid. The procedures necessary to answer such questions are beyond the scope of this book and will be encountered by post-graduate students in advanced animal breeding and statistics courses.

Quantitative traits of livestock

All breeders undoubtedly have a personal list of traits which they feel are important to improve. The relative importance of the traits is also likely to be different from breeder to breeder. Various segments of each livestock industry will also have different ideas as to which traits are most important. For example, the meat packer is most interested in the backfat thickness and dressing percentage of pigs at market weight. The owner of the herd of sows is more likely to be concerned with the number of healthy pigs weaned and how many litters per year the sows will produce. The purchaser of feeder pigs is greatly concerned with growth rate. Lists of most of the traits that would be considered important by many breeders are given in Tables 16-2 to 16-7 for several classes of animals.

The *beef cattle* industry illustrates the typical conflicts which occur along the chain of production (Table 16-2). The calf producer sells calves. Weaning weight and number of calves per cow determine gross income. Maternal ability (primarily milk production) helps determine weaning weight. Reproductive rate and calving difficulty are the main factors affecting number of calves per cow. Maintenance of the weight of the mature cow is the primary cost.

The feeder is mostly concerned with gain from weaning until market weight and somewhat with muscling and dressing percentage.

The packer is interested in the quality traits—dressing percentage, fat depth, rib eye area, and the consumer trait of marbling (fat distributed within the muscles).

Crossbreeding has become important partly because more segments of the industry can be satisfied. Heterosis for maternal ability results in more and larger calves at weaning. A terminal sire breed can be chosen to produce calves with rapid growth and desirable carcasses.

Dairy cattle breeders are perhaps the luckiest. One trait—milk yield per lactation—has overwhelming importance (Table 16-3). In addition, the eco-

Table 16-2 Quantitative traits of possible economic importance—beef cattle

Trait	Heritability	Variation	Economic Value
Weight			
Birth	Low	Moderate	Indirect
Weaning (205-day)	Moderate	Moderate	Direct
Yearling	Moderate	Moderate	Indirect
Gain to market	Moderate	Moderate	Direct
Age at market	Moderate	Moderate	Direct
Mature	High	Moderate	Direct
Maternal ability	Moderate	Moderate	Indirect
Carcass			
Fat depth	Moderate	Moderate	Indirect
Rib eye area	High	Moderate	Indirect
Dressing percentage	Moderate	Low	Direct
Marbling	Moderate	Low	Indirect
Yield grade	Moderate	Moderate	Direct
Reproduction			
Age first breeding	Low	Low	Indirect
Calving			
Difficulty (first parity)	Low	Moderate	Direct
Live births	Low	Low	Direct
Calving interval	Low	Low	Indirect
Services per conception	Low	Low	Direct

Table 16-3 Quantitative traits of possible economic importance—dairy cattle

Trait	Heritability	Variation	Economic value
Yield per lactation (year)(lifetime)			
Milk	Moderate	Moderate	Direct
Fat	Moderate	Moderate	Direct
Protein	Moderate	Low	Direct
Milk constituents			
Fat %	High	Moderate	Indirect
Protein %	High	Low	Indirect
Somatic cell count (mastitis indicator)	Moderate	Moderate	Indirect
Reproduction			
Age first calving	Low	Moderate	Direct
Services per conception	Very Low	Low	Direct
Days open	Very Low	Low	Direct
Calving interval	Low	Low	Direct
Calving difficulty (first parity)	Low	Moderate	Direct
Conformation			
Soundness	Low	Low	Indirect

nomic values of the protein and fat content can be easily incorporated into evaluations for gross income. Thus selection decisions can be greatly simplified. Reproduction is mostly a management problem and is not as critical as with meat animals where the offspring provide the only income and, for management reasons, must often be born within a limited time period.

The most significant development in dairy cattle breeding has been the widespread use, through artificial insemination, of highly superior bulls. The main factors in finding and using the best bulls have been

1. a set of national milk records collected by Dairy Herd Improvement cooperatives,
2. sampling of bulls with a few daughters in representative herds, and
3. selection of the best bulls for heavy use by artificial insemination.

Dairy farmers at first quickly accepted artificial insemination because of the dangers of keeping dairy bulls on the farm and because technicians would arrive quickly to inseminate a cow in heat. Later the popularity increased because sire summaries derived from the DHI records and frozen semen gave them easy access to the best bulls in the country. Because of daily contact

Table 16-4 Quantitative traits of possible economic importance—swine

Trait	Heritability	Variation	Economic value
Reproduction			
Litter size	Low	Moderate	Direct
Litters per year	Low	Low	Direct
Viability	Low	Low	Direct
Weight			
Weaning	Low	Moderate	Indirect
Market	Moderate	Moderate	Direct
Age at market	Moderate	Moderate	Direct
Carcass			
Fat depth	Moderate	Moderate	Indirect
Rib eye area (volume)	High	Moderate	Indirect
Dressing percentage	Moderate	Low	Direct

between the herd manager and the cows, detection of animals in heat is not as great a problem as with some types of animals.

Swine producers face essentially the same problem as beef producers (Table 16-4). Crossbreeding has been widely used to increase litter size, litter survival, and growth rate. Number of offspring sold per female per year is probably more important with pig producers than with producers of other livestock. Growth rate has been increased greatly because of the selection intensity that is possible with many pigs per litter and because of the high heritability. Because of low heritabilities, litter size and survival have required heterosis arising from breed crosses to make improvement. Unfortunately for the consumer, most meat packers have not bought market hogs based on carcass yield. Therefore muscling and dressing percentage have not improved very much. Backfat thickness, however, was reduced dramatically from between 2 and 3 inches to $\frac{1}{2}$ inch or less at market weights in relatively few years.

Sheep provide not only meat and fiber but also, in some countries, milk. Producers of lamb are interested primarily in numbers of lambs weaned and growth rate to market size (Table 16-5). Lamb is often thought of as a seasonal product because of the seasonal nature of breeding in temperate zones. A steady supply of lamb would require out-of-season breeding. Thus, many breeders are interested in obtaining more than one crop of lambs per year which would also reduce the overhead cost of production. With some breeds the fleece is only a byproduct of producing lamb. For other breeds the fleece is the major product with fleece weight and spinning quality of primary importance.

The *horse* is a multipurpose animal (Table 16-6). Most are recreation or

Table 16-5 Quantitative traits of possible economic importance—sheep

Trait	Heritability	Variation	Economic value
Weight			
Birth	Moderate	Low	Indirect
Weaning	Moderate	Moderate	Direct
Age at market	Moderate	Moderate	Indirect
Mature	Moderate	Moderate	Direct
Maternal ability	Low	Moderate	Indirect
Reproduction			
Multiple births	Low	Moderate	Direct
Age at first breeding	Low	Low	Indirect
Carcass			
Muscling—quality grade	Moderate	Low	Direct
Dressing percentage	Moderate	Low	Direct
Fleece			
Weight	Moderate	Moderate	Direct
Fiber diameter	Moderate	Low	Direct
Staple length	Moderate	Low	Direct
Face cover score	Moderate	Moderate	Indirect

pleasure animals. Most intensive selection, however, has been for performance—primarily racing in the Thoroughbred, Standardbred, and Quarter Horse breeds. There have been no coordinated efforts to produce better pleasure horses or racing horses. Because of the moderate heritability of racing speed, selection of sires based on their track performance has been

Table 16-6 Quantitative traits of possible economic importance—horse (blanks indicate a lack of research results)

Trait	Heritability	Variation	Economic value
Performance			
Racing	Moderate	High	Direct
Jumping	Low	Moderate	Indirect
Pulling	Low	Moderate	Indirect
Reproduction			
Foaling rate	Low	Low	Direct
Conformation			
Soundness			
Show ring			
Gaits			
Action			

Table 16-7 Quantitative traits of possible economic importance—poultry

Trait	Heritability	Variation	Economic value
Mortality			
Hatchability	Low	Low	Direct
Fertility	Very low	Low	Direct
Egg production			
Rate of lay	Moderate	Moderate	Direct
Age at first egg	Moderate	Moderate	Indirect
Weight	Moderate	Moderate	Direct
Quality			
Firmness of albumen	Low	Low	Indirect
Yolk color	Low	Moderate	Indirect
Shell thickness	Low	Moderate	Indirect
Meat production			
Weight			
At market	Moderate	Moderate	Direct
Growth rate	Moderate	Moderate	Direct
Carcass			
Dressing percentage	High	Moderate	Direct
Breast angle	Moderate	Moderate	Indirect

successful. Little effort has been made to select the best brood mares on their own performances. The racing industry has either resisted or limited the potential of artificial insemination. Thus only a few mare owners can afford the services of the best stallions.

Poultry breeding has become the most commercialized of all livestock species (Table 16-7). A very few large companies supply the seedstock for most of the United States and the world. Their breeding programs are well-kept secrets. The chicken industry is divided into two distinct segments—egg layers and broilers. Both have made spectacular progress because of the high selection intensity that is possible. Both groups are concerned with viability and resistance to disease as only a live hen or broiler produces any income.

Egg producers are interested primarily in rate of lay and age at first egg. The weight of the egg and quality traits present a problem since these may be negatively related to rate of lay. Some markets require eggs with white shells and others prefer eggs with brown shells.

Meat producers want a broiler at market weight as soon as possible. A young market weight broiler is a much more efficient converter of feed to a market product than an older broiler.

The turkey industry has been so successful in increasing the production of white meat in the breast (broad breasted) that artificial insemination must be used to obtain fertilized eggs.

16-3 Genetic Engineering

Some technologies that manipulate the genetic material have already had an impact on the science of animal breeding, while others have potential that is as yet untapped. The most notable example of a technology that has been used to advantage in the dairy industry, but is yet to be fully exploited with other species, is artificial insemination. This technology has a tremendous impact on sire evaluation, as it permits the mating of large numbers of females, in many different environments, to individual bulls. The result has been a vast improvement in the reliability and accuracy of sire evaluations, thus increasing the rate of genetic progress.

Embryo transfer is another manipulation of genotypes that is being exploited primarily by the beef and dairy industries. This technique allows an increase in the number of progeny produced by individual females, thus improving genetic evaluations for this sex also. The sexing of embryos before transfer gives yet another new tool to the animal breeder, by allowing the production of offspring of the desired sex.

The manipulation of embryos includes the technique of splitting embryos to produce several genetically identical progeny from one fertilized egg. This provides the animal breeder with clones that could be employed in studies of the effect of the environment, that hitherto have been restricted by the low natural rate of occurrence of twins.

Techniques as yet unexplored by, or unavailable to, the animal breeder include manipulations of the chromosomes themselves and of individual genes. The insertion of genes into the genotypes of domestic livestock is entering the realm of feasibility but is not yet practical on a commercial scale.

With any new technology, time must elapse before it is widely accepted, and it must be economically attractive to be adopted by producers. Also, the industry must be structured in such a way that the potential benefit of the new technology can be realized. A prime example is artificial insemination which has unquestioned advantages for animal improvement but which is not being widely used with all species.

16-4 Some Last Thoughts

The first key to genetic improvement is obtaining accurate records. Breed associations for some classes of livestock have provided pedigree information but generally have not provided a means of collecting records of performance. Many breed societies are now correcting this problem. Cooperative testing associations have been very successful in the dairy industry in providing both production testing and pedigree information. Artificial insemination organizations have often been forced to collect records which have sometimes been suspect because of the potential relationship between genetic evaluations and sale price of semen.

The second key is to define as accurately as possible the relative economic values of economic traits. Progress may be limited because of too much emphasis on unimportant traits.

The third key is to make proper use of records on the animal and its relatives using the principles outlined in this book. A famous animal several generations back in the pedigree is unlikely to have much influence on the production of its descendents. Overemphasis on famous but distant relatives has held back progress.

The fourth key is to use the genetic evaluations. If the evaluation properly puts together the records and economic values, then progress will be rapid, but only if selection is based on that combined evaluation. Too often, the physical features of a herdmate with a poor evaluation prove more influential than a high evaluation of a less attractive animal.

With this mixture of positive and negative admonitions, the authors wish the reader success in attempting to apply genetic principles and good sense to selecting better and better animals.

Further Readings

American Sheep Producers Council. 1984. *The Sheepman's Production Handbook.* Sheep Industry Development Program, Denver, Colo.

Everett, R. W., chmn. 1984. *Proceedings of the National Invitational Workshop on Genetic Improvement of Dairy Cattle.* Department of Animal Science. Cornell University, Ithaca, N.Y.

Johansson, I., and J. Rendel. 1968. *Genetics and Animal Breeding.* W. H. Freeman Company, New York.

Turner, H. W., and S. S. Y. Young. 1969. *Quantitative Genetics in Sheep Breeding.* Cornell University Press, Ithaca, N.Y.

U.S. Department of Agriculture. 1981. *Guidelines for Beef Improvement Programs.* Program aid 1020. Extension Service, Science and Education Administration.

U.S. Department of Agriculture. 1981. *Guidelines for Uniform Swine Improvement Programs.* Program aid 1157. Extension Service, Science and Education Administration.

See also the following technical journals:

Animal Breeding Abstracts
Animal Production
Journal of Animal Science
Journal of Dairy Science
Livestock Production Science
Poultry Science

Index